WS.

FIBER OPTICS
Devices and Systems

PRENTICE-HALL SERIES
IN SOLID STATE PHYSICAL ELECTRONICS

Nick Holonyak, Jr., Editor

FIBER OPTICS
Devices and Systems

Peter K. Cheo

United Technologies Research Center
East Hartford, CT

and

The Hartford Graduate Center
Hartford, CT

Prentice-Hall, Inc., Englewood Cliffs, NJ 07632

Library of Congress Cataloging in Publication Data

CHEO, P. K. (Peter K.)   (date)
  Fiber optics.

  (Prentice-Hall series in solid state physical
electronics)
  Includes bibliographical references and index.
  1. Fiber optics. I. Title. II. Series.
TA1800.C48 1985   621.36′92   84-4922
ISBN 0-13-314204-3

Editorial/production supervision and
   chapter opening design: *Gretchen K. Chenenko*
Cover design: *Diane Saxe*
Manufacturing buyer: *Gordon Osbourne*

Printed in the United States of America

10 9 8 7 6 5 4 3 2

ISBN 0-13-314204-3

PRENTICE-HALL INTERNATIONAL, INC., *London*
PRENTICE-HALL OF AUSTRALIA PTY. LIMITED, *Sydney*
EDITORA PRENTICE-HALL DO BRASIL, LTDA., *Rio de Janeiro*
PRENTICE-HALL CANADA INC., *Toronto*
PRENTICE-HALL OF INDIA PRIVATE LIMITED, *New Delhi*
PRENTICE-HALL OF JAPAN, INC., *Tokyo*
PRENTICE-HALL OF SOUTHEAST ASIA PTE. LTD., *Singapore*
WHITEHALL BOOKS LIMITED, *Wellington, New Zealand*

Dedicated
to Dorothy and my children

Contents

Preface

This book comprises an accumulation of three years of lecture notes that I prepared for a course offered at the Hartford Graduate Center to a group of first-year graduate students who are employed by aerospace, electronics, and telecommunications industries. One of the problems that confronted me was to find an appropriate textbook that would satisfy the needs of my students, most of whom obtained their bachelor degrees in electrical engineering or applied sciences several years ago and needed an extensive review of the fundamentals. There are many excellent reference books on fiber optics and on topics related to this field, but they are not suitable for use as a text. The emerging field of fiber optics actually involves several independent technologies, such as guided-wave optics, semiconductor light sources and photodetectors, and digital communications. There are excellent books available that contain detailed treatments of each of these topics. However, an introductory textbook that covers the entire field is lacking at present. On the other hand, many books provide a broad coverage of this field by offering a collection of papers in which the fundamentals are often left to references. In these cases, various aspects of fiber optics are presented by experts, each with a different writing style and approach. The lack of coherence and fundamental concepts make these books difficult for students to comprehend unless they have already acquired sufficient background knowledge and expertise in the field.

The goal of this book is to present the above-mentioned topics in a self-consistent manner so that students can follow the development of this text with a clear understanding of the subject matter. The presentation has been kept at an introductory level, with emphasis on the physical concepts underlying the interpretation of the properties of fiber optical waveguides, semiconductor light emission, and detection devices. I have made an effort to remain as rigorous and up to date as possible within the constraint of the presentation level.

This book is intended as a text for a course to be taught at the first-year graduate level to students enrolled in departments of electrical engineering, physics, and applied sciences. In this book I have treated in a pedagogic manner topics that form the basics for all optical fiber systems. The fundamental principles and theories are introduced and developed to the extent that the student can gain a better insight into each topic without losing track of the basic concepts. When mathematics becomes cumbersome, approximate methods or intuitive approaches are introduced to provide students with a semiquantitative picture and a physical interpretation of the process. It is assumed that students have already taken introductory courses in electromagnetic theory, solid-state physics, and quantum mechanics. However, a certain amount of background material has been included so that students who have not taken some of these courses should still be able to follow this text by putting in some extra efforts. Therefore, advanced undergraduate students in engineering and physical sciences, who have already fulfilled sufficient prerequisites, should be able to take this course without much difficulty. It is also my intention to provide general readers with an easy-to-read text, which can help them become acquainted with this subject through self-study.

The first part of the text deals with the principles and applications of optical fibers as data transmission media. Both ray and wave approaches have been employed to explain the mode structures of optical fibers. Special emphasis is given to their interactions and propagation characteristics. The second part reviews and treats in detail the properties and operating characteristics of optical sources and photoreceivers with special emphasis on the emission and regeneration processes in semiconductors and their noise characteristics, which have a great impact on the system performance. The last part is devoted to several applications of these components for data transmission and telecommunication purposes. Efforts are made to present the materials as explicitly as possible. Occasionally, some details are intentionally left out as exercises for students, solely for the purpose of stimulating the learning process.

Fabrication techniques for fibers and other optical components are introduced together with the specific subject matter. A list of problems is included at the end of each chapter. Designs of fiber optical systems are discussed only briefly, because it is difficult to elaborate on topics that are highly specialized and in many cases are still in the development stage. The field that covers the current research on most sophisticated optical devices and circuits is known as integrated optics. It is omitted because its relevance to fiber optical systems is still not clear at this time. Furthermore, no attempt has been made to provide a bank of references or to acknowledge the original work by various contributors. References listed in this text are only those directly related to the subject matter which can provide students with more detailed information. There are many review articles available, most of which provide comprehensive lists of original research papers.

I would like to express my appreciation to Professor W. R. Kolk for his continuing interest and encouragement in this endeavor and to Dr. E. Snitzer, Dr. T. Li, Professor W. S. C. Chang, and to my students, who have made contributions by reducing the number of errors and ambiguities occurring in this text. I wish to express my deep gratitude to my wife, Dorothy, for her unfailing patience in proofreading this text in its entirety. I am also grateful to United Technologies Research Center for providing me with excellent word-processing and illustration services and to the Alcoa Foundation for its interest and support of this project.

Peter K. Cheo
West Hartford, CT

1

Introduction

Fiber optics has gained prominence in telecommunications, instrumentation, cable TV network, and data transmission and distribution. The major application, however, is in the area of telecommunications. Within this decade, there will be a significant changeover from wires and coaxial cables to optical fibers for telecommunication systems and information services. This anticipated change is dictated almost entirely by economics. The increasing cost and demand for high-data-rate or large-bandwidth-per-transmission channels and the lack of available space in already congested conduits in every metropolitan area are the reasons for this changeover. Furthermore, fiber optical devices interface well with digital data-processing equipment, and their technology is compatible with modern microelectronic technology. For these reasons, it is anticipated that in the future most telephones, television receivers, bank machines, computers, and to a lesser extent, medical and industrial instruments will be linked by optical fibers.

Since 1960 the availability of laser sources has stimulated research into optical communication. However, optical communication was not considered to be practical until 1970, when optical fiber technology had advanced to the point where relatively low-loss (<20 dB/km) fiber could be drawn routinely. Today, fibers with an absorption coefficient a (λ) as low as 0.5 dB/km can be manufactured for optical transmission at wavelengths $\lambda \geq 1.2$ μm. For a complete and up-to-date reference list on fiber optics development, readers should consult the review article by Li (Ref. 1.1). General information can be found in two other articles (Refs. 1.2 and 1.3).

A typical optical fiber system linkage is shown in Figure 1.1. The input

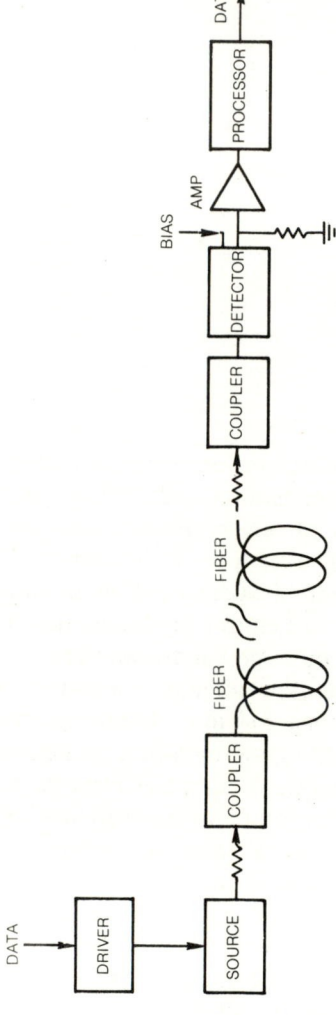

Figure 1.1 Schematic of a typical optical fiber data transmission link.

data are usually coded by using a current pulse network which can directly modulate the light source. The output of light pulses is coupled into a fiber by using a lens or simply a butt joint. The optical power received by the photodiode through a certain length of fiber is always substantially reduced from its initial value due to losses through coupling, absorption, scattering, leakage, dispersion, and mode conversion. To maintain reliable, high-fidelity system operation, the power must be sufficient to overcome system losses. The fidelity of the signals transmitted depends on the detectable level of the signal-to-noise (S/N) ratio, which can be estimated from the detection probability function of a given distribution. For a telecommunication system, the minimum bit error rate (BER) is 10^{-9}, which corresponds to an optical S/N ratio of about 12 dB. The noise equivalent power of an optical receiver used in a typical fiber optical circuit depends on the data rate. Therefore, the minimum required power must be determined by also taking into account the data rate. For an avalanche photodiode, the minimum detectable power, which is equivalent to a S/N ratio of unity, can be as low as -45 dBm at a data rate of 400 megabits per second (Mb/s). System analysis of this type requires knowledge of various parameters that govern the performance of the light source and the detector, the system noise, and propagation characteristics of light in a fiber transmission channel. Various signal processing techniques are required to deal with problems of signal distortion and interference, and statistical methods are often employed to determine system error-detection probability.

This book consists of four main topics: (1) the theory of optical fiber waveguides and pulse propagation phenomena in fibers; (2) the emission process, structure, and performance characteristics of semiconductor light sources; (3) optical receivers and noise characteristics of semiconductor photodiodes; and (4) telecommunication and data transmission systems via optical fibers. It provides a self-contained treatment of these topics so that readers can reach a reasonable level of understanding of the fundamentals without relying too much on other sources for information.

One of the most important components in an optical fiber system is the optical fiber, which is discussed in the next four chapters. In most cases, it is made of glass material (SiO_2) mixed with various dopants primarily to control the refractive index and reduce the softening point. Most fibers have a cylindrical core with an index n_0 of slightly higher value than that of the cladding material n_c. However, some fibers, primarily those made for optical imaging applications, have a square cross section. As shown in Figure 1.2, the radii of the core and the cladding are denoted by a and b, respectively. For a step-index fiber the refractive index is expressed by

$$n(r) = \begin{cases} n_0 & (r < a) \\ n_c & (a \leq r \leq b) \end{cases} \qquad (1.1)$$

and for a graded-index fiber with a parabolic profile for its core, the refractive index is expressed by

(a)

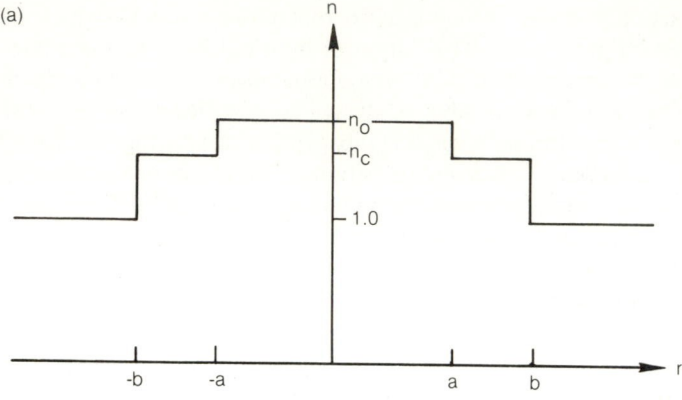

(b)

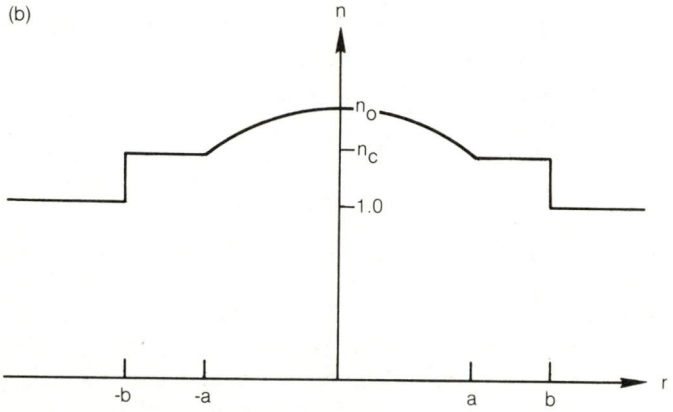

Figure 1.2   Index profile for (a) a step-index fiber having a core diameter $2a$ and clad diameter of $2b$ and (b) a graded-index fiber.

$$n(r) = \begin{cases} n_0 \left[1 - 2\Delta \left(\dfrac{r}{a} \right)^\alpha \right]^{1/2} & (r < a) \\ n_c & (a \le r \le b) \end{cases} \qquad (1.2)$$

where

$$\Delta = \frac{(n_0 - n_c)}{n_c} \qquad (1.3)$$

For a variety of glass fibers with different dopants, $\alpha \simeq 2$.

A single-mode fiber has a core radius typically of the order of one optical wavelength λ. A multimode fiber is one whose core radius is substantially larger than λ and is about 25 to 50 μm. In this case there are hundreds or even thousands of allowable modes propagating in the guide. As a rule, the number of modes in a step-index fiber is about twice that in a graded-index fiber of the same dimension. Because each mode possesses a unique group velocity, a short light pulse of energy distributed among these modes will be broadened as it travels through a certain length of the fiber. This phenomenon, known as **modal dispersion**, is treated in greater detail in Chapter 5 because of its impact on the information-carrying capability of a fiber. To reduce the modal dispersion, a fiber with a graded index has been introduced to compensate for the differences in group velocities, so that all modes travel in a graded-index fiber at nearly the same speed. Another alternative for reducing modal dispersion, and thereby increasing the information-carrying capability, is to use a single-mode step-index fiber. The penalty paid for using a single-mode fiber can be very high because it is considerably more difficult to splice and couple single-mode fibers than multimode fibers, because the physical dimensions of single-mode fibers are too small ($r \simeq \lambda$) to be handled conveniently.

Propagation in a fiber can be viewed from a simplified ray picture such as that shown in Figure 1.3. Let θ_M be the maximum angle beyond which the rays that enter the fiber will no longer be confined within the fiber. Therefore, θ'_M is the critical angle beyond which rays will not be bounded. Using Snell's law for rays at the air–fiber interface and at the core–cladding interface, we obtain

$$\sin \theta_M = n_0 \sin \theta'_M \qquad (1.4)$$

and

$$n_c = n_0 \cos \theta'_M \qquad (1.5)$$

Combining Equations (1.4) and (1.5), we obtain

$$\sin \theta_M = \sqrt{n_0^2 - n_c^2} \equiv \text{NA} \qquad (1.6)$$

where NA denotes the numerical aperture of the fiber. Since the difference between the refractive indices of the core and the cladding is usually very small,

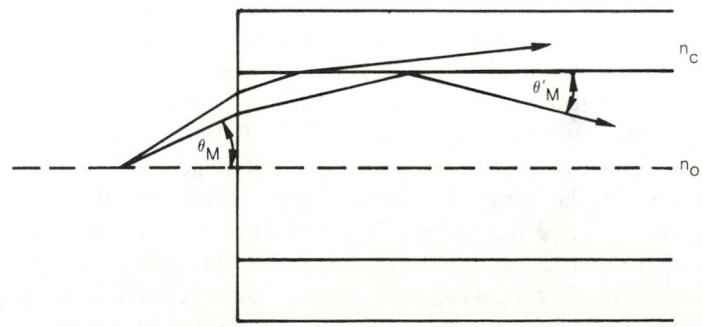

Figure 1.3   Ray picture showing total reflected and radiated rays in a glass fiber.

we can express Equation (1.6) by the expression

$$NA \simeq n_0 \sqrt{2\Delta} \tag{1.7}$$

where Δ is given by Equation (1.3). If we let $\sin \theta_M \simeq \theta_M$ and from Equation (1.6), we see that NA represents the maximum acceptance angle within which all rays will be guided by the fiber. For $NA = 0.2$, the maximum angle to capture rays is about $11°$. Rays that make larger angles with the fiber axis within the acceptance angle correspond to high-order modes and travel in longer paths than do those propagating along the axis of the fiber. The delay difference between the axial rays and the rays traveling at the maximum angle is given by

$$\delta\tau = \frac{n_0}{c} \frac{L}{\cos \theta'_M} - \frac{n_0 L}{c} \tag{1.8}$$

Substituting Eq. (1.5) into (1.8) we can write

$$\delta\tau = \frac{n_0 L (n_0/n_c - 1)}{c} \tag{1.9}$$

Substituting Equation (1.3) into (1.9) for $n_0/n_c - 1$, and noting that $n \simeq n_0 \simeq n_c$, we obtain

$$\delta\tau = \frac{nL\Delta}{c} \tag{1.10}$$

For $\Delta = 0.01$, Equation (1.10) yields a delay time of 50 ns/km. In a graded-index fiber, the rays traveling away from the axis can gain speed and arrive with a very small delay difference. With careful control of the index profile, one can obtain roughly two orders of magnitude of reduction in the delay difference among all modes. These effects and the effects of material dispersion and mode coupling are discussed in Chapter 5. Fiber fabrication and measurement techniques are discussed in Chapter 6.

Two types of sources commonly used as transmitters in optical fiber systems are semiconductor light-emitting diodes (LEDs), which are incoherent sources, and laser diodes (LDs). The group of semiconducting materials used in making these sources are GaAs, InAs, InP, AlGaAs, InGaAsP, and so on, with direct bandgap energies extending from 0.7 to 1.6 μm in the spectral region. For most glass materials the spectral region of negligible material dispersion is in the range 1.2 to 1.3 μm. Therefore, the use of longer wavelengths leads to a reduction in material dispersion. Because the effects of material dispersion are at least one order of magnitude smaller than that of modal dispersion, the modal dispersion must first be reduced in order to reach a certain level of data rate. This can be accomplished by employing either a single-mode or graded-index fiber. A further increase in data rate can be

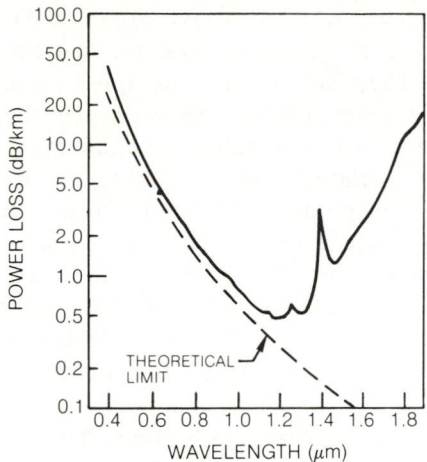

Figure 1.4 Loss spectrum of a phosphosilicate glass fiber with a borosilicate cladding having a NA value of 0.18. [After M. Horiguchi and H. Osanai, *Electron. Lett., 12,* 310 (1976).]

achieved by using a longer-wavelength source. There are also other advantages to selecting longer wavelengths, because as λ increases, the scattering losses decrease as $1/\lambda^4$, as shown in Figure 1.4. However, as λ increases beyond 1.3 μm, the absorption losses arising from the first overtone of the fundamental infrared band structures of the OH radical in the fiber begin to dominate. For this particular fiber, as shown in Figure 1.4, the OH content was estimated to be in the range of 50 parts in 10^9, which is an exceptionally low level for all practical purposes. Because of low losses and low material dispersion at longer wavelengths, most of the recent research activity on sources has centered around the InGaAsP and AlGaAsSb quaternary systems, which emit in the wavelength range 1.3 to 1.8 μm. As a result, very low threshold double-heterostructure InGaAsP lasers with improved surface morphology and reproducibility have been produced. The relatively low threshold current of 1 kA/cm^2 has been achieved for these lasers. This is only about a factor of 2 to 5 greater than that achieved for the more advanced AlGaAs ternary system. Even though the quaternary system is more complex than the ternary system, present results indicate that quaternary semiconductor sources can be made as efficient and reliable as those made from ternary compounds. Besides wavelength considerations, other factors, such as component cost, reliability, output power, and coupling efficiency, are also important in formulating the criteria for selecting a source. More details on sources and transmitters are given in Chapter 7, 8, 9, and 10.

The photodetector is another important component in an optical fiber system. The type of detector commonly used in optical fiber systems is the semiconductor photodiode, which is a reverse-biased *pn*-junction device. The two most commonly used photodetectors are the *pin* and the avalanche photodiode (APD). They are actually modified *pn*-junction devices with additional layers at slightly different doping levels to provide either more efficient quantum conversion or avalanche gain through ionization. Basically, a photon is absorbed in a relatively high *E*-field region, where an electron–hole pair is created. This will produce current in the detector circuit. To obtain higher quantum efficiency, a device such as a *pin* can provide adequate absorption in the relatively high-resistivity central *i* region. Another approach for obtaining higher detector currents is to create an avalanche gain effect, as in the case of APD. In this case an electron–hole pair may generate tens or hundreds more secondary electron–hole pairs. Because these events occur at random and are of a statistical nature, the noise generated in these devices can be a limiting factor on the detectivity. A trade-off between the gain or quantum efficiency and noise exists for these devices. Techniques are available for processing and regeneration of digital postdetection signals through amplification, pulse shaping, equalization, timing extraction, decision, and error detection. More details on detectors and receivers are given in Chapter 11.

In addition to the discussion of various optical fiber components, standard techniques for signal coding, modulation, and some basic circuits for transmitters and receivers are presented in Chapters 10 and 11, and some examples of commonly used digital and analog fiber systems are presented in Chapter 12. A system designer must choose various components that are best suited for a specific application. These choices should be made based on a trade-off analysis among various system parameters involving optical power, fiber loss, receiver noise, signal type, data rate or bandwidth, bit error rate or signal-to-noise ratio, and the length between terminals or repeater spacing. Once an optimized system configuration is established, the designer must then consider other factors, including environmental conditions, cost, reliability, flexibility, size, weight, installation, and maintenance. The procedure must be iterative because most of these factors are interrelated. An added complication is that the cost of fiber optical components is changing rapidly with time. All indications are that the cost of optical fiber components will decrease significantly in time and that the rate of decrease will be determined primarily by the growth rate or production rate.

One of the attractive features of an optical fiber system is the potentially large repeater spacing for large-capacity data transmission, which can bring about a substantial system cost reduction. Figure 1.5 shows the existing and the projected repeater spacing as a function of data rate. The solid curve represents the results obtained for existing systems for which AlGaAs lasers emitting at 0.9 μm have been used. The average system loss at this wavelength is assumed to be about 4.5 dB/km. If future systems can be built in the region 1.2 to 1.6 μm, a marked increase in repeater spacing is expected, as shown in Figure 1.5 by the dashed curves. These curves are obtained by assuming a total loss of 0.7 dB/

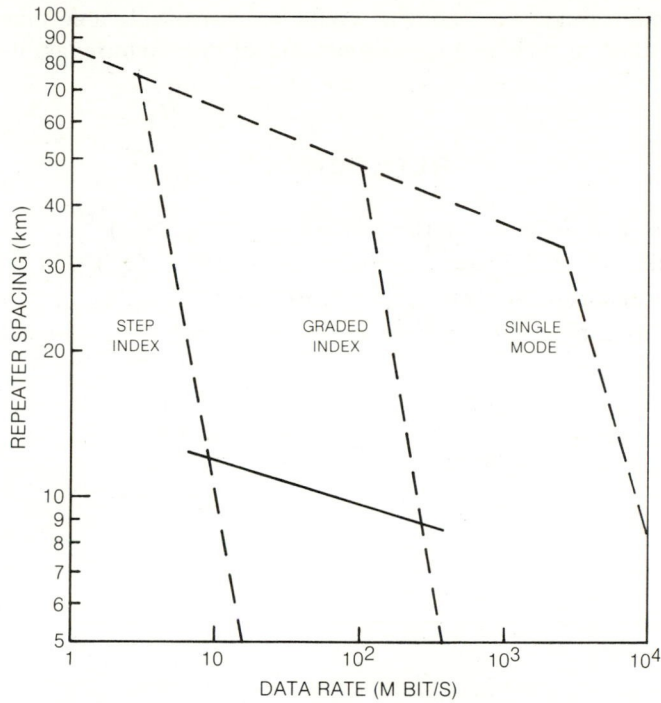

Figure 1.5   Measured (solid curve) and estimated (dashed curves) repeater spacing for various types of fibers as a function of data rate. The solid curve represents results obtained from optical fiber systems operating at wavelengths near 0.9 μm.

km, which includes both fiber and splicing losses. Other components used in this estimate are state-of-the-art InGaAsP double-heterostructure lasers and germanium APD detectors with a low ionization factor, $k \simeq 0.1$. The results, as shown in Figure 1.5, indicate that it is possible to achieve a repeater spacing greater than 30 km at 1 gigabit per second by using a single-mode fiber with a total dispersion of 1 ps/km and a source that has a spectral width of 2 Å. Most recent results (Ref. 1.4) indicate that a repeater spacing greater than 115 km at 0.43 Gb/s can be achieved by using a single-mode fiber and a cleaved coupled-cavity (C^3) laser operating at 1.55 μm. With the possibility of a substantial increase in repeater spacing at high data transmission rates, not only the cost of system components but also the installation, operation, and maintenance costs could be reduced significantly. If the system were not limited by its transmission-line length, it would then be possible to install repeaters in substations instead of manholes. In this case many advantages could be obtained. For example, the power supply would be readily available at each substation, thus eliminating the need for transmitting electric power along the fiber cable. Also, the convenience of maintaining and repairing optical repeaters at substations, instead of in a manhole environment, could lead to considerable savings in the cost of operation as well as a large improvement in system reliability.

REFERENCES

1.1. T. Li, *IEEE J. Select. Areas Commun., SAC-1,* 356 (1983).

1.2. S. E. Miller, E. A. J. Marcatili, and T. Li, *Proc. IEEE, 61,* 1703 (1973).

1.3. D. Botez and G. J. Herskowitz, *Proc. IEEE, 68,* 689 (1980).

1.4. W. T. Tsang, N. A. Olsson, R. A. Logan, *Appl. Phys. Lett., 42,* 650 (1983).

2

Field Relations for Dielectric Waveguides

2.1 REVIEW OF BASIC LAWS OF ELECTROMAGNETICS

Maxwell's equations are the embodiment of four phenomenological laws of electromagnetics. They are:

1. Gauss's Law. Gauss's law is a direct consequence of Coulomb's law of electrostatics and can be expressed in a variety of ways. A differential formulation of Gauss's law is given by

$$\nabla \cdot \mathbf{E} = \frac{\rho}{\varepsilon} \tag{2.1}$$

where $\mathbf{E}$ is the electric field vector, ρ the charge density, and ε the permittivity, which is related to the dielectric constant of the medium. Equation (2.1) indicates that the electric field diverging from an arbitrarily chosen surface is equal to the charge density enclosed by that surface divided by the permittivity. It indicates further that the electric field must originate from electrical charges with a radially outward line of force. Since the displacement field vector $\mathbf{D}$ is related to $\mathbf{E}$ as

$$\mathbf{D} = \varepsilon \mathbf{E} \tag{2.2}$$

Equation (2.1) can also be written as

$$\nabla \cdot \mathbf{D} = \rho \tag{2.3}$$

2. Magnetostatic Law. Since there are no free magnetic charges, a counterpart of Gauss's law for magnetostatics can be written as

$$\nabla \cdot \mathbf{B} = 0 \qquad (2.4)$$

where $\mathbf{B}$ is a magnetic field vector that defines the magnetic flux diverging from a closed surface and is related to the magnetic field intensity vector $\mathbf{H}$ through the permeability or susceptivity μ of the medium by the relation

$$\mathbf{B} = \mu \mathbf{H} \qquad (2.5)$$

3. Ampère's Law.   Ampère's law describes the relationship between $\mathbf{H}$ and electric current I in a circuit, a phenomenon first observed by Hans C. Oersted. An integral formulation of this law can be given in terms of a line integral of the magnetic field $\mathbf{H}$ along any closed path encircling a circuit having a value equal to the total current I carried by the circuit. Specifically, we can write

$$\oint \mathbf{H} \cdot d\mathbf{l} = I$$

By Stokes's theorem, we can rewrite the equation above in differential form as

$$\nabla \times \mathbf{H} = \mathbf{J} \qquad (2.6)$$

where $\mathbf{J}$ is the current density vector pointing in a direction normal to the surface that is enclosed by the integration path. For a medium that contains displacement currents, Equation (2.6) must also include an induced current term as given by

$$\nabla \times \mathbf{H} = \mathbf{J} + \frac{\partial \mathbf{D}}{\partial t} \qquad (2.7)$$

The second term on the right-hand side of Equation (2.7) is obtained by using the law of conservation of charge with the help of a very useful continuity equation:

$$\frac{\partial \rho}{\partial t} + \nabla \cdot \mathbf{J} = 0 \qquad (2.8)$$

4. Faraday's Law of Induction.   The most important law in electro-dynamics is Faraday's law, which relates electric field to time-dependent magnetic field. Specifically, it describes the behavior of an induced current in a circuit that has been subjected to a time-varying magnetic field. The integral form of this law can be expressed by

$$\oint \mathbf{E} \cdot d\mathbf{l} = -\frac{\partial}{\partial t} \iint \mathbf{B} \cdot \mathbf{n} \, dS$$

Again using Stokes's theorem, we obtain from the equation above the following differential form of Faraday's law:

$$\nabla \times \mathbf{E} + \frac{\partial \mathbf{B}}{\partial t} = 0 \qquad (2.9)$$

2.2 MAXWELL'S EQUATIONS

Optical power propagating in either a planar or a cylindrical dielectric waveguide is described by the field vectors $\mathbf{E}$, $\mathbf{D}$ and $\mathbf{H}$, $\mathbf{B}$, satisfying Maxwell's equations for a linear homogeneous and nonconducting medium with no sources. They comprise a set of four relations:

$$\nabla \times \mathbf{E} = -\frac{\partial \mathbf{B}}{\partial t} \tag{2.10}$$

$$\nabla \times \mathbf{H} = \frac{\partial \mathbf{D}}{\partial t} \tag{2.11}$$

$$\nabla \cdot \mathbf{D} = 0 \tag{2.12}$$

$$\nabla \cdot \mathbf{B} = 0 \tag{2.13}$$

By substituting Equations (2.5) and (2.2) into (2.10) and (2.11) we obtain

$$\nabla \times \mathbf{E} = -\mu \frac{\partial \mathbf{H}}{\partial t} \tag{2.14}$$

$$\nabla \times \mathbf{H} = \varepsilon \frac{\partial \mathbf{E}}{\partial t} \tag{2.15}$$

By taking the curl of Equation (2.14) and using Equation (2.15) to eliminate the time derivative of $\mathbf{H}$, we obtain

$$\nabla \times (\nabla \times \mathbf{E}) = \nabla (\nabla \cdot \mathbf{E}) - \nabla^2 \mathbf{E} = -\mu \varepsilon \frac{\partial^2 \mathbf{E}}{\partial t^2}$$

Since $\nabla \cdot \mathbf{E} = 0$, the equation above reduces to

$$\nabla^2 \mathbf{E} = \mu \varepsilon \frac{\partial^2 \mathbf{E}}{\partial t^2} \tag{2.16}$$

and similarly, for $\mathbf{H}$ we obtain the other wave equation

$$\nabla^2 \mathbf{H} = \mu \varepsilon \frac{\partial^2 \mathbf{H}}{\partial t^2} \tag{2.17}$$

Equations (2.16) and (2.17) are the equations of motion of electromagnetic waves in dielectric waveguides. These waves are represented by the coupled $\mathbf{E}$ and $\mathbf{H}$ vectors and propagate with a phase velocity v_p, which is determined by parameters of the medium μ and ε as given by the expression $1/\sqrt{\mu \varepsilon}$.

2.3 SOLUTIONS OF THE GENERAL FORM

We shall develop the field relations first for a planar dielectric waveguide and then for a cylindrical dielectric waveguide. We take this approach because the field components for a planar waveguide involve only elementary functions, which are simpler to manipulate than those describing the fields in cylindrical fibers. However, the mathematical procedures are similar and very instructive when dealing with both types of waveguides.

Figures 2.1 and 2.2 define the coordinate systems used for the two waveguides. In both cases, we have chosen the z axis as the direction for wave propagation. Therefore, in the case of a planar waveguide, we look for solutions of the form

$$\mathbf{E} = \mathbf{E}\,(x,\,y)\,\exp\left[i(\omega t - \beta z\,)\right]$$
$$\mathbf{H} = \mathbf{H}\,(x,\,y)\,\exp\left[i(\omega t - \beta z\,)\right]$$

(2.18)

and in the case of a cylindrical waveguide,

$$\mathbf{E} = \mathbf{E}\,(r,\,\theta)\,\exp\left[i(\omega t - \beta z\,)\right]$$
$$\mathbf{H} = \mathbf{H}\,(r,\,\theta)\,\exp\left[i(\omega t - \beta z\,)\right]$$

(2.19)

where β is the propagation constant of the field and its values are subject to the boundary conditions imposed on $\mathbf{E}(x,\,y)$ or $\mathbf{E}(r,\,\theta)$, and others, and ω is the angular frequency of the wave, which is related to the phase velocity v_p:

$$v_p = \frac{\omega}{\beta}$$

(2.20)

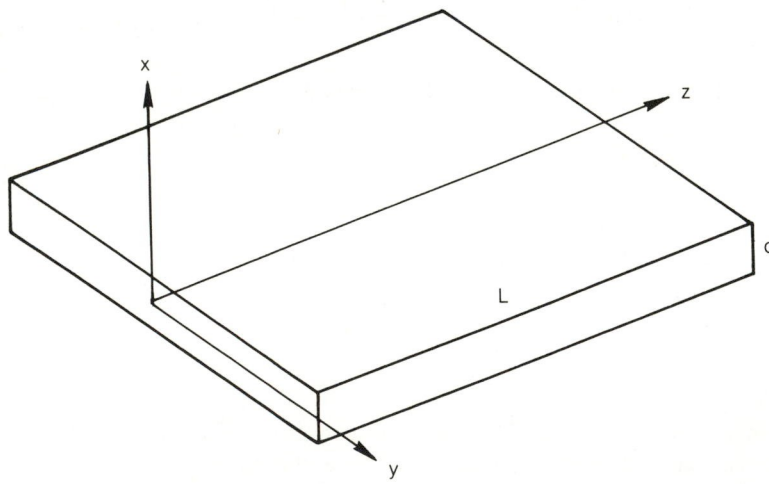

Figure 2.1   Planar waveguide in the yz plane having thickness d and length L.

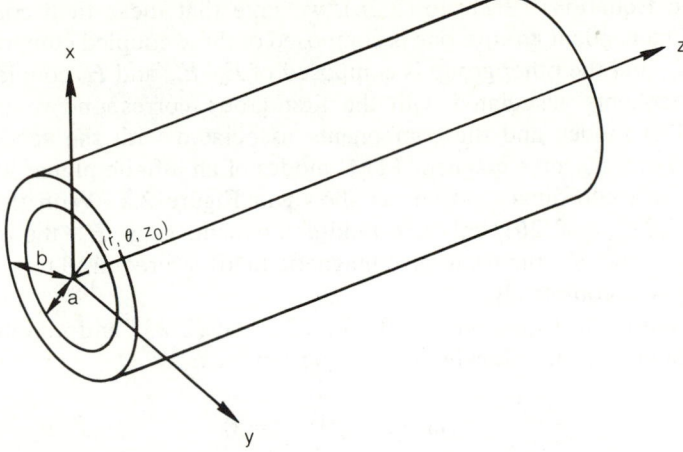

Figure 2.2   Cylindrical waveguide with core radius a and clad radius b.

In free space, $\varepsilon = \varepsilon_0$, $\mu = \mu_0$, and $v_p = c$, and the ratio of the velocity in free space to that in a dielectric medium, where μ is also equal to μ_0, defines the refractive index, $n = \sqrt{\varepsilon/\varepsilon_0}$.

2.4 RELATIONS FOR PLANAR WAVEGUIDES

For an infinite planar waveguide lying in the yz plane, we can assume that $\partial/\partial y = 0$. Substituting Equation (2.18) into (2.14) and (2.15) and resolving into component forms, we have

TE waves
$(E_y,\ H_x,\ H_z)$
$$\begin{cases} \beta E_y = -\mu \omega H_x & (2.21) \\[2ex] -i\beta H_x - \dfrac{\partial H_z}{\partial x} = i\varepsilon \omega E_y & (2.22) \\[2ex] \dfrac{\partial E_y}{\partial x} = -i\mu \omega H_z & (2.23) \end{cases}$$

TM waves
$(E_x,\ E_z,\ H_y)$
$$\begin{cases} \beta H_y = \varepsilon \omega E_x & (2.24) \\[2ex] i\beta E_x + \dfrac{\partial E_z}{\partial x} = i\omega \mu H_y & (2.25) \\[2ex] \dfrac{\partial H_y}{\partial x} = i\varepsilon \omega E_z & (2.26) \end{cases}$$

From Equations (2.21) to (2.26), we note that these field components form two independent groups: one is composed of three coupled components E_y, H_x, and H_z; and the other group is composed of E_x, E_z, and H_y coupled waves. Field components associated with the first group correspond to transverse electric (TE) modes and the components associated with the second group correspond to transverse magnetic (TM) modes of an infinite planar waveguide described in a coordinate system as shown in Figure 2.1. With the help of relations (2.21) to (2.26), only two field components (e.g., E_y, the transverse electric field, and H_y, the transverse magnetic field) are required to specify TE and TM waves completely.

By combining Equations (2.21), (2.22), and (2.23) and eliminating H_x and H_z, we obtain the following wave equation for E_y:

$$\frac{\partial^2 E_y}{\partial x^2} + (\omega^2 \varepsilon \mu - \beta^2) E_y = 0 \tag{2.27}$$

Similarly, by combining Equations (2.24), (2.25), and (2.26) and eliminating E_x and E_z, we obtain an identical wave equation for H_y:

$$\frac{\partial^2 H_y}{\partial x^2} + (\omega^2 \varepsilon \mu - \beta^2) H_y = 0 \tag{2.28}$$

Equations (2.27) and (2.28) indicate, as expected, that E_y and H_y are plane waves of the form

$$E_y = \exp(ik_x x)$$
$$H_y = \exp(ik_x x) \tag{2.29}$$

where k_x is the propagation constant in the x direction with an amplitude defined by the equation

$$k_x = \sqrt{\omega^2 \varepsilon \mu - \beta^2} \tag{2.30}$$

From Equation (2.30) we can write the amplitude of the resultant wave vector $\mathbf{k}$ as

$$|\mathbf{k}| = \sqrt{k_x^2 + \beta^2} = \omega \sqrt{\varepsilon \mu} \tag{2.31}$$

Often $\varepsilon \mu$ is expressed in terms of $K_e K_m \varepsilon_0 \mu_0$, where K_e, the dielectric constant, is approximately equal to n^2, and K_m is approximately equal to unity; and $\varepsilon_0 \mu_0 = 1/c^2$, where c is the velocity of light in vacuum. Therefore, we can rewrite Equation (2.31) as

$$\omega \sqrt{\varepsilon \mu} = nk_0 \tag{2.32}$$

where $k_0 = 2\pi/\lambda_0$, which is the amplitude of the propagation vector in free space.

2.5 RELATIONS FOR CYLINDRICAL WAVEGUIDES

By substituting Equations (2.19) into (2.14) and (2.15) and using cylindrical coordinates for $\nabla \times \mathbf{E}$ and $\nabla \times \mathbf{H}$, we obtain the following expressions:

$$\frac{1}{r}\left(\frac{\partial E_z}{\partial \theta} + i\beta r E_\theta\right) = -i\omega\mu H_r \tag{2.33}$$

$$i\beta E_r + \frac{\partial E_z}{\partial r} = i\omega\mu H_\theta \tag{2.34}$$

$$\frac{1}{r}\left(\frac{\partial r E_\theta}{\partial r} - \frac{\partial E_r}{\partial \theta}\right) = -i\omega\mu H_z \tag{2.35}$$

$$\frac{1}{r}\left(\frac{\partial H_z}{\partial \theta} + i\beta r H_\theta\right) = i\omega\varepsilon E_r \tag{2.36}$$

$$i\beta H_r + \frac{\partial H_z}{\partial r} = -i\omega\varepsilon E_\theta \tag{2.37}$$

$$\frac{1}{r}\left(\frac{\partial r H_\theta}{\partial r} - \frac{\partial H_r}{\partial \theta}\right) = i\omega\varepsilon E_z \tag{2.38}$$

From Equations (2.33) to (2.38) we can solve for r and θ components in terms of z components to obtain the following relationships:

$$E_r = -\frac{i}{k_r^2}\left(\beta\frac{\partial E_z}{\partial r} + \omega\mu\frac{1}{r}\frac{\partial H_z}{\partial \theta}\right) \tag{2.39}$$

$$E_\theta = -\frac{i}{k_r^2}\left(\frac{\beta}{r}\frac{\partial E_z}{\partial \theta} - \omega\mu\frac{\partial H_z}{\partial r}\right) \tag{2.40}$$

$$H_r = -\frac{i}{k_r^2}\left(\beta\frac{\partial H_z}{\partial r} - \omega\varepsilon\frac{1}{r}\frac{\partial E_z}{\partial \theta}\right) \tag{2.41}$$

$$H_\theta = -\frac{i}{k_r^2}\left(\frac{\beta}{r}\frac{\partial H_z}{\partial \theta} + \omega\varepsilon\frac{\partial E_z}{\partial r}\right) \tag{2.42}$$

where k_r is the radial component of the $\mathbf{k}$ vector in the guide and has an amplitude given by

$$k_r = \sqrt{\omega^2\varepsilon\mu - \beta^2} \tag{2.43}$$

From Equations (2.39) to (2.42) we note that as in the case of an optical fiber waveguide, field components in cylindrical coordinates are generally not separable. They are considerably more complex and represented not only by the linearly polarized TE and TM modes introduced previously, but also by hybrid HE and EH modes to be discussed in Chapter 4. However, when the difference in refractive index between core and cladding is very small, we can use a simplified approach in which these coupled relations are ignored. In this way a set of solutions are obtained that describe the modes in terms of a set of linearly polarized fields—the LP modes. This problem is discussed further in Chapter 4. A thorough review of the electromagnetic theory can be found in the first eight chapters of the book by Jackson (Ref. 2.1). A similar treatment for planar and cylindrical waveguides can be found in the book by Midwinter (Ref. 2.2).

PROBLEMS

2.1. Derive Equations (2.21) to (2.26).

2.2. Construct a vector diagram of **k** in both rectangular and cylindrical coordinate systems.

2.3. Derive Equations (2.33) to (2.38).

2.4. Derive Equations (2.39) and (2.41).

REFERENCES

2.1. J. D. Jackson, *Classical Electrodynamics,* John Wiley & Sons, Inc., New York, 1962.

2.2. J. E. Midwinter, *Optical Fibers for Transmission,* John Wiley & Sons, Inc., New York, 1979.

3

Planar Dielectric
Waveguides

3.1 INTRODUCTION

The conditions necessary for establishing guided waves in a planar waveguide are that (1) the waves must be totally reflected at upper and lower boundaries, and (2) the total phase shift after two consecutive reflections must be an integer multiple of 2π. To illustrate the effect of these conditions, we shall first treat the TE and TM propagating waves at a planar boundary and then extend the analysis to a planar waveguide, which consists of two boundaries. One of the pioneering papers on the theory and techniques for the excitation and propagation of guided-wave modes in a planar waveguide was the work by Tien and Ulrich (Ref. 3.1).

3.2 TOTAL INTERNAL REFLECTION

The two TE components parallel to the interface are E_y and H_z, as shown in Figure 3.1. From Equation (2.27) we can express E_y in the form $A_0 e^{-ik_1 x}$ as the incident wave, $B_0 e^{ik_1 x}$ as the reflected wave in the medium with an index of refraction denoted by n_1, and $C_0 e^{-ik_2 x}$ as the transmitted wave into the medium with an index denoted by n_2. By using Equation (2.23) we can write H_z in the form $(k_1 A_0/\mu\omega)e^{-ik_1 x}$ as the incident wave and $-(k_1 B_0/\mu\omega)e^{+ik_1 x}$ as the reflected wave and $(k_2 C_0/\mu\omega)\, e^{-ik_2 x}$ as the transmitted wave. The principle of continuity requires that at the boundary $x = 0$, the total transverse field in one

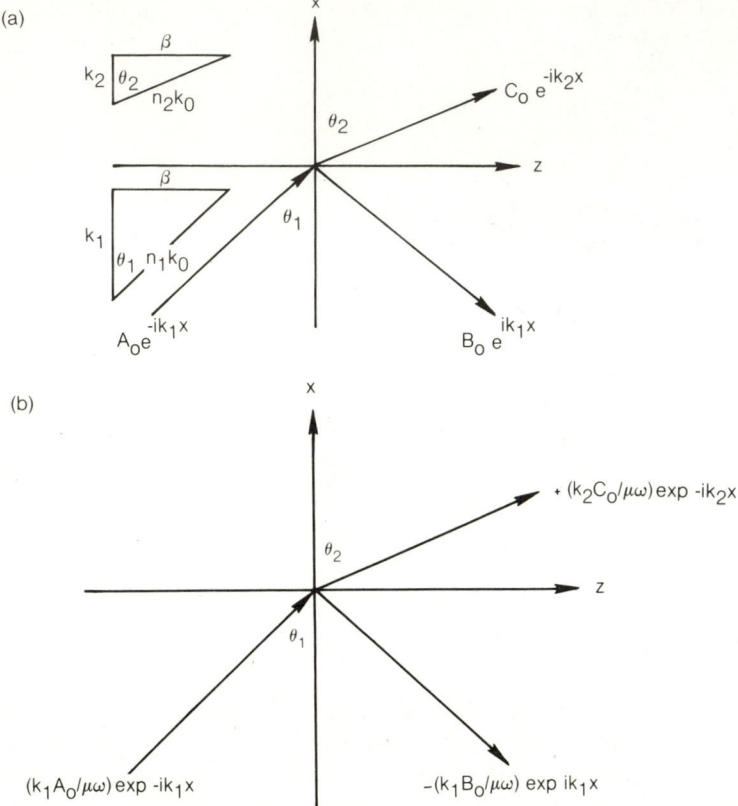

Figure 3.1 TE wave at a planar interface: (a) E_y component; (b) H_z component.

medium must be equal to that in the other medium. We obtain two equations containing three constants A_0, B_0, and C_0 as follows:

$$A_0 + B_0 = C_0 \tag{3.1}$$

$$k_1 A_0 - k_1 B_0 = k_2 C_0 \tag{3.2}$$

Expressing B_0 and C_0 in terms of A_0, we have

$$B_0 = A_0 \frac{k_1 - k_2}{k_1 + k_2} = A_0 r_E \tag{3.3}$$

and

$$C_0 = A_0 \frac{2k_1}{k_1 + k_2} = A_0 t_E \tag{3.4}$$

where r_E and t_E are reflection and transmission coefficients associated with this component. The following relationships between k_1, k_2 and $n_1, n_2, \theta_1, \theta_2$ can be obtained from Figure 3.1(a):

$$k_1 = n_1 k_0 \cos \theta_1 \tag{3.5}$$

$$k_2 = n_2 k_0 \cos \theta_2 \tag{3.6}$$

and

$$k_1^2 - k_2^2 = (n_1^2 - n_2^2)k_0^2 \tag{3.7}$$

Total internal reflection, at which $k_2 = 0$, occurs as the angle of incidence θ_1 is increased to a critical value θ_c. As θ_1 increases further, Equation (3.7) indicates that k_2 must take on imaginary values; therefore, for the case of a guided wave, we let

$$k_2 = -i\gamma_2 \tag{3.8}$$

Substituting Equation (3.8) into (3.3) and letting $Z = k_1 + i\gamma_2$, we get

$$B_0 = A_0 \frac{Z}{Z^*} = A_0 \frac{Z^2}{ZZ^*} = A_0 \frac{Z^2}{|Z|^2} = A_0 e^{i2\phi_E} \tag{3.9}$$

where

$$\phi_E = \tan^{-1} \frac{\gamma_2}{k_1} \tag{3.10}$$

Equation (3.9) indicates that when a wave is totally internally reflected, the phase angle of the reflected wave differs from that of the incident wave by $2\phi_E$. This phenomenon, known as the Goos–Haenchen shift, is caused by the penetration of the wave into the less dense medium before it is totally reflected. To incur a phase shift of $2\phi_E$, it can be shown that the wave has to extend into medium 2 with a penetration depth d equal to $1/\gamma_2$.

We now extend the treatment to TM waves by examining the behavior of the H_y and E_z components at the planar interface. For a dielectric medium, we use $n_1 k_0 = \omega \sqrt{\varepsilon_1 \mu}$ and $n_2 k_0 = \omega \sqrt{\varepsilon_2 \mu}$ in the expressions for H_y and E_z, as shown in Figure 3.2. By equating the incident, reflected, and refracted H_y and E_z waves at $x = 0$, we get

$$A_0 + B_0 = C_0 \tag{3.11}$$

$$\frac{k_1 A_0}{n_1^2} - \frac{k_1 B_0}{n_1^2} = \frac{k_2 C_0}{n_2^2} \tag{3.12}$$

From Equations (3.11) and (3.12), we can express B_0 and C_0 in terms of A_0 as

$$B_0 = A_0 \frac{k_1 n_2^2 - k_2 n_1^2}{k_1 n_2^2 + k_2 n_1^2} = A_0 r_M \tag{3.13}$$

and

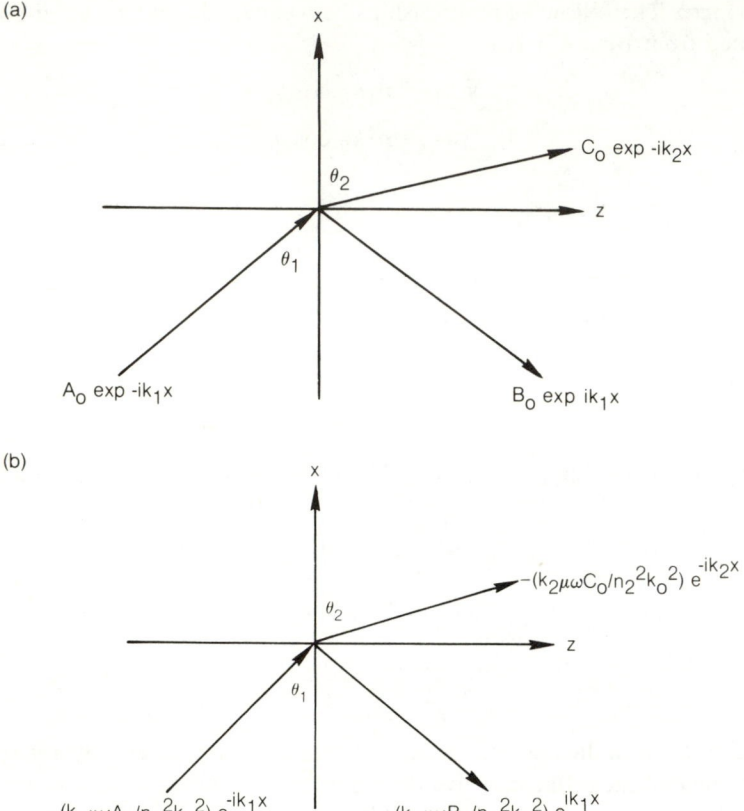

Figure 3.2   TM wave at a planar interface: (a) H_y component; (b) E_z component.

$$C_0 = A_0 \frac{2n_2^2 k_1}{k_1 n_2^2 + k_2 n_1^2} = A_0 t_M \qquad (3.14)$$

In the case of total internal reflection, which occurs if we substitute $k_2 = i\gamma_2 n_1^2$ into Equation (3.13), we obtain the result for TM waves that

$$B_0 = A_0 e^{i2\phi_M} \qquad (3.15)$$

where

$$\phi_M = \tan^{-1} \frac{\gamma_2/n_2^2}{k_1/n_1^2} \qquad (3.16)$$

From Equations (3.15) and (3.16) we see that there will again be a phase shift of $2\phi_M$ for the TM waves when the angle of incidence is increased beyond the critical angle.

3.3 GUIDED-WAVE MODES

The treatment described above can be extended to include both upper and lower boundaries of a planar waveguide. In the case of TE waves, we shall determine the phase shift for E_y and H_z components after reaching the condition satisfying total internal reflection at both upper and lower boundaries. Total internal reflection implies that the amplitude A of the incident E_y wave must be equal to the amplitude E of the reflected wave [see Figure 3.3(a)], provided that there is no loss in the media. As shown in Figure 3.3, expressions for the incident, reflected, and evanescent waves are obtained with the help of Equations (2.23) and (2.27). The term "evanescent" rather than "refracted" is used here, because under the condition that $\theta > \theta_c$, k_2 and k_3 become imaginary and must be replaced by $-i\gamma_2$ and $-i\gamma_3$, respectively. Therefore, the waves that extend beyond the boundaries switch from the oscillatory form to the exponential decaying form. For simplicity we choose a coordinate system such that the lower boundary of the waveguide lies in the $x = 0$ plane and the upper boundary lies in the $x = d$ plane as shown in Figure 3.3.

By equating the fields in medium 1 to those in medium 3 at $x = 0$ and the fields in medium 1 to those in medium 2 at $x = d$, we obtain a set of four equations:

$$B + E = D \tag{3.17}$$

$$-k_1 B + k_1 E = i\gamma_3 D \tag{3.18}$$

$$A e^{-ik_1 d} + B e^{ik_1 d} = C e^{-\gamma_2 d} \tag{3.19}$$

$$k_1 A e^{-ik_1 d} - k_1 B E^{ik_1 d} = -i\gamma_2 C e^{-\gamma_2 d} \tag{3.20}$$

By eliminating D from Equations (3.17) and (3.18), we obtain

$$E = B \frac{k_1 + i\gamma_3}{k_1 - i\gamma_3} = B e^{i2\phi_{13}} \tag{3.21}$$

where

$$\phi_{13} = \tan^{-1} \frac{\gamma_3}{k_1} \tag{3.22}$$

is the phase shift of the guided wave after total internal reflection at the boundary between 1 and 3.

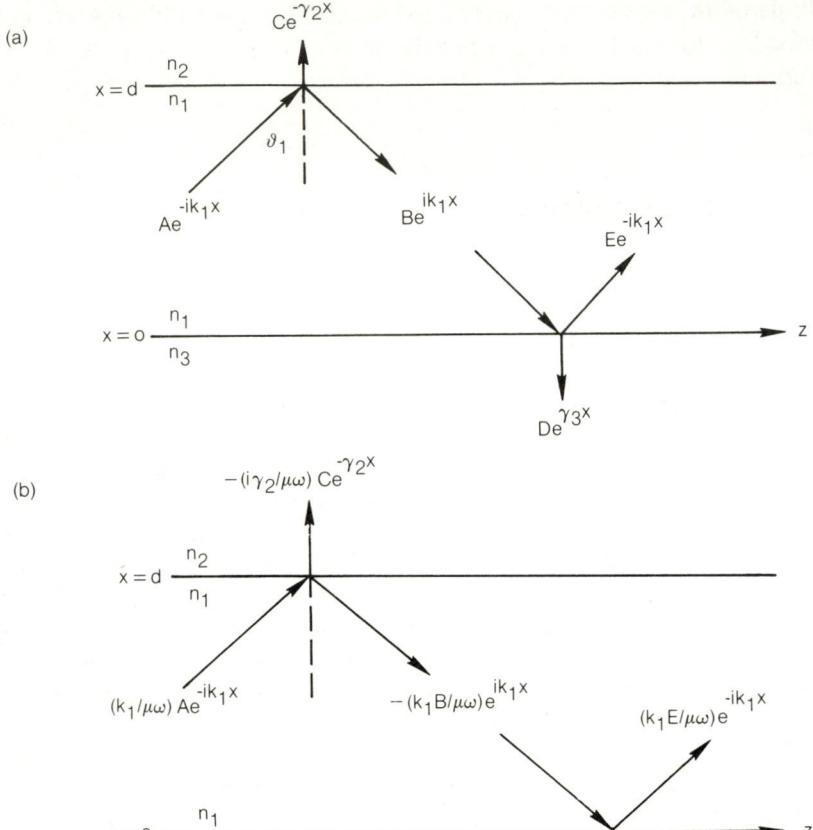

Figure 3.3 Guided TE mode in an asymmetric waveguide with thickness d: (a) E_y component; (b) the H_z component.

By eliminating C from Equations (3.19) and (3.20), we obtain

$$B = A e^{-i2k_1 d} \frac{k_1 + i\gamma_2}{k_1 - i\gamma_2} = A e^{i(2\phi_{12} - 2k_1 d)} \qquad (3.23)$$

where

$$\phi_{12} = \tan^{-1} \frac{\gamma_2}{k_1} \qquad (3.24)$$

The phase shift of the wave after total internal reflection at the boundary between 1 and 2 is equal to $2\phi_{12} - 2k_1 d$. By applying the condition that the total phase shift after two consecutive reflections must be equal to an integer multiple of 2π, we can write the eigenvalue equation for k_1 of the guided-wave TE modes as

$$k_1 = \frac{1}{d}(\phi_{12} + \phi_{13} - m\pi) \qquad (3.25)$$

where $m = 0, 1, 2, \ldots$. Substituting Equation (3.25) into (2.30), where $k_x = k_1$, we obtain β values for various propagating TE modes.

By the same analogy, we can derive an expression for the k_1 values of TM modes. In this case the analysis of H_y and E_z components has to be extended to include both boundaries, as shown in Figure 3.2. The major difference between the results of TE and TM modes is due to the susceptivity associated with these fields. In the TE case, E_y and H_z are related by the coupling coefficient $k_1/\mu\omega$, where μ is a constant; in the TM case, H_y and E_z are related by $k_1/\varepsilon\omega$, where ε takes on a different value for the different medium. Therefore, the eigenvalue equation for TM modes has a slightly modified form:

$$k_1 = \frac{1}{d}\left(\tan^{-1}\frac{\gamma_2 n_1^2}{k_1 n_2^2} + \tan^{-1}\frac{\gamma_3 n_1^2}{k_1 n_3^2} - m\pi \right) \qquad (3.26)$$

where $m = 0, 1, 2, \ldots$.

The number of modes as given by Equation (3.25) for TE waves and Equation (3.26) for TM waves is limited and depends on the thickness of the guide. For a very thin guide, only a finite number of modes are allowed to propagate, while others fall beyond cutoff. This can be explained by the fact that only a few angles of incidence $\theta > \theta_c$ satisfy the phase-shift condition. As d approaches zero, the lowest-order mode is always allowed because its phase-shift angle also approaches zero as both θ and θ_c approach $90°$.

The modal dispersion relationships and their cutoff conditions as a function of waveguide thickness can best be illustrated using a graphic method. We shall present only the graphical solution for TE modes of a symmetric waveguide ($n_2 = n_3$; $\gamma_2 = \gamma_3$) with a thickness d. We shall see in Chapter 8 that such a waveguide structure is commonly used for semiconductor lasers. For convenience, we set the origin of the coordinate system at the center of the planar waveguide. The eigenvalue equation (3.25) can be separated into two independent groups (Ref. 3.3):

$$\frac{k_1 d}{2} \tan \frac{k_1 d}{2} = \frac{\gamma_2 d}{2} \qquad \text{(even } m\text{)} \qquad (3.27)$$

and

$$\frac{k_1 d}{2} \cot \frac{k_1 d}{2} = -\frac{\gamma_2 d}{2} \qquad \text{(odd } m\text{)} \qquad (3.28)$$

From Equations (3.7) and (3.8), we obtain an equation of a circle in the $k_1 d/2$ and $\gamma_2 d/2$ plane as given by

$$\left(\frac{k_1 d}{2}\right)^2 + \left(\frac{\gamma_2 d}{2}\right)^2 = (n_1^2 - n_2^2)\left(\frac{k_0 d}{2}\right)^2 \tag{3.29}$$

where k_1 and γ_2, in accordance with the vector diagram shown in Figure 3.1(a), are related to β in the following ways:

$$k_1^2 = n_1^2 k_0^2 - \beta^2 \tag{3.30}$$

$$\gamma_2^2 = \beta^2 - n_2^2 k_0^2 \tag{3.31}$$

Equations (3.30) and (3.31) show that the requirement for guided modes is

$$n_2^2 k_0^2 < \beta^2 < n_1^2 k_0^2 \tag{3.32}$$

From Equation (3.29) we obtain a family of circles, as shown in Figure 3.4, for three different waveguide thicknesses: 0.2, 1.0, and 1.5 μm. These values are representative of the active-layer thickness of single- and double-heterostructure AlGaAs lasers, emitting at $\lambda = 0.9$ μm. Typical values for n_1 and n_2 of these devices are 3.590 and 3.385, respectively. In Figure 3.4, the dashed curves are plots of Equations (3.27) and (3.28). The points of intersection uniquely define the values of k_1 and γ_2 and also determine the value of β for various modes of the waveguide. For $d = 0.2$ μm, there is only one point of intersection between the circle and the $m = 0$ curve, corresponding to only one eigenvalue for the lowest-order TE_0 mode, while all other modes are beyond cutoff. There is no cutoff for the TE_0 mode no matter how small d becomes. For $d = 1$ μm, there exist TE_0, TE_1, and TE_2 modes with corresponding $k_1 d/2$ values 1.3, 2.5, and 3.6, respectively. For $d = 1.5$ μm there are four modes in the guide and the fifth

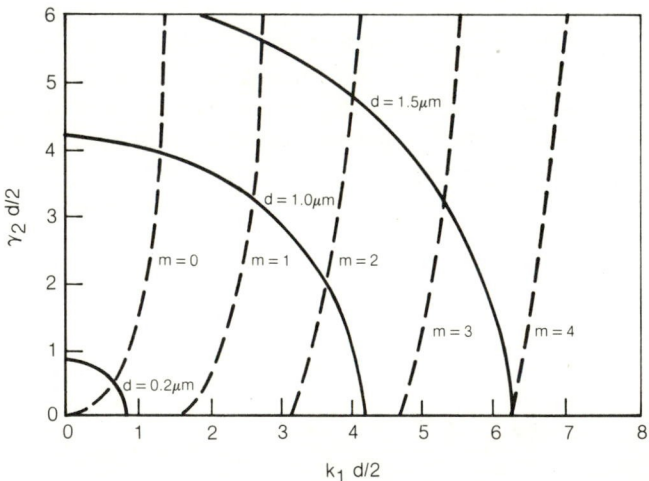

Figure 3.4 Graphical solution of the eigenvalue equation for TE modes in a symmetric $Al_{0.3}Ga_{0.7}As$ double-heterostructure laser. λ is assumed to be 0.9 μm.

TE$_4$ mode is just at cutoff, as indicated by the intersection point occurring at $\gamma_2 d/2 = 0$. From the intersections of the $(k_1 d/2) \tan (k_1 d/2)$ curves with the $k_1 d/2$ axis that occur at $m\pi/2$, we obtain the cutoff condition for the waveguide thickness:

$$d_c = \frac{1}{2} \frac{m\lambda}{(n_1^2 - n_2^2)^{1/2}} \tag{3.33}$$

where $m = 0, 1, 2, \ldots$.

3.4 FIELD EXPRESSIONS FOR PLANAR WAVEGUIDES

There are two independent modes, TE and TM, in a planar waveguide as described by the wave equations (2.27) and (2.28), respectively. The TE mode has its electric polarization transverse to the direction of propagation and is specified by the E_y component. The TM mode has its magnetic polarization transverse to the propagation direction and is specified by the H_y component. Two other components, H_z, H_x for the TE mode and E_x, E_z for the TM mode, can be generated from E_y and H_y. Since the same exponential factor, exp $[i(\omega t - \beta z)]$, appears in all components, it will be omitted from all expressions in the following treatment. We consider here only a symmetric waveguide. Again, for this case it is more convenient to set the origin of the coordinate system at the center of the waveguide and deal with even and odd modes separately.

Within the thickness of the guide $|x| < d/2$, the standing wave of the TE mode has a form that satisfies the wave equation (2.27), as given by

$$E_y(x) = \begin{cases} A_e \cos k_1 x & m = 0, 2, \ldots \\ A_o \sin k_1 x & m = 1, 3, \ldots \end{cases} \tag{3.34}$$

Using Equation (2.23), we write

$$H_z(x) = \begin{cases} -\dfrac{ik_1 A_e}{\mu\omega} \sin k_1 x & m = 0, 2 \ldots \\ \dfrac{ik_1 A_o}{\mu\omega} \cos k_1 x & m = 1, 3 \ldots \end{cases} \tag{3.35}$$

For the field outside the guide, $|x| > d/2$,

$$E_y(x) = \begin{cases} A_e \cos \dfrac{k_1 d}{2} \exp\left[-\gamma_2 \left(|x| - \dfrac{d}{2} \right) \right] & m = 0, 2, \ldots \\ \dfrac{x}{|x|} A_o \sin \dfrac{k_1 d}{2} \exp\left[-\gamma_2 \left(|x| - \dfrac{d}{2} \right) \right] & m = 1, 3, \ldots \end{cases} \tag{3.36}$$

and

$H_z(x) =$

$$\begin{cases} -i\dfrac{x}{|x|} \dfrac{\gamma_2 A_e}{\mu\omega} \cos\dfrac{k_1 d}{2} \exp\left[-\gamma_2\left(|x| - \dfrac{d}{2}\right)\right] & m = 0, 2, \ldots \\[4mm] -i\dfrac{\gamma_2 A_o}{\mu\omega} \sin\dfrac{k_1 d}{2} \exp\left[-\gamma_2\left(|x| - \dfrac{d}{2}\right)\right] & m = 1, 3, \ldots \end{cases}$$

$$(3.37)$$

The variation of the electric field E_y as a function x is shown in Figure 3.5(a) for a symmetric waveguide of thickness 1 μm. For even modes, the maximum field occurs at the center. For odd modes, the field amplitude always has a zero at the center. All field amplitudes are modulated by the exponential term $\exp[i(\omega t - \beta z)]$. Figure 3.5(b) shows the intensity distribution for these modes. These curves, shown in Figure 3.5(b), correspond to the variation of the square of the electric field as a function of x.

3.5 POWER DISTRIBUTION AND CONFINEMENT FACTOR

Figure 3.5 shows that the electric field and power of a guided-wave mode can extend beyond the geometric boundary of the guide and the amount is dependent on the thickness and the mode number. As we shall see in Chapter 8, it is important to determine the fraction of an optical mode within the guiding layer of a semiconductor laser in which strong interaction exists between the field and the gain medium. In this section we evaluate the spatial extent of the light intensity within the guide. It is measured by the confinement factor Γ, which is defined by the ratio of light intensity within the layer to the total light intensity.

The light intensity I is proportional to the magnitude of the Poynting vector $\mathbf{P}$, defined by

$$\mathbf{P} = \mathbf{E} \times \mathbf{H} \qquad (3.38)$$

where $\mathbf{E}$ and $\mathbf{H}$ are usually expressed in the form of complex functions as given by Equation (2.18). However, one must realize that actual fields are the real part of the complex quantity with sinusoidally varying functions oscillating in time. The energy density associated with these fields can be obtained by taking the time average of the product of the real parts of $\mathbf{E}$ and $\mathbf{H}$ vectors. This is equivalent to one-half the real part of the product of one vector and the complex conjugate of the other. Therefore, the time average of the Poynting vector along the z-axis is given by

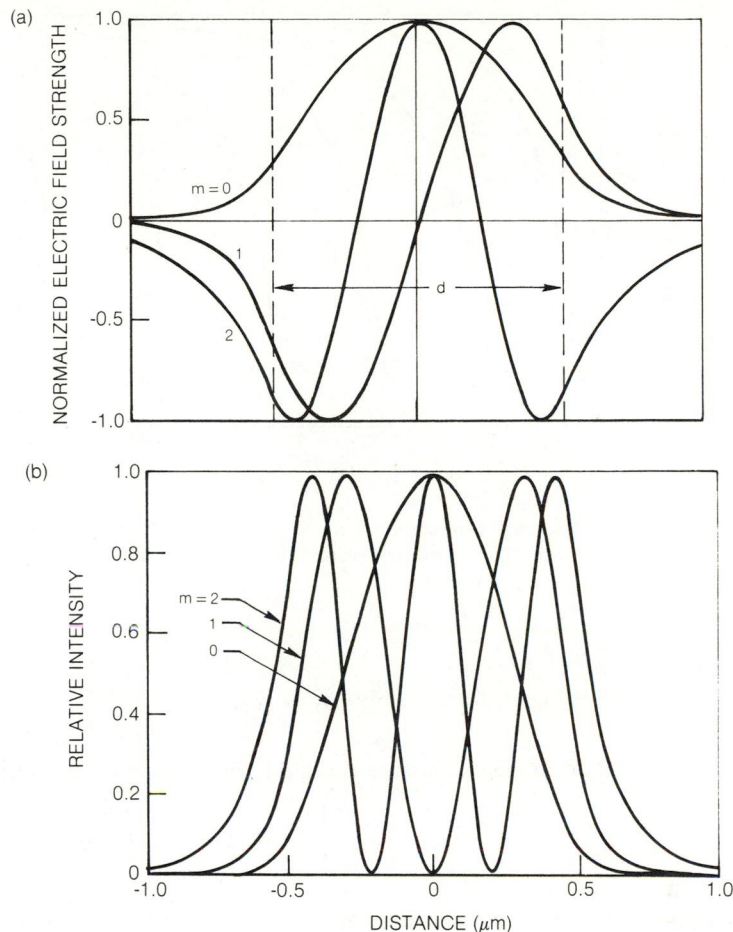

Figure 3.5 (a) Plot of normalized electric field for the first three modes of a symmetric waveguide with parameters $n_1 = 3.590$, $n_2 = 3.385$, and $\lambda = 0.9\ \mu m$; (b) corresponding intensity distribution, which is the square of electric field strength.

$$P_z = \tfrac{1}{2} \int (E_x H_y - H_x E_y)\,dx \qquad (3.39)$$

For TE modes, E_x and H_y vanish and using Equation (2.21), we obtain

$$P_z = \frac{1}{2}\ \frac{\beta}{\mu\omega} \int E_y^2\,dx \qquad (3.40)$$

Now, taking the real part of E_y for even modes of TE waves as given by Equations (3.34) and (3.36), we obtain the fraction of light intensity inside the guide as denoted by the integral I_{in}:

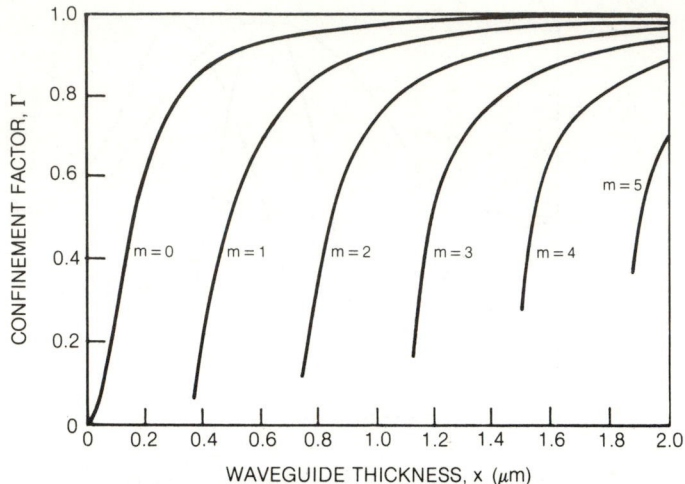

Figure 3.6 Variations of confinement factor Γ as a function of waveguide thickness for various mode orders in a symmetric planar waveguide, having $n_1 = 3.590$ and $n_2 = 3.385$. λ is assumed to be 0.9 μm.

$$I_{\text{in}} = \frac{\beta A_e^2}{\mu \omega} \int_0^{d/2} \cos^2 k_1 x \, dx \tag{3.41}$$

The fraction of light intensity outside the guide is denoted by the integral I_{out}:

$$I_{\text{out}} = \frac{\beta A_e^2}{\mu \omega} \int_{d/2}^{\infty} \cos^2 k_1 \frac{d}{2} \exp\left[-2\gamma_2 \left(x - \frac{d}{2}\right)\right] dx \tag{3.42}$$

By definition, we obtain the confinement factor Γ for even TE modes:

$$\Gamma = \left\{ 1 + \frac{\cos^2(k_1 d/2)}{\gamma_2 \left[d/2 + \left(\dfrac{1}{k_1}\right) \sin(k_1 d/2) \cos(k_1 d/2)\right]} \right\}^{-1} \tag{3.43}$$

which is the ratio of light intensity within the waveguide layer to the total light intensity.

From Equation (3.43) it is interesting to note that when $\gamma_2 = 0$, Γ approaches zero; therefore, the entire power resides in the cladding. This is consistent with the cutoff condition for modes. Because of the difference in k_1 and γ_2 values among various modes, Γ values vary significantly with m values; however, the differences in Γ values for TE and TM waves of the same m value remain negligibly small. As the guiding-layer thickness increases from zero, the Γ value for the lowest-order mode increases rapidly and reaches a saturation level. A further increase in thickness allows the growth of the next-higher-order mode. This behavior is shown in Figure 3.6. The confinement factor is plotted for various m values as a function of the guiding-layer thickness of a symmetric planar waveguide, with $n_1 = 3.590$, $n_2 = 3.385$, and $\lambda = 0.9\ \mu$m.

PROBLEMS

3.1. Show that if $\theta_1 + \theta_2 = \pi/2$, r_M defined in Equation (3.13) is zero or t_M is unity. This result is known as Brewster's condition for the TM wave.

3.2. Derive Equations (3.21) and (3.23).

3.3. Establish the eigenvalue equation for TE modes of a symmetric planar waveguide (e.g., $n_2 = n_3$).

3.4. Obtain the eigenvalue equation for TM modes of a asymmetric planar waveguide.

3.5. Construct the vector diagram of **k** for a guided wave both inside and outside the guide.

3.6. From the curves of Figure 3.4, calculate the β values for the 1-μm guide.

3.7. Calculate the fraction of the light intensity inside the guiding layer for the even TE modes.

3.8. Calculate the fraction of the light intensity outside the guiding layer for the even TE mode.

3.9. Calculate the Γ factor for the odd TE modes.

REFERENCES

3.1. P. K. Tien, and R. Ulrich, *J. Opt. Soc. Am., 60,* 1325 (1970).

3.2. J. E. Midwinter, *Optical Fibers for Transmission,* John Wiley & Sons, Inc., New York, 1979.

3.3. N. C. Casey, Jr. and M. B. Panish, *Heterostructure Lasers,* Part A: *Fundamental Principles,* Academic Press, New York, 1978.

4

Cylindrical Dielectric Waveguides

4.1 INTRODUCTION

Exact solutions of Maxwell's equations can be obtained for a cylindrical dielectric waveguide with a step-index profile as shown in Figure 1.2(a), by using appropriate boundary conditions applied to the transverse field components. However, the results are considerably more complex than those obtained for planar waveguides, because in this case all the field components are coupled, and for each axial propagation constant β, there are two sets of nearly degenerate modes. To illustrate this degeneracy, one can visualize the situation from a ray picture by following two opposite helical paths in the fiber. Because of circular symmetry, these modes with opposite helicity have almost the same β value. These helical waves are called hybrid EH or HE modes. The notation EH implies that the z component of the E field is larger than that of the H field. The converse is true for HE modes. It should be noted that both E_z and H_z are usually very small compared with other components. The polarization associated with these modes is circular and rotating either clockwise or counterclockwise in accordance with the helicity. Other groups of modes associated with meridional rays are the TE and TM linearly polarized waves. In this case the field components have no angular or azimuthal dependence. Besides guided-wave modes there are radiation and leaky-wave modes, which are also solutions of Maxwell's equations, but will not be treated. We start with a simplified analysis to establish a simple eigenvalue equation for a weakly guiding fiber ($n_0 \simeq n_c$). With these results we can obtain approximate solutions for two cases: one is close to cutoff and the other is far from cutoff. We then proceed to establish the exact eigenvalue equation by which a complete characterization of hybrid modes can be achieved with considerable complexity. In all cases the results indicate that there exist an enormous number of modes in a multimode fiber, each of which propagates at a distinct group velocity. If a short pulse of light is launched at one end of a multimode fiber, the width of this pulse will be broadened primarily as a result of the modal dispersion. However, a graded-index fiber, if made with a correct index profile, can greatly suppress the modal dispersion and consequently reduce the group delay. An alternative approach is, of course, to use a single-mode step-index fiber. But the disadvantage of using such a fiber is the difficulty encountered in efficiently splicing and optically coupling a fiber that has a core diameter of only a few micrometers. Dispersion, coupling, and splicing will be discussed later. In this chapter we introduce some important properties related to the propagation characteristics of light in both step-index and graded-index fibers. Because of the complexity of wave equations, which contain a term functionally dependent on the core radius, it is impossible to obtain exact solutions for the case of graded-index fibers. Instead, we treat this problem using simple ray analysis. With this approach we can trace light paths precisely for either meridional or skew rays inside the fiber with a given graded-index profile. Conceptually, we can visualize a bundle of rays that strike the cylindrical boundary at different angles, each of which represents a mode in the wave description. Lower-order modes strike the boundary at angles near grazing and therefore travel very close to the core of the fiber. If the core index is constant, lower-order modes will travel at faster velocities than will higher-order modes. To slow down lower-order modes or speed up higher-order modes, a parabolic index profile such as the one shown in Figure 1.2(b) can be used. However, the exact shape of this profile has to be determined for the least dispersion. We shall introduce a procedure for selecting a proper index profile in ray analysis by requiring that all rays propagate with nearly the same period. Following the ray analysis, we introduce a method known as the Wentzel–Kramers–Brillouin (WKB) approximation by which we can determine β values for graded-index fibers. An explicit expression for β is needed to evaluate the effects of modal dispersion on the information-carrying capacity of a fiber.

4.2 SCALAR FIELD SOLUTIONS FOR STEP-INDEX FIBERS

For simplicity we first solve the wave equation for a cylindrical waveguide with a step-index profile in the approximation of a scalar field by neglecting the complications of coupled fields. In this approximation, each transverse component ψ of the electric field obeys the scalar Helmholtz equation, which can be derived directly from Equation (2.16) or (2.17). By eliminating the time dependence, we write

$$[\nabla^2 + k_0^2 n^2(r)]\psi = 0 \tag{4.1}$$

where ∇^2 is the Laplacian operator. For cylindrical coordinates, we write

$$\nabla^2 = \frac{d^2}{dr^2} + \frac{1}{r}\frac{d}{dr} + \frac{1}{r^2}\frac{d^2}{d\theta^2} + \frac{d^2}{dz^2} \tag{4.2}$$

Because of the axial and circular symmetry of the fiber, we assume a solution of Equation (4.1) of the form

$$\psi = \psi(r)\exp[i(l\theta + \beta z)] \tag{4.3}$$

where l is the azimuthal eigenvalue and β is the propagation wave number along the z axis of a fiber with a core radius a. Substituting Equation (4.3) into (4.1), we obtain for the case of a step-index profile:

$$\frac{d^2\psi}{dr^2} + \frac{1}{r}\frac{d\psi}{dr} + \left(k_0^2 n_0^2 - \beta^2 - \frac{l^2}{r^2}\right)\psi = 0 \qquad r \le a \tag{4.4}$$

$$\frac{d^2\psi}{dr^2} + \frac{1}{r}\frac{d\psi}{dr} + \left(k_0^2 n_c^2 - \beta^2 - \frac{l^2}{r^2}\right)\psi = 0 \qquad r > a \tag{4.5}$$

To simplify the equations above, we define

$$u^2 \equiv (k_0^2 n_0^2 - \beta^2)a^2 \tag{4.6}$$

$$\gamma^2 \equiv (\beta^2 - k_0^2 n_c^2)a^2 \tag{4.7}$$

We get an important parameter V for the fiber:

$$V = (u^2 + \gamma^2)^{1/2} = k_0 a(n_0^2 - n_c^2)^{1/2}$$
$$= \frac{2\pi a}{\lambda}(NA) \tag{4.8}$$

where NA is known as the numerical aperture of a step-index fiber.

The solutions of Equations (4.4) and (4.5) are well known. For bounded solutions, we choose the first kind of Bessel function of order l for Equation (4.4) and the second kind of modified Bessel function of order l for Equation (4.5). They are

$$\psi(r) = AJ_l\left(\frac{ur}{a}\right) \qquad r < a \tag{4.9}$$

$$\psi(r) = BK_l\left(\frac{\gamma r}{a}\right) \qquad r > a \tag{4.10}$$

where J_l is given in Problem 4.2 and $K_l(x) = (\pi/2)i^{-(l+1)}H_l(-ix)$. $H_l(-ix)$ is known as the Hankel function, which is a linear combination of Bessel functions of the first kind J_l and of the second kind Y_l. The choice of these functions is clear from the following asymptotic forms. For $x \ll 1$,

$$J_l(x) = \frac{1}{l!} \left(\frac{x}{2} \right)^l \qquad l = 0, 1, \ldots \qquad (4.11)$$

$$K_l(x) = (l - 1)! 2^{l-1} x^{-l} \qquad l \geq 1 \qquad (4.12)$$

and for $x \gg 1$,

$$J_l(x) = \sqrt{\frac{2}{\pi x}} \cos \left[x - \frac{\pi(2l + 1)}{4} \right] \qquad (4.13)$$

$$K_l(x) = \sqrt{\frac{2}{\pi x}} e^{-x} \left(1 + \frac{4l^2 - 1}{8x} \right) \qquad (4.14)$$

These results indicate that J_l and K_l are well-behaved functions inside and outside the boundary of a cylindrical waveguide, respectively. Other Bessel functions are not suitable for representing this waveguide. For further analysis, the following recurrence relations for these functions (with argument x) and some asymptotic forms that are very useful are

$$J_{-l} = (-1)^l J_l \qquad (4.15)$$

$$J_l' = \tfrac{1}{2} (J_{l-1} - J_{l+1}) = \pm J_{l\mp 1} \mp \frac{l J_l}{x} \qquad (4.16)$$

$$J_{l\mp 1} = \frac{2l J_l}{x} - J_{l\pm 1} \qquad (4.17)$$

$$J_{l\mp 2} = \frac{2(l \mp 1) J_{l\mp 1}}{x} - J_l \qquad (4.18)$$

$$K_l = K_{-l} \qquad (4.19)$$

$$K_l' = -\tfrac{1}{2} (K_{l-1} + K_{l+1}) = \mp \frac{l K_l}{x} - K_{l\mp 1} \qquad (4.20)$$

$$K_{l\mp 1} = \mp \frac{2l K_l}{x} + K_{l\pm 1} \qquad (4.21)$$

$$K_{l\mp 2} = \mp \frac{2(l \mp 1) K_{l\mp 1}}{x} + K_l \qquad (4.22)$$

For $x \ll 1$,

$$\frac{K_0}{K_1} = x \ln \frac{2}{1.782x} \qquad (4.23)$$

$$\frac{K_{l-1}}{K_l} = \frac{x}{2(l-1)} \qquad l \geq 2 \tag{4.24}$$

$$\frac{K_{l+1}}{K_l} = \frac{2l}{x} \qquad l \geq 1 \tag{4.25}$$

For $x \gg 1$,

$$\frac{K_{l\mp 1}}{K_l} = 1 + \frac{1 \mp 2l}{2x} \tag{4.26}$$

The continuity of ψ and its derivative at the boundary of the core and the cladding (e.g., $r = a$) leads to a simple eigenvalue equation of the form

$$\frac{u J_l'(u)}{J_l(u)} = \frac{\gamma K_l'(\gamma)}{K_l(\gamma)} \tag{4.27}$$

Substituting Equations (4.16) and (4.20) into (4.27), we obtain two equivalent eigenvalue equations of the form

$$\frac{u J_{l\pm 1}(u)}{J_l(u)} = \pm \frac{\gamma K_{l\pm 1}(\gamma)}{K_l(\gamma)} \tag{4.28}$$

For various values of l, we can obtain from Equation (4.28) the corresponding values of u and γ, either of which leads to the corresponding value for the propagation wave number β, lying within the range

$$n_c < \frac{\beta}{k_0} < n_0 \tag{4.29}$$

As we shall see in the next section, this is a special case of a cylindrical dielectric waveguide for which $n_0 \simeq n_c$. Due to the oscillatory nature of Bessel functions, for every value of l there exists m allowed solutions for β. Therefore, each allowed β value is characterized by two integers l and m. The first integer, l, associated with two circular functions, $\cos l\theta$ and $\sin l\theta$, and the second integer, m, corresponds to the mth root of the eigenvalue equation. By convention, $HE_{l,m}$ modes are those whose longitudinal electric field dominates and $EH_{l,m}$ modes are those whose longitudinal magnetic field dominates. If the longitudinal fields are zero, as in the case of very weakly guiding fibers, we have linearly polarized LP modes. It is to be noted that the eigenvalue equation (4.28) contains $\pm$ signs. The plus sign usually goes with $HE_{l+1,m}$ modes and the minus sign goes with $EH_{l-1,m}$ modes.

Equation (4.28) implies that $HE_{l+1,m}$ modes are degenerate with $EH_{l-1,m}$ modes, because the same equation applies for both modes. However, for $l = 0$, the $HE_{1,m}$ modes have special significance because $EH_{-1,m}$ modes do not exist. Therefore, $HE_{1,m}$ modes are nondegenerate. In the following, we calculate the β value for the HE_{11} mode for only two limiting cases. For the first case, we

TABLE 4.1   First Four Zeros of J_0 and J_1 Functions

		m		
l	1	2	3	4
0	0	3.832	7.016	10.173
1	2.405	5.520	8.654	11.790

establish a simple relationship between β and V near cutoff. This is the case when $\gamma = 0$. For the second case, we let $\gamma = \infty$. This is the situation when the mode is considered to be far above cutoff.

If $\gamma = 0$, we obtain from Equation (4.28) that $J_{l\pm1}(u) = 0$. If $l = 0$, the lowest root of $J_{\mp1}(u)$ is that for $a = 0$. This means that there is no cutoff for the lowest-order HE_{11} mode in a fiber. The cutoff for the next-order mode in this group ($l = 0$) is at $u_m = 3.832$. However, before this mode, the first cutoff of another group ($l = 1$) will occur at $u_m = 2.405$, which is the first root of $J_1(u)$. Table 4.1 is a list of cutoff frequencies for a few lower-order modes. Figure 4.1 shows the oscillatory behavior of J_0 and J_1 functions and the roots of these two functions.

From Equation (4.28), we can write for $l = 0$,

$$\frac{uJ_1(u)}{J_0(u)} = \frac{\gamma K_1(\gamma)}{K_0(\gamma)}$$

Using Equation (4.23) for the limiting value near cutoff, we can write

$$\frac{J_0(V)}{VJ_1(V)} = -\ln\frac{1.782\gamma}{2}$$

where we have used the approximation $u \simeq V$. Therefore, we obtain

$$\gamma = 1.122 \exp\left[-\frac{J_0(V)}{VJ_1(V)}\right] \tag{4.30}$$

This approximation is as good as the exact solution (Ref. 4.1) for values of V ranging from zero to 1, and is off by only about 10% as V approaches 2.5.

For the case in which the mode is far from cutoff, we let $\gamma \to \infty$. From Equation (4.26) we have

$$\lim_{\gamma \to \infty}\frac{K_0(\gamma)}{K_1(\gamma)} \simeq 1$$

Substituting this limiting value into Equation (4.28) and letting $\gamma \simeq V$, we get

$$VJ_0(u) = uJ_1(u) \tag{4.31}$$

Further simplification can be made by eliminating J_0 and J_1 from Equation

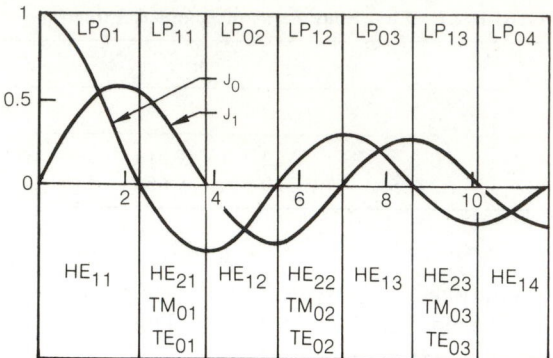

Figure 4.1   Plot of the Bessel J_0 and J_1 functions.

(4.31). We first differentiate Equation (4.31) with respect to V and make use of the identities

$$J_1'(x) = J_0 - \frac{1}{x} J_1(x)$$

$$J_0'(x) = - J_1(x)$$

We get

$$\frac{du}{dV} \left[u + V \frac{J_1(u)}{J_0(u)} \right] = 1$$

We can now eliminate $J_1(u)/J_0(u)$ from the equation above by again using Equation (4.31) and obtain

$$\frac{du}{u} (u^2 + V^2) = dV$$

In the limit as $\gamma \to \infty$, $u \ll V$; therefore, we can neglect the u^2 term in the equation above and obtain a simple expression for u after integrating the equation above from (u_{0m}, V) to (u_{0m}^∞, ∞), giving

$$u_{0m} = u_{0m}^\infty \exp\left(-\frac{1}{V} \right) \tag{4.32}$$

where u_{0m}^∞ is the root of $J_0(u)$ for V values far from cutoff. Similar relationships can be obtained for higher-order modes by extending the foregoing analysis for $l > 1$. In the limit as $\gamma \to \infty$, we obtain from the eigenvalue equation (4.28)

$$\gamma J_l(u) = u J_{l+1}(u)$$

We again take the derivative of the equation above with respect to V and

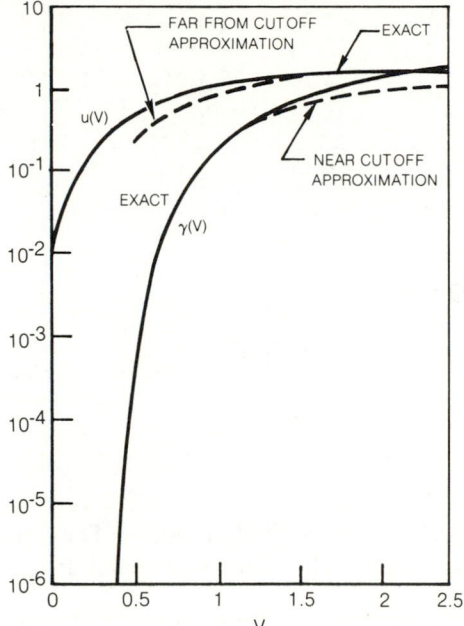

Figure 4.2 Plot of u and γ as a function of V for the HE_{11} mode.

eliminate the Bessel functions from the equation by using the same approximation. We obtain

$$\frac{du}{dV} = \frac{u}{V[V - 2(l - 1)]}$$

Upon integration we obtain the approximate solution in the limit as $\gamma \to \infty$ for HE modes ($l > 1$),

$$u_{lm} = u_{lm}^{\infty} \left[1 - \frac{2(l - 1)}{V} \right]^{1/2(l-1)} \tag{4.33}$$

where u_{lm}^{∞} is the mth root of $J_l(u)$, and for large m, the value of u_{lm}^{∞} can be approximated by the expression

$$u_{lm}^{\infty} \simeq (l + 2m)\frac{\pi}{2} \tag{4.34}$$

Figure 4.2 is a plot of u and γ as a function of V for the HE_{11} mode. The approximate solutions are shown by dashed curves. We see from Figure 4.2 that the results given by the approximation near cutoff coincide with the exact solution for $V \leq 1$, and the approximation far from cutoff is very good for $V \geq 1.5$. The next two higher-order modes are the transverse electric (TE_{01})

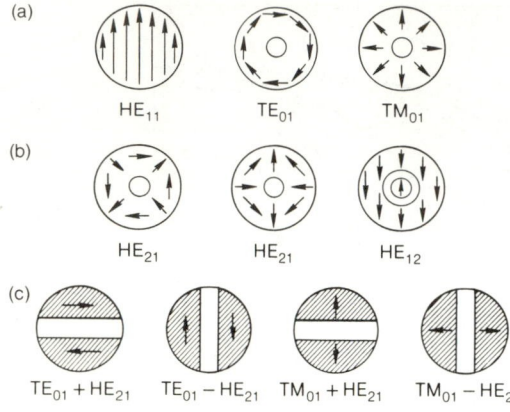

Figure 4.3   Electric vectors of several lowest-order modes of a step-index fiber.

and transverse magnetic (TM_{01}) modes. They are the two lowest cylindrically symmetric modes. The electric vectors are oriented as shown in Figure 4.3(a).

In general, TE_{0m} and TM_{0m} are two linearly polarized sets of modes with polarization vectors perpendicular to each other. Therefore, they can easily be distinguished from each other by using a polarizer to analyze the orientation of the electric vectors. The next higher-order mode is HE_{21}. It is a hybrid mode with a twofold degeneracy, which is a consequence of circular symmetry as shown in Figure 4.3(b). There are two possible field configurations for the HE_{21} mode as shown in Figure 4.3(b) and they can be obtained simply by rotating one 90° with respect to the other. Because TE_{01}, TM_{01}, and HE_{21} modes have nearly the same β-values and cutoff characteristics (see Figure 4.5), they usually occur simultaneously. Figure 4.3(c) shows four independent linear combinations of these three modes. The first two combinations have their resultant electric vector parallel to the null line, whereas the second two combinations have their resultant electric vector perpendicular to the null line. Again it is possible to distinguish TE_{0m} and TM_{0m} modes from HE_{2m} modes by rotating an analyzer placed at the entrance to the fiber. As the analyzer rotates, the null line will rotate in the same direction for the $l = 0$ modes but in the opposite direction for the HE_{2m} modes.

The third group of higher-order modes are HE_{12}, EH_{11}, and HE_{31} modes with the same cutoff value at $u_c = 3.832$. As the diameter of the core increases further ($V > 3.832$), more and more modes are introduced with values of β much closer to each other. As a result, spatial resolution of these modes becomes extremely difficult to achieve. Figure 4.4 shows a few typical lower-order mode patterns, some of which are observed with the help of polarizers.

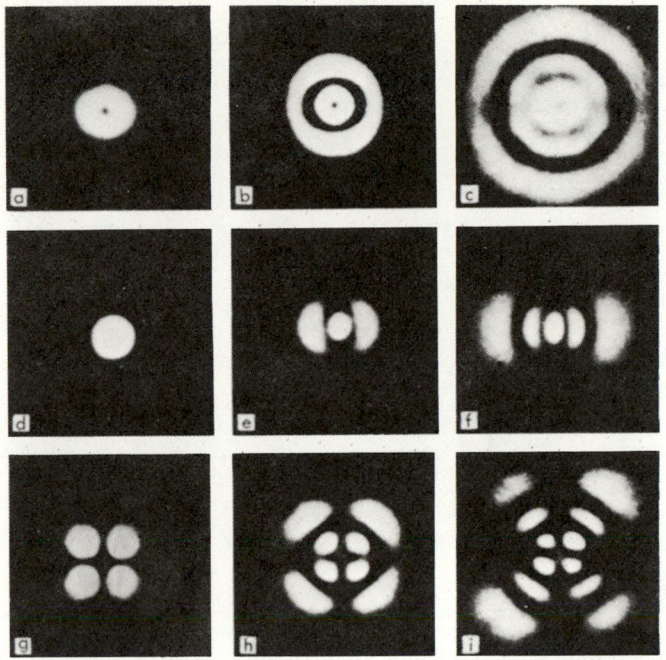

Figure 4.4   Images of several lower-order modes: (a), (b), and (c) are either TE_{0m}, TM_{0m}, or HE_{2m} ($m = 1,2,3$); (d) is the HE_{11} mode; (e) is a linear combination of HE_{12} with either HE_{31} or EH_{11}; (f) is a linear combination of HE_{13} with either HE_{32} or EH_{12}; (g), (h), and (i) are either EH_{1m} or HE_{3m} ($m = 1,2,3$). (Courtesy of Dr. Elias Snitzer, United Technologies Research Center.)

4.3 APPROXIMATION
FOR WEAKLY GUIDED STEP-INDEX FIBERS

Next, we carry out a more rigorous analysis of the modes in a cylindrical dielectric waveguide with a very small refractive index difference between the cladding and the core by following the treatment of Gloge (Ref. 4.2). He recognized the fact that the field expressions are considerably simpler in appearance if they are expressed in Cartesian coordinates rather than in cylindrical coordinates. The eigenvalue equation obtained in this treatment is identical to Equation (4.28) if the assumption $\Delta \ll 1$ is made. This approximation is valid for most fibers of practical interest.

In this approximation, transverse field components are assumed to be essentially linearly polarized. These transverse field components for the LP modes can be expressed in the following form:

$$E_y = \begin{cases} \dfrac{E_l}{J_l(u)} \, J_l\left(\dfrac{ur}{a}\right) \cos l\theta & \text{for } r < a \qquad (4.35) \\[4mm] \dfrac{E_l}{K_l(\gamma)} \, K_l\left(\dfrac{\gamma r}{a}\right) \cos l\theta & \text{for } r > a \qquad (4.36) \end{cases}$$

where E_l is the electric field strength at the core–cladding interface. Using Equation (2.21), we write

$$H_x \simeq \begin{cases} -\dfrac{n_0}{Z_0} E_y & \text{for } r < a \qquad (4.37) \\[4mm] -\dfrac{n_c}{Z_0} E_y & \text{for } r > a \qquad (4.38) \end{cases}$$

where

$$Z_0 = \frac{\omega\mu}{k_0} \qquad (4.39)$$

which is the plane-wave impedance in vacuum. For Equations (4.37) and (4.38) an approximation has been made by letting $\beta = n_0 k_0 = n_c k_0$. Furthermore, the choice of cos $l\theta$ instead of sin $l\theta$ for Equations (4.35) and (4.36) is entirely arbitrary. In general, there are two orthogonal states of polarization, and two orthogonal circular functions that form four sets of modes for $l \neq 0$. For $l = 0$, we have only two sets of modes polarized orthogonally with respect to each other.

In this approximation ($\Delta \ll 1$), E_x and H_y are very small compared to E_y and H_x. Therefore, the longitudinal components can be obtained from Equations (2.14) and (2.15) by a coordinate transformation as follows:

$$E_z = -\frac{iE_l u}{2k_0 a n_0} \left[\frac{J_{l+1}(ur/a)}{J_l(u)} \sin(l+1)\theta \right. \\ \left. + \frac{J_{l-1}(ur/a)}{J_l(u)} \sin(l-1)\theta \right] \qquad \text{for } r < a \qquad (4.40)$$

$$E_z = -\frac{iE_l \gamma}{2k_0 a n_c} \left[\frac{K_{l+1}(\gamma r/a)}{K_l(\gamma)} \sin(l+1)\theta \right. \\ \left. - \frac{K_{l-1}(\gamma r/a)}{K_l(\gamma)} \sin(l-1)\theta \right] \qquad \text{for } r > a \qquad (4.41)$$

$$H_z = -\frac{iE_l u}{2k_0 Z_0 a}\left[\frac{J_{l+1}(ur/a)}{J_l(u)}\cos(l+1)\theta\right.$$

$$\left. -\frac{J_{l-1}(ur/a)}{J_l(u)}\cos(l-1)\theta\right] \qquad \text{for } r < a$$

(4.42)

$$H_z = -\frac{iE_l \gamma}{2k_0 Z_0 a}\left[\frac{K_{l+1}(\gamma a/r)}{K_l(\gamma)}\cos(l+1)\theta\right.$$

$$\left. +\frac{k_{l-1}(\gamma a/r)}{K_l(\gamma)}\cos(l-1)\theta\right] \qquad \text{for } r > a$$

(4.43)

In Equations (4.40) to (4.43), the amplitude of the factors u/k_0 and γ/k_0 are both of the order $\sqrt{\Delta}$. From these longitudinal components we can also generate the transverse components in the cylindrical coordinate system by using Equations (2.40) and (2.42), as given by

$$E_\theta = \begin{cases} \frac{1}{2}E_l\dfrac{J_l(ur/a)}{J_l(u)}[\cos(l+1)\theta + \cos(l-1)\theta] & \text{for } r < a \quad (4.44) \\[4mm] \frac{1}{2}E_l\dfrac{K_l(\gamma r/a)}{K_l(\gamma)}[\cos(l+1)\theta + \cos(l-1)\theta] & \text{for } r > a \quad (4.45) \end{cases}$$

$$H_\theta = \begin{cases} -\frac{1}{2}\dfrac{E_l n_0}{Z_0}\dfrac{J_l(ur/a)}{J_l(u)}[\sin(l+1)\theta - \sin(l-1)\theta] & \text{for } r < a \\[2mm] & (4.46) \\[2mm] -\frac{1}{2}\dfrac{E_l n_0}{Z_0}\dfrac{K_l(\gamma r/a)}{K_l(\gamma)}[\sin(l+1)\theta - \sin(l-1)\theta] & \text{for } r > a \\[2mm] & (4.47) \end{cases}$$

By equating the tangential components H_z at the boundary $r = a$ and letting $n_0 = n_c$, we get

$$u\frac{J_{l+1}(u)}{J_l(u)}\cos(l+1)\theta - u\frac{J_{l-1}(u)}{J_l(u)}\cos(l-1)\theta$$

$$= \gamma\frac{K_{l+1}(\gamma)}{K_l(\gamma)}\cos(l+1)\theta + \gamma\frac{K_{l-1}(\gamma)}{K_l(\gamma)}\cos(l-1)\theta$$

From the equation above, we obtain

$$u\,\frac{J_{l\pm1}(u)}{J_l(u)} = \pm\,\gamma\,\frac{K_{l\pm1}(\gamma)}{K_l(\gamma)} \tag{4.48}$$

which is identical to the two eigenvalue equations given by Equation (4.28). Other transverse components lead to the same eigenvalue equations in this approximation.

If $n_0 \neq n_c$, this degeneracy ceases to exist. Each LP_{lm} mode breaks up into modes with terms $(l + 1)\theta$, which can be identified as $HE_{l+1,m}$, and modes with terms $(l - 1)\theta$, which are labeled $EH_{l-1,m}$ or TE_{0m} and TM_{0m}. For all practical purposes Equation (4.28) is sufficiently accurate for calculating the propagation constant β of these LP modes. Again, we should remember that the subscripts on HE_{lm} and EH_{lm} refer to the lth order and mth rank, where l is an integer and is associated with the circular functions $\sin l\theta$ or $\cos l\theta$; m is also an integer, which identifies the successive roots of $J_l = 0$. Physically, the values l and m represent the number of angular and radial antinodes in the field pattern, respectively.

It is useful to establish an explicit functional relationship between u and V for higher order modes as we have done previously for the HE_{11} mode. We shall follow the analysis of Snyder (Ref. 4.3) by taking the derivative of Equation (4.48) with respect to V and using the lower case $(l - 1)$. We have

$$\frac{du}{dV} = \frac{-1}{J_{l-1}K_l}\left[\frac{d}{dV}(\gamma K_{l-1}J_l) + u\,\frac{d}{dV}K_lJ_{l-1}\right] \tag{4.49}$$

Using Equation (4.8), we write

$$\frac{d\gamma}{dV} = \left(V - u\,\frac{du}{dV}\right)\Big/\gamma$$

The derivatives in Equation (4.49) can be expressed as

$$\frac{d}{dV}(\gamma K_{l-1}J_l) = \frac{du}{dV}\left[\gamma K_{l-1}J_l' - \frac{uJ_l(\gamma K_{l-1}' + K_{l-1})}{\gamma}\right]$$
$$+ \frac{V(\gamma K_{l-1}' + K_{l-1})J_l}{\gamma}$$

and

$$\frac{d}{dV}(J_{l-1}K_l) = \left(K_lJ_{l-1}' - \frac{uJ_{l-1}K_l'}{\gamma}\right)\frac{du}{dV} + \frac{VJ_{l-1}K_l'}{\gamma}$$

After considerable algebraic manipulation and using Bessel functional relationships, we have

$$\frac{du}{dV} = \frac{X}{Y} \tag{4.50}$$

where

$$X = -\frac{lV^2 J_{l-1} K_l}{\gamma^2} - \frac{V^2 J_{l-1} K_{l-1}}{\gamma} + \frac{lV^2 K_{l-1} J_l}{u\gamma} \tag{4.51}$$

$$Y = \frac{V}{\gamma J_{l-1}} \left[u K_{l-1} \left(\frac{2lJ_l}{uJ_{l-1}} - 1 \right) - \frac{\gamma K_l J_l}{J_{l-1}} \right] \tag{4.52}$$

With the help of the eigenvalue equation (4.48), we can simplify Equations (4.51) and (4.52) and get

$$X = -\frac{V^2 J_{l-1} K_{l+1}}{\gamma} \tag{4.53}$$

$$Y = -\frac{uV}{\gamma} \left(K_{l+1} - \frac{K_l^2}{K_{l-1}} \right) \tag{4.54}$$

Substituting Equations (4.53) and (4.54) into (4.50), we have

$$\frac{du}{dV} = \frac{u}{V} \left[1 - \frac{K_l^2(\gamma)}{K_{l+1}(\gamma) K_{l-1}(\gamma)} \right] \tag{4.55}$$

Again, we shall evaluate the equation above for two cases: near cutoff and far from cutoff. For $\gamma \to \infty$ we have

$$\frac{K_l^2(\gamma)}{K_{l+1}(\gamma) K_{l-1}(\gamma)} \simeq 1 - \frac{1}{V} \tag{4.56}$$

Equation (4.56) is valid for all l values. In the case that $\gamma \to 0$, the values of $K_l^2(\gamma)/K_{l+1}(\gamma) K_{l-1}(\gamma)$ are listed below for a few lower-order LP modes.

$$\lim_{\gamma \to 0} \frac{K_l^2(\gamma)}{K_{l+1}(\gamma) K_{l-1}(\gamma)} = \begin{cases} 0 & \text{for } l = 0, l = 1 \\ \dfrac{l-1}{l} & \text{for } l > 1 \end{cases}$$

It has been shown by Gloge (Ref. 4.2) that the approximation

$$\frac{K_l^2}{K_{l+1} K_{l-1}} \simeq 1 - (\gamma^2 + l^2 + 1)^{-1/2} \tag{4.57}$$

provides a reasonable fit for all values of V. If we replace γ^2 by $V^2 - u_c^2$, where u_c is the cutoff value, we can solve Equation (4.55) for u as a function of V by using this approximation, and obtain the following expression:

$$u(V) = \frac{u_c}{s} \exp\left[\arcsin\frac{s}{u_c} - \arcsin\left(\frac{s}{V}\right)^2 \right] \quad (4.58)$$

where

$$s = (u_c^2 - l^2 - 1)^{1/2} \quad (4.59)$$

Equation (4.58) is good for all LP modes except the LP_{01} (HE_{11}) mode. In the case of the HE_{11} mode, the functional relationship is given by (Ref. 4.2)

$$u(V) = \frac{(1 + \sqrt{2})V}{1 + (4 + V^4)^{1/4}} \quad (4.60)$$

Figure 4.5 is a plot of the propagation constant β for a few lower-order modes in a step-index fiber as a function of V. A comparison with the exact solutions shows that the difference between the two is too small to be displayed in this figure.

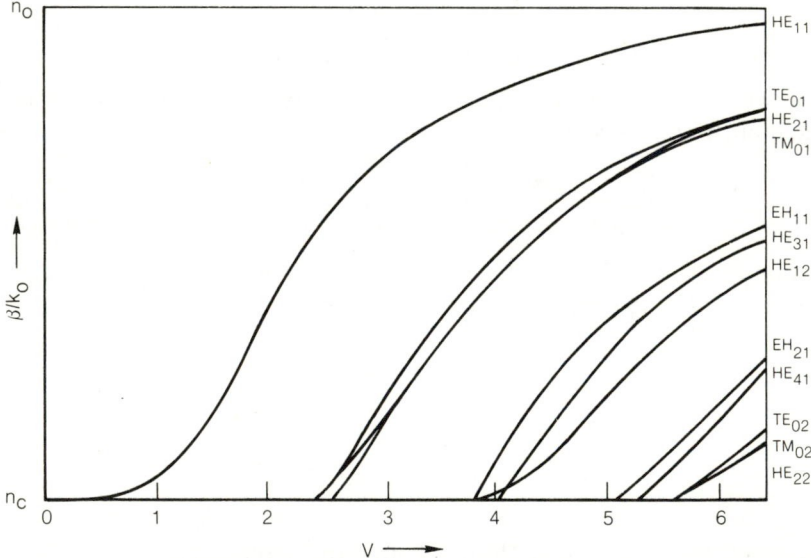

Figure 4.5   Plot of axial wave number for several lower-order modes in a step-index fiber as a function of the parameter V. (From Ref. 4.2.)

4.4 POWER DISTRIBUTION

Using the scalar approximation, we now calculate the modal power distribution in the core and in the cladding by carrying out the following integrals:

$$P_{\text{core}} = 2\pi \int_0^a \psi^2 r \, dr \tag{4.61}$$

$$P_{\text{cladding}} = 2\pi \int_a^\infty \psi^2 r \, dr \tag{4.62}$$

where ψ is assumed to be normalized to unity at the boundary and can be expressed as follows:

$$\psi = J_l^{-1}(u) J_l\left(\frac{ur}{a}\right) \qquad r < a \tag{4.63}$$

$$\psi = K_l^{-1}(\gamma) K_l\left(\frac{\gamma r}{a}\right) \qquad r > a \tag{4.64}$$

Using the identity for integrals of Bessel functions, we obtain

$$P_{\text{core}} = \pi a^2 [1 - \bar{J}_l(u)] \tag{4.65}$$

$$P_{\text{cladding}} = \pi a^2 [\bar{K}_l(\gamma) - 1] \tag{4.66}$$

where

$$\bar{J}_l(u) = \frac{J_{l-1}(u) J_{l+1}(u)}{J_l^2(u)} \tag{4.67}$$

$$\bar{K}_l(\gamma) = \frac{K_{l-1}(\gamma) K_{l+1}(\gamma)}{K_l^2(\gamma)} \tag{4.68}$$

We can express the eigenvalue equation (4.28) in terms of $\bar{J}_l$ and $\bar{K}_l$ as

$$u^2 \bar{J}_l(u) = -\gamma^2 \bar{K}_l(\gamma) \tag{4.69}$$

By adding Equations (4.65) and (4.66) and using Equation (4.69), we obtain the expression for the total power P_T:

$$P_T = \pi a^2 \frac{V^2}{u^2} \bar{K}_l(\gamma) \tag{4.70}$$

The fractional power in the cladding can be obtained from Equations (4.66) and (4.70) and expressed by

$$\frac{P_{\text{cladding}}}{P_T} = \frac{u^2}{V^2} \left[1 - \frac{1}{\bar{K}_l(\gamma)} \right] \tag{4.71}$$

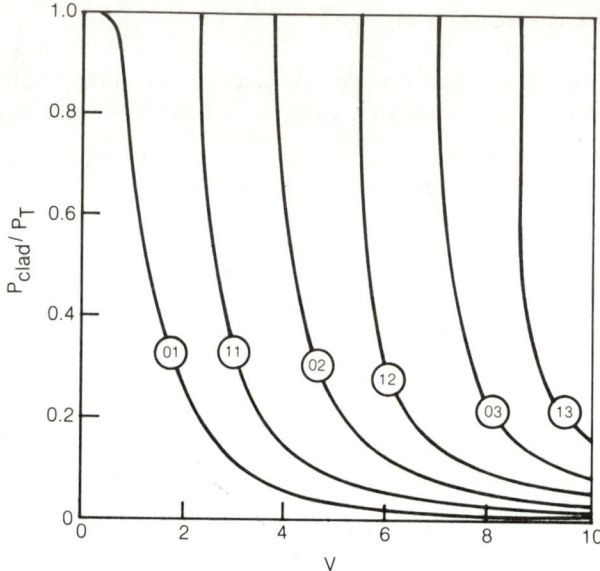

Figure 4.6 Plot of normalized optical power in the cladding as a function of V for a few LP modes. (From Ref. 4.2.)

and the fractional power in the core is given simply by

$$\frac{P_{core}}{P_T} = 1 - \frac{P_{cladding}}{P_T} \tag{4.72}$$

It is interesting to examine the power distribution for different modes near cutoff. In doing so, we use the calculated results for the LP mode as given in Equations (4.58) and (4.60). Figure 4.6 is a plot of the fractional power inside the cladding as a function of V for a few lower-order LP modes in a step-index fiber. We see that for the first two lowest-order modes, the power flow is mostly in the cladding near cutoff. For very large values of l, however, the power remains in the core even at or just beyond cutoff.

4.5 EXACT SOLUTIONS FOR STEP-INDEX FIBERS

The treatment discussed above turns out to be valid only for the special case that $n_0 \simeq n_c$. To treat this problem exactly, one must start with appropriate E_z and H_z field expressions of the form given by Equations (4.40) to (4.43) and develop appropriate expressions for other field components by using the relationships as given by Equations (2.39) to (2.42).

For those components inside the core, we let

$$E_z = \frac{AJ_l(ur/a)e^{il\theta}}{J_l(u)} \tag{4.73}$$

$$H_z = \frac{BJ_l(ur/a)e^{il\theta}}{J_l(u)} \tag{4.74}$$

and for those components outside the core, we write

$$E_z = \frac{AK_l(\gamma r/a)e^{il\theta}}{K_l(\gamma)} \tag{4.75}$$

$$H_z = \frac{BK_l(\gamma r/a)e^{il\theta}}{K_l(\gamma)} \tag{4.76}$$

These E_z and H_z field components can be expressed in the forms given by Equations (4.73) to (4.76), because we have incorporated the boundary condition that E_z and H_z are continuous at $r = a$. In doing so, we have reduced the number of constants to A and B. To determine A and B, we apply the continuity requirement on the other tangential components E_θ and H_θ at the boundary. From Equations (2.39) to (2.42), we obtain appropriate expressions for the field components in the core of a cylindrical fiber as

$$E_r = -i\frac{a^2}{u^2 J_1(u)}\left[\frac{\beta uA}{a}J'_l(ur/a) + \frac{i\omega\mu B}{r}J_l(ur/a)\right]e^{il\theta} \tag{4.77}$$

$$E_\theta = -i\frac{a^2}{u^2 J_l(u)}\left[-\frac{u\omega\mu B}{a}J'_l(ur/a) + i\frac{\beta lA}{r}J_l(ur/a)\right]e^{il\theta} \tag{4.78}$$

$$H_r = -i\frac{a^2}{u^2 J_l(u)}\left[\frac{u\beta B}{a}J'_l(ur/a) - i\frac{\omega\varepsilon_1 lA}{r}J_l(ur/a)\right]e^{il\theta} \tag{4.79}$$

$$H_\theta = -i\frac{a^2}{u^2 J_l(u)}\left[\frac{\varepsilon_1 u\omega A}{a}J'_l(ur/a) + i\frac{\beta lB}{r}J_l(ur/a)\right]e^{il\theta} \tag{4.80}$$

For the field components in the cladding of a cylindrical fiber, we have

$$E_r = i\frac{a^2}{\gamma^2 K_l(\gamma)}\left[\frac{\beta\gamma A}{a}K'_l(\gamma r/a) + i\frac{\omega\mu lB}{r}K_l(\gamma r/a)\right]e^{il\theta} \tag{4.81}$$

$$E_\theta = i\frac{a^2}{\gamma^2 K_l(\gamma)}\left[-\frac{\gamma\omega\mu B}{r}K'_l(\gamma r/a) + i\frac{\beta lA}{r}K_l(\gamma r/a)\right]e^{il\theta} \tag{4.82}$$

$$H_r = i\frac{a^2}{\gamma^2 K_l(\gamma)}\left[\frac{\gamma\beta B}{a}K'_l(\gamma r/a) - i\frac{\omega\varepsilon_2 lA}{r}K_l(\gamma r/a)\right]e^{il\theta} \tag{4.83}$$

$$H_\theta = i \frac{a^2}{\gamma^2 K_l(\gamma)} \left[\frac{\gamma \varepsilon_2 A}{a} K_l'(\gamma r/a) + i \frac{\beta l B}{r} K_l(\gamma r/a) \right] e^{il\theta} \qquad (4.84)$$

By equating Equations (4.78) and (4.82) for E_θ and similarly Equations (4.80) and (4.84) for H_θ at $r = a$, we obtain two homogeneous equations containing two unknown constants A and B:

$$i \frac{l\beta}{a} \left(\frac{1}{u^2} + \frac{1}{\gamma^2} \right) A - \frac{\omega\mu}{a} \left[\frac{1}{u} \frac{J_l'(u)}{J_l(u)} + \frac{1}{\gamma} \frac{K_l'(\gamma)}{K_l(u)} \right] B = 0 \qquad (4.85)$$

$$\frac{\omega\varepsilon_1}{a} \left[\frac{n_0^2}{u} \frac{J_l'(u)}{J_l(u)} + \frac{n_c^2}{\gamma} \frac{K_l'(\gamma)}{K_l(\gamma)} \right] A + i \frac{l\beta}{a} \left(\frac{1}{u^2} + \frac{1}{\gamma^2} \right) B = 0 \qquad (4.86)$$

A nontrivial solution exists only if the determinant of the coefficient vanishes. From this determinant, we obtain the desired eigenvalue equation:

$$\left[\frac{n_0^2}{n_c^2} \frac{\gamma^2}{u^2} \frac{J_l'(u)}{J_l(u)} + \gamma \frac{K_l'(\gamma)}{K_l(\gamma)} \right] \left[\frac{\gamma^2}{u^2} \frac{J_l'(u)}{J_l(u)} + \gamma \frac{K_l'(\gamma)}{K_l(\gamma)} \right]$$
$$= \left[l \left(\frac{n_0^2}{n_c^2} - 1 \right) \beta n_c k_0 \left(\frac{a}{u} \right)^2 \right]^2 \qquad (4.87)$$

For a given waveguide, this equation will give a set of discrete values for β falling within the range

$$n_c \leq \frac{\beta}{k_0} \leq n_0$$

We note that when n_0 is very close to n_c, Equation (4.87) reduces to the simple form given by Equation (4.28). More detailed treatment can be found in the book by Marcuse (Ref. 4.1).

4.6 RAY ANALYSIS FOR GRADED-INDEX FIBERS

It has been pointed out in Chapter 1 that a graded-index fiber having a nearly parabolic index profile is a good choice for optical data transmission because it reduces significantly the modal dispersion that exists in multimode step-index fibers. To demonstrate this fact, we adopt here the simple approach of geometrical ray analysis. In this analysis we seek an appropriate index profile which can simultaneously satisfy the synchronization condition for all rays. The path of a light ray in a medium of varying index can be described by the Eikonal equation, which can be derived directly from Maxwell's equations:

$$|\nabla S| = n(r) \qquad (4.88)$$

where S is a surface of constant phase along a ray path s. The vector ∇S specifies the direction of the energy flow and is equivalent to the Poynting vector as introduced previously. Let **r** be a position vector. Then $d\mathbf{r}/ds$ is a unit vector normal to the surface S. In a vector form, Equation (4.88) can be written as

$$\frac{d\mathbf{r}}{ds} = \frac{1}{n} \nabla S \tag{4.89}$$

If we differentiate Equation (4.89) with respect to s and again make use of the Eikonal equation, we obtain an equivalent form of the ray equation in terms of **r** as follows:

$$\frac{d}{ds}\left(n\frac{d\mathbf{r}}{ds}\right) = \frac{d}{ds}\nabla S$$

Since

$$\frac{d}{ds} = \frac{d\mathbf{r}}{ds} \cdot \nabla$$

We can write that

$$\frac{d}{ds}\nabla S = \frac{d\mathbf{r}}{ds} \cdot \nabla(\nabla S)$$

Substituting Equation (4.89) into the equation above, we have

$$\frac{1}{n}\nabla S \cdot \nabla(\nabla S) = \frac{1}{2n}\nabla(\nabla S)^2$$

Substituting Equation (4.88) into the equation above, we have

$$\frac{d}{ds}\left(n\frac{d\mathbf{r}}{ds}\right) = \frac{1}{2n}\nabla n^2 = \nabla n \tag{4.90}$$

It is desirable to express Equation (4.90) in cylindrical coordinates. Let $\mathbf{e}_r$, $\mathbf{e}_\theta$, and **k** be three unit vectors in the cylindrical coordinate system. We write that $\mathbf{r} = r\mathbf{e}_r + z\mathbf{k}$, and

$$\frac{d\mathbf{r}}{ds} = \frac{dr}{ds}\mathbf{e}_r + r\frac{d\mathbf{e}_r}{ds} + \frac{dz}{ds}\mathbf{k} \tag{4.91}$$

and

$$\frac{d\mathbf{e}_r}{ds} = \frac{d\theta}{ds}\mathbf{e}_\theta \tag{4.92}$$

$$\frac{d\mathbf{e}_\theta}{ds} = -\frac{d\theta}{ds}\mathbf{e}_r \tag{4.93}$$

We can now rewrite Equation (4.90) by using the relationships above as follows:

$$\left[\frac{d}{ds}\left(n\frac{dr}{ds}\right) - nr\left(\frac{d\theta}{ds}\right)^2\right]\mathbf{e}_r + \left[n\frac{dr}{ds}\frac{d\theta}{ds} + \frac{d}{ds}\left(nr\frac{d\theta}{ds}\right)\right]\mathbf{e}_\theta$$

$$+ \left[\frac{d}{ds}\left(n\frac{dz}{ds}\right)\right]\mathbf{k} = \nabla n \tag{4.94}$$

We obtain from Equation (4.94) three equations of motion, of which the θ and z components offer two time-invariant relationships as a result of the fact that n is only a function of r:

$$z \text{ component:} \quad n\frac{dz}{ds} = \text{constant} \equiv E \tag{4.95}$$

$$\theta \text{ component:} \quad nr^2\frac{d\theta}{ds} = \text{constant} \equiv I \tag{4.96}$$

A combination of these two relationships yields another time-invariant quality:

$$r^2\frac{d\theta}{dz} = \frac{I}{E} \equiv l \tag{4.97}$$

The quantities dr/ds, $d\theta/ds$, and dz/ds are directly related to the directional cosines of the incident and refracted rays as defined in Figure 4.7. The r component, on the other hand, gives the equation

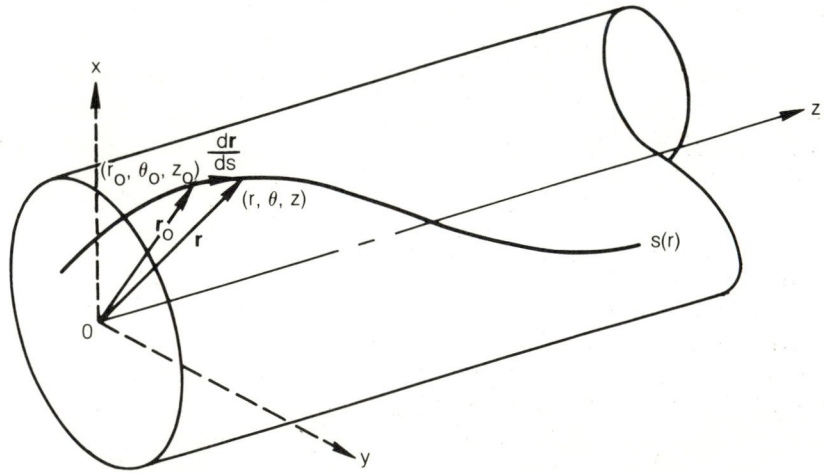

Figure 4.7 Path of a skew ray passing through the point (r_0, θ_0, z_0) and the point (r, θ, z) in a cylindrical graded-index fiber.

$$\frac{d}{ds}\left(n\,\frac{dr}{ds}\right) - nr\left(\frac{d\theta}{ds}\right)^2 = \frac{dn}{dr} \tag{4.98}$$

Substituting the invariant relationships above into Equation (4.98), we obtain

$$\frac{d^2r}{dz^2} - \frac{l^2}{r^3} - \frac{1}{2E^2}\frac{d}{dr}\,n^2 = 0 \tag{4.99}$$

Equation (4.99) can further be reduced to a first-order differential equation:

$$\frac{dr}{dz} = \sqrt{\frac{n^2}{E^2} - \frac{l^2}{r^2} - 1} \tag{4.100}$$

From Equations (4.100) and (4.97) we obtain two integrals, which describe the ray path completely for a given $n(r)$, if the initial direction of the ray is specified. They are

$$z = z_0 + \int_{r_0}^{r}\frac{E\,dr}{\sqrt{n^2 - l^2E^2/r^2 - E^2}} \tag{4.101}$$

$$\theta = \theta_0 + \int_{r_0}^{r}\frac{El\,dr}{r^2\sqrt{n^2 - l^2E^2/r^2 - E^2}} \tag{4.102}$$

We first make the following observation. Equations (4.101) and (4.102) both involve a quadratic form

$$n^2 - \frac{l^2E^2}{r^2} - E^2 \tag{4.103}$$

where n is a slowly varying function of r and has a maximum value of n_0 at $r = 0$ and a minumum value of n_c at $r = a$. As shown in Figure 4.8, these integrals exist only within a range defined by R_{min} and R_{max}, which are the roots of the quadratic form (4.103). We now define two integrals:

$$P \equiv 2E \int_{R_{min}}^{R_{max}}\frac{dr}{\sqrt{n^2 - l^2E^2/r^2 - E^2}} \tag{4.104}$$

$$\Theta \equiv 2E \int_{R_{min}}^{R_{max}}\frac{l\,dr}{r^2\sqrt{n^2 - l^2E^2/r^2 - E^2}} \tag{4.105}$$

where P is the period and Θ is the total angular change over one period for a ray specified by the initial values E and l. Therefore, the functions P and Θ are very important for the study of time dispersion of the ray path. The usual way (Ref.

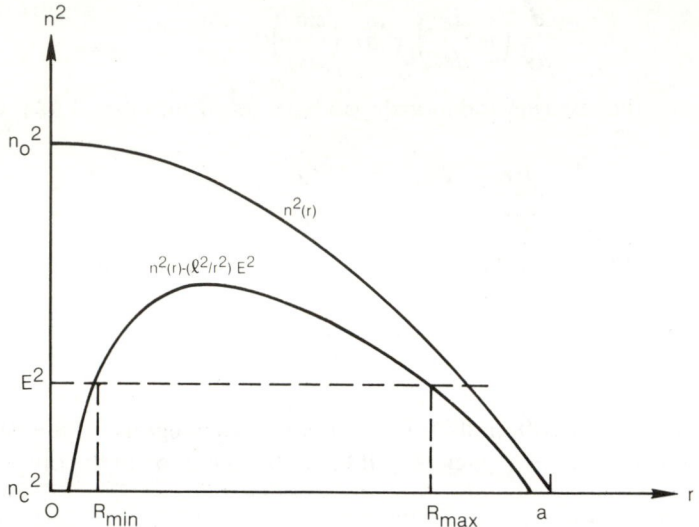

Figure 4.8 Region in which skew rays are confined in a graded-index fiber.

4.4) to treat this problem is to apply Fermat's principle, which requires that the optical paths for the minimum time dispersion are of the same length. This implies that

$$\frac{dP}{dE} = 0, \qquad \frac{d\Theta}{dE} = 0$$

and

$$\frac{dP}{dl} = 0, \qquad \frac{d\Theta}{dl} = 0$$

These conditions can be realized if we can find an index profile $n(r)$ such that both P and Θ can be independent of E and l.

We shall examine the following three types of index profile:

Case 1—Parabolic profile:

$$n(r) = n_0 \left[1 - \varepsilon^2 r^2\right]^{1/2} \qquad (4.106)$$

Case 2—α profile:

$$n(r) = n_0 \left[1 - 2\Delta \left(\frac{r}{a}\right)^{\alpha}\right]^{1/2} \qquad (4.107)$$

Case 3—Hyperbolic or "selfoc" profile:

$$n(r) = n_0 \operatorname{sech}\frac{\Delta r}{a} \qquad (4.108)$$

For case 1 we obtain upon integration that

$$P = \frac{2\pi\varepsilon E}{n_0} \qquad \text{and} \qquad \Theta = \pi$$

These results indicate that Θ is independent of both E and l, and P is now independent of l but is still proportional to E. Therefore, we can conclude that the parabolic profile is not the best choice. Because P is proportional to E, the parabolic profile does not yield the least modal dispersion. In fact, the period of a ray propagating in this fiber increases with decreasing angle defined by that ray and the z axis.

For case 2 we consider only the meridional rays by letting $l = 0$, and $R_{\min} = 0$. We have

$$P = 2E \int_0^{R_{\max}} \frac{dr}{\sqrt{1 - 2\Delta(r/a)^\alpha - E^2}} = \frac{1 + \sqrt{1 - 2\Delta}}{2\sqrt{1 - E^2}} \left(\frac{1 - E^2}{2\Delta}\right)^{1/\alpha} C(\alpha)$$

$$(4.109)$$

where

$$C(\alpha) = \int_0^1 \frac{dx}{\sqrt{1 - x^\alpha}}$$

It is to be noted that $C(\alpha)$ is independent of E. To find the α value such that P is independent of E, we simply let dP/dE equal zero. In this way, we obtain from Equation (4.109) the following condition:

$$1 - \frac{2(1 - \Delta)}{\alpha} = 0$$

From this we obtain

$$\alpha = 2 - 2\Delta \qquad (4.110)$$

Therefore, the path of all rays in such an index profile is periodic with a period independent of the angle of incidence. From the wave description, we have seen that there is a direct correspondence between the mode angle and the mode number. The result of this ray analysis reflects that all modes traveling in a fiber with an appropriate α profile as given by $2 - 2\Delta$ should have nearly the same group velocity. Even though a graded-index fiber is multimode, it has a minimum modal dispersion, and a maximum data-carrying capacity.

For case 3 it can be shown that all ray paths are periodic in z with a period as given by

$$P = \frac{\pi}{\Delta} \qquad (4.111)$$

Equation (4.111) indicates that the period of all ray paths in a hyperbolic index fiber is also independent of the angle of incidence.

4.7 CALCULATIONS OF GUIDED-WAVE MODES IN THE WKB APPROXIMATION

The WKB (Wentzel–Kramers–Brillouin) method, which is well known in quantum mechanics, can be applied to obtain some useful solutions of otherwise very complex dielectric waveguide problems for graded-index fibers. This approximation takes advantage of the fact that wave functions in the guide change very slowly. We shall use this method to determine (1) the total number of modes, and (2) the propagation constant β_{lm} associated with the corresponding mode. The establishment of an explicit expression for β_{lm} is necessary for studies of modal dispersion characteristics in Chapter 5.

Recall that the wave equation for a graded-index fiber is of the form

$$\frac{d^2\psi}{dr^2} + \frac{1}{r}\frac{d\psi}{dr} + \left[n^2(r)k_0^2 - \beta^2 - \frac{l^2}{r^2} \right]\psi = 0 \qquad (4.112)$$

where $n(r)$ in general is a very slowly varying function of r. Therefore, we can assume a general solution, which is a superposition of plane waves, of the form

$$\psi(r) = e^{ik_0\phi(r)} \qquad (4.113)$$

where $\phi(r)$ can be expanded in a power series in terms of $1/k_0$ as

$$\phi(r) = \phi_0 + \frac{1}{k_0}\phi_1 + \cdot\cdot\cdot \qquad (4.114)$$

Substituting Equation (4.113) into (4.112) and collecting terms in accordance with the power of $1/k_0$, we obtain two equations belonging to the zeroth and the first order:

$$k_0^2 \left(\frac{d\phi_0}{dr}\right)^2 - \left[k_0^2 n^2(r) - \beta^2 - \frac{l^2}{r^2} \right] = 0 \qquad (4.115)$$

and

$$i\left(\frac{d^2\phi_0}{dr^2} + \frac{1}{r}\frac{d\phi_0}{dr}\right) - 2\frac{d\phi_0}{dr}\frac{d\phi_1}{dr} = 0 \qquad (4.116)$$

We obtain from Equation (4.115) that

$$k_0 \phi_0(r) = \int_0^r \left[k_0^2 n^2(r) - \beta^2 - \frac{l^2}{r^2} \right]^{1/2} dr \qquad (4.117)$$

Substituting Equation (4.117) into (4.116), we obtain

$$\phi_1(r) = \frac{i}{4} \ln \left[r^2 n^2(r) - \frac{\beta^2 r^2}{k_0^2} - \frac{l^2}{k_0^2} \right] \qquad (4.118)$$

We observe that the real limits of the integral in Equation (4.117) are two points inside the fiber at which the integrand vanishes and they represent the "caustic" or turning points, which separate regions of oscillatory and evanescent field variation, as illustrated by Figure 4.9. In other words, within the region bound by r_1 and r_2, as shown in Figure 4.9, guided-wave modes exist and can have the allowed β values. At the caustic point, the function ϕ_1 in Equation (4.118) possesses a pole and hence the first-order WKB approximation fails at these points.

Since we are interested in establishing an expression for β, in the following we deal only with Equation (4.117). As we have seen before, one of the conditions for establishing guided-wave modes is that on two consecutive reflections, the total phase angle must be an integer multiple of 2π. By integrating Equation (4.117) from r_1 to r_2, we get only one-half of the cycle for a skew ray ($l \neq 0$); therefore, we write

$$m\pi = \int_{r_1}^{r_2} \left[k_0^2 n^2(r) - \beta^2 - \frac{l^2}{r^2} \right]^{1/2} dr \qquad (4.119)$$

The integer m is associated with the mth mode number. Equation (4.119) is useful for determining the number of modes in a given range. However, some comments and corrections to this equation must be made. First, the validity of the zeroth WKB approximation is good only for the ray optics picture, in which the phase changes at the caustic or turning points r_1 and r_2 are ignored. Second, there are twofold degeneracies associated with each lm mode: one goes with the clockwise or counterclockwise rotation and the other one goes with the orientations of the linear polarization. These degeneracies increase the mode number by a factor of 4 from what has been already accounted for by Equation (4.119). Third, for each m value, there exist a set of l numbers with an upper limiting value l_{max} at which the wave is no longer bound. Since the largest l value for a given m occurs for the mode near its cutoff point, we shall replace β by $k_0 n_c$. Therefore, we can write

$$l_{max} = k_0 r [n^2(r) - n_c^2]^{1/2} \qquad (4.120)$$

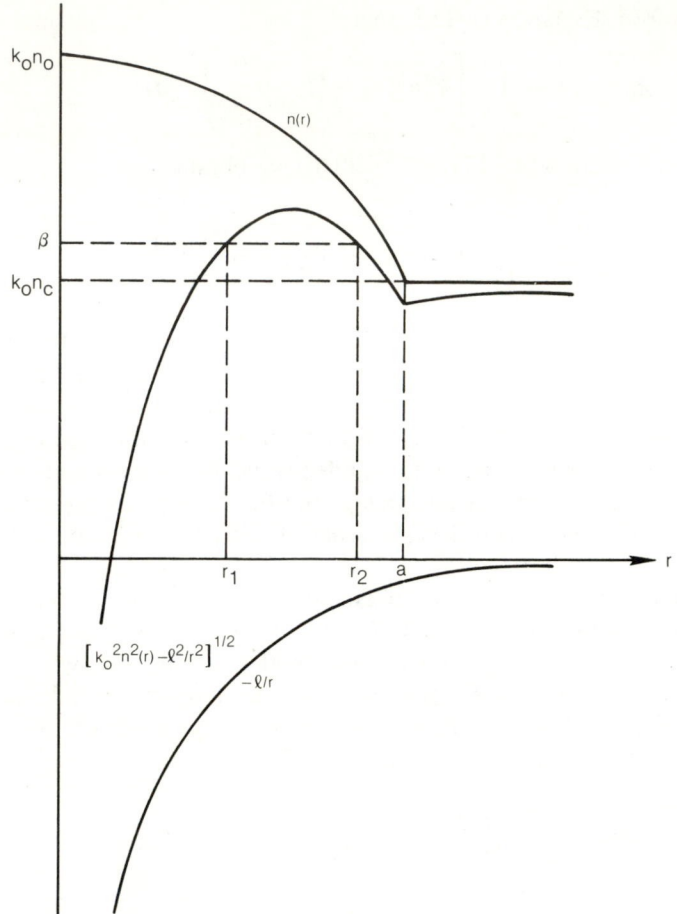

Figure 4.9   Region in which hybrid modes are confined in a graded-index fiber.

By taking these corrections into account and treating l as a continuous variable, we can replace the sum over l by an integral and thus obtain from Equation (4.119) the total number of modes as expressed by

$$M = \frac{4k_0}{\pi} \int_0^a \int_0^{l_{max}} \left[n^2(r) - n_c^2 - \frac{l^2}{k_0^2 r^2} \right]^{1/2} dl \; dr$$

After integrating over l and using Equation (4.120), we get

$$M = k_0^2 \int_0^a [n^2(r) - n_c^2] r \; dr \qquad\qquad (4.121)$$

In Equation (4.121), β has been replaced by $k_0 n_c$ and the integral has been extended over the entire core.

If we substitute

$$n^2(r) = n_0^2 \left[1 - 2\Delta \left(\frac{r}{a} \right)^\alpha \right]$$

$$n_c^2 = n_0^2 (1 - 2\Delta)$$

into Equation (4.121), we obtain after integrating over r that

$$M = \frac{\alpha}{\alpha + 2} n_0^2 k_0^2 a^2 \Delta \tag{4.122}$$

Equation (4.122) gives an expression for calculating the total number of modes in a graded-index fiber with an α profile. This expression can be extended to a step-index fiber by letting $\alpha = \infty$. We obtain from Equation (4.122)

$$M = n_0^2 k_0^2 a^2 \Delta = \frac{V^2}{2} \tag{4.123}$$

an expression for calculating the total number of modes in a step-index fiber with a given V value. Since $\alpha \simeq 2$, the results of Equations (4.122) and (4.123) indicate that the number of modes in a graded-index fiber is only about one-half of that in a step-index fiber of identical core diameter.

The number of modes having the corresponding β values falling within the limit from $n_0 k_0$ to β_m is given by

$$m = \int_0^{r_m} [n^2(r)k_0^2 - \beta_m^2]^{1/2} \, r \, dr \tag{4.124}$$

where

$$r_m = a \left[\frac{1 - (\beta_m/n_0 k_0)^2}{2\Delta} \right]^{1/\alpha} \tag{4.125}$$

Upon integration of Equation (4.124), we obtain

$$m = a^2 \Delta n_0^2 k_0^2 \frac{\alpha}{\alpha + 2} \left(\frac{n_0^2 k_0^2 - \beta_m^2}{2\Delta k_0^2 n_0^2} \right)^{(\alpha+2)/\alpha} \tag{4.126}$$

By combining the results of Equations (4.126) and (4.123), we obtain an expression for β_m in terms of total number of m-fold degenerated groups

$$\beta_m = k_0 n_0 \left[1 - 2\Delta \left(\frac{m}{M} \right)^{2\alpha/(\alpha+2)} \right]^{1/2} \tag{4.127}$$

where values for m have been obtained by summing the degeneracy from zero to the mth-level (e.g., $\Sigma\, m$ is approximately equal to m^2).

The expression for β_m, which will be used in Chapter 5 for the calculation of modal dispersion, is useful in gaining some understanding of physical properties of fibers. First, we can calculate the modal spacing in a fiber by differentiating β_m with respect to m and obtain

$$\frac{d\beta_m}{dm} = \left(\frac{\alpha}{\alpha + 2} \right)^{1/2} \frac{2\sqrt{\Delta}}{a} \left(\frac{m}{M} \right)^{(\alpha-2)/(\alpha+2)} \tag{4.128}$$

Equation (4.128) indicates that, for a parabolic index profile ($\alpha = 2$) the modal spacing is independent of the mode number. In the case of a step-index fiber ($\alpha = \infty$), Equation (4.128) yields

$$\frac{d\beta_m}{dm} = \frac{2\sqrt{\Delta}}{a} \left(\frac{m}{M} \right) \tag{4.129}$$

This result indicates that the modal spacing in a step-index fiber increases linearly with increasing mode order, which is one of the major distinctions between these two types of fiber. Another interesting application of Equation (4.127) is to relate β_m with the z component of $\mathbf{k}$. We get

$$\cos \theta_m = \left[1 - 2\Delta \left(\frac{m}{M} \right)^{2\alpha/(\alpha+2)} \right]^{1/2} \tag{4.130}$$

where θ_m is the maximum angle that the wave vector $\mathbf{k}$ makes with the z axis at $r = 0$. From Equation (4.130) we obtain

$$\sin \theta_m = \sqrt{2\Delta} \left(\frac{m}{M} \right)^{\alpha/(\alpha+2)} \tag{4.131}$$

The expression (4.131) indicates that there is definitely a direct correspondence between the mode angle and mode number. In the case of a parabolic index fiber ($\alpha = 2$)

$$\sin \theta_m = \sqrt{\frac{2\Delta m}{M}} \tag{4.132}$$

whereas in the case of a step-index fiber

$$\sin \theta_m = \sqrt{2\Delta} \left(\frac{m}{M} \right) \qquad (4.133)$$

A comparison of these results indicates that modes in a graded-index fiber are more confined to the core of the guide than those in a step-index fiber. At the cutoff, $m = M$ so that $n_0 \sin \theta_c = n_0 \sqrt{2\Delta}$, which is precisely the definition of numerical aperture of a step-index fiber. Expressions of (4.132) and (4.133) can be used to calculate the field patterns radiated from the ends of these fibers.

PROBLEMS

4.1. Give geometric interpretations of u and γ as defined by Equations (4.6) and (4.7).

4.2. Derive the identity given by Equation (4.16), using the series definition of $J_l(x)$ given by

$$J_l(x) = \sum_{n=0}^{\infty} \frac{(-1)^n}{n!\Gamma(n+l+1)} \left(\frac{x}{2} \right)^{2n+l}$$

where $\Gamma(x)$ is the gamma function.

4.3. Using the scalar field approximation, derive the eigenvalue equation given by Equation (4.28).

4.4. Using recurrence relations for j_l and K_l, show that the two eigenvalue equations given by Equation (4.28) are equivalent.

4.5. Summarize the cutoff conditions for TM_{om}, HE_{1m}, and HE_{lm} modes.

4.6. For large V, show that

$$u(V) = u(\infty)e^{-1/V}$$

4.7. Derive the expressions for the fractional power residing in the core and in the cladding as given by Equations (4.65) and (4.66). *Hint:*

$$\int_0^r r J_{l\mp 1}(ar) J_{l\mp 1}(ar)\, dr = \frac{r^2}{2} [J_{l\mp 1}^2(ar) - J_l(ar)J_{l\mp 2}(ar)]$$

and

$$\int_r^\infty r K_{l\mp 1}(ar) K_{l\mp 1}(ar)\, dr = \frac{r^2}{2} [K_{l\mp 1}(ar)K_l(ar) - K_{l\mp 2}(ar)]$$

4.8. Show that the period P of a ray propagating in a parabolic index fiber is equal to $2\pi E/n_0 \varepsilon$.

4.9. Derive Equations (4.121) and (4.122).

4.10. Derive Equation (4.126).

REFERENCES

4.1. D. Marcuse, *Theory of Dielectric Waveguides,* Academic Press, Inc., New York, 1974.

4.2. D. Gloge, *Appl. Opt., 10,* 2252 (1971).

4.3. A. W. Snyder, *IEEE Trans. Microwave Theory Tech., MIT-17,* 1130 (1969).

4.4. M. Eve, *Opt. Quantum Electron., 8,* 285 (1976).

5

Dispersion,
Mode Coupling,
and Loss Mechanisms

5.1 INTRODUCTION

Many limiting factors that originate from the geometric and physical nature of glass fibers have a profound effect on the information-carrying capacity of optical fiber waveguides. In this chapter we deal with various effects of fiber dispersions and imperfections. We show that in addition to modal dispersion, other factors, such as the dispersive properties of the glass and the spectral distribution of the source, can lead to a significant change in α values for graded-index fibers and can cause further pulse broadening. In practice, there exist many structural imperfections in fibers that introduce losses through scattering and absorption of optical power. Imperfections in a multimode fiber also create a random coupling of modes that in effect can produce pulse narrowing at the cost of power reduction through leakage into unguided radiative modes. To take advantage of this pulse-narrowing effect, the core–cladding interface must be prepared very carefully to minimize the radiation loss.

5.2 GROUP VELOCITY AND GROUP DELAY

In Chapter 4 we derived the conditions for obtaining β values associated with the modes in a fiber of a given index profile. Each mode propagates with a phase velocity as defined by Equation (2.20). In reality the phase velocity has no physical significance because there is no monochromatic source. Even for a laser source there is a finite spectral width or a spread of frequencies associated with the output. If the transmission medium is dispersive, the phase velocity is not the same for each frequency component of the wave; therefore, the wave must be represented by a group velocity, v_g. To show the above-mentioned effects, we consider a special case for which the pulse energy propagates in a planar dispersive waveguide. It is necessary to express this wave packet as a sum of plane waves characterized by different β values, each of which travels at a different phase velocity ω/β. Within the spectral range of interest, we must express ω in terms of β [e.g., $\omega = \omega(\beta)$]. Therefore, the simple plane-wave solution $E = A \exp[-i(\omega t - \beta z)]$ must be modified by the form

$$E(z, t) = \frac{1}{\sqrt{2\pi}} \int_{-\infty}^{\infty} A(\beta)e^{i[\beta z - \omega(\beta)t]}d\beta \tag{5.1}$$

where the amplitude $A(\beta)$ is determined by a linear superposition of different frequency components. If the distribution of $A(\beta)$ is sharply peaked around a value β_0, one can expand $\omega(\beta)$ around β_0 in a power series as

$$\omega(\beta) = \omega_0 + \frac{d\omega}{d\beta}(\beta - \beta_0) + \cdots \tag{5.2}$$

where $d\omega/d\beta$ is evaluated at $\beta = \beta_0$. Substituting Equation (5.2) into (5.1), we obtain, by neglecting the higher-order terms in the expansion,

$$E(z, t) \simeq \frac{1}{\sqrt{2\pi}} e^{i[\beta_0(d\omega/d\beta) - \omega_0]t} \int_{-\infty}^{\infty} A(\beta)e^{i[z - (d\omega/d\beta)t]\beta} d\beta \tag{5.3}$$

The inverse transform of Equation (5.3) evaluated at $t = 0$ is

$$A(\beta) = \frac{1}{\sqrt{2\pi}} \int_{-\infty}^{\infty} E(z', 0)e^{-i\beta z'} dz' \tag{5.4}$$

where $z' = z - (d\omega/d\beta)t$. Apart from an overall phase factor, the pulse travels along the waveguide undistorted in shape with a group velocity

$$v_g = \frac{d\omega}{d\beta} \tag{5.5}$$

To analyze the propagation characteristic of a pulse in a fiber, we must deal with the group delay time for a given mode β_{lm}. Consider that all modes are excited by a short pulse at the input. Each mode transports an equal amount of energy to the end of the fiber. As they recombine, the short pulse is expected to suffer a certain distortion, depending on the $\beta-\omega$ characteristics of each mode and the dispersion in the fiber. The power profile of the inpulse response can be measured by several methods (see Chapter 6). To compute the impulse response, we define the group delay τ_g in terms of the group velocity v_g as given by Equation (5.5):

$$\tau_g = L \frac{d\beta_{lm}}{d\omega} = \frac{L}{c} \frac{d\beta_{lm}}{dk_0} \tag{5.6}$$

where β_{lm} depends on both waveguide parameters and wavelength λ of the source. Once τ_g is known, the impulse response can be established by a superposition of the energy of all lm modes that arrive between τ and $\tau + d\tau$.

From Equations (4.6), (4.7), and (4.8) we obtain an expression for a multimode step-index fiber in the form

$$\frac{\beta^2/k_0^2 - n_c^2}{n_0^2 - n_c^2} = 1 - \frac{u^2}{V^2} \equiv b \tag{5.7}$$

where we have defined a quantity b in terms of V. For small index differences, Equation (5.7) becomes

$$b \simeq \frac{\beta/k_0 - n_c}{n_0 - n_c} \tag{5.8}$$

Since β and b are proportional, the quantity b can be regarded as a normalized propagation constant. From Equation (5.8), we can express β in terms of b as

$$\beta = n_c k_0 (b\Delta + 1) \tag{5.9}$$

Substituting Equation (5.9) into (5.6) and keeping in mind that both b and the refractive indices are functions of k_0, we obtain an expression for the group delay, given by

$$\tau_g = \frac{L}{c} \left(\frac{d}{dk_0} n k_0 + n\Delta \frac{d}{dV} Vb \right) \tag{5.10}$$

In Equation (5.10) we have ignored the difference in the dispersive effect between the core and cladding materials by omitting the subscript. Also the products of Δ with $(k_0/n(dn/dk_0)$ and $d\Delta/dk_0$ terms are omitted because they are small compared with all terms.

The first part of Equation (5.10) characterizes the material dispersion for all modes. The second term, which represents the group delay caused by waveguide dispersion, is governed by the derivative $d(Vb)/dV$.

Using Equation (5.8), we write

$$\frac{d}{dV} Vb = 1 - \frac{u^2}{V^2} \left(1 - 2 \frac{V}{u} \frac{du}{dV} \right) \tag{5.11}$$

Substituting the result of Equation (4.55) into (5.11), we obtain

$$\frac{d}{dV} Vb = 1 - \frac{u^2}{V^2} \left[1 - 2 \frac{K_l^2(\gamma)}{K_{l+1}(\gamma)K_{l-1}(\gamma)} \right] \tag{5.12}$$

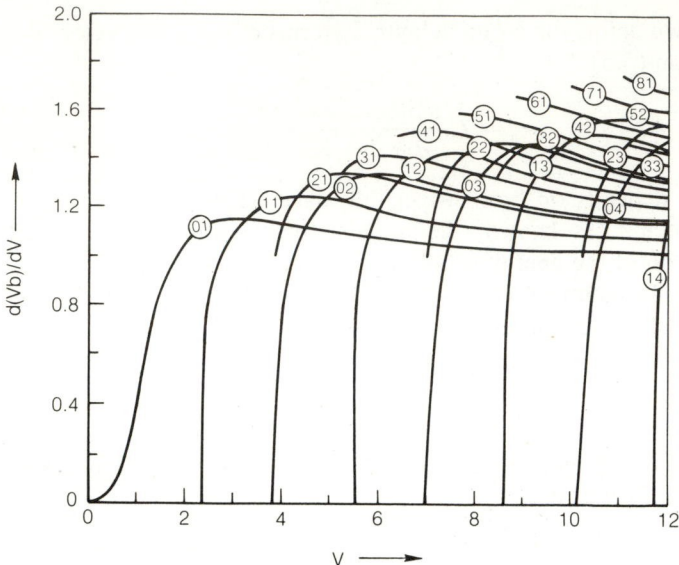

Figure 5.1 Relative group delay among various guided-wave modes in a weakly guided step-index fiber is plotted as a function of V. (From Ref. 5.1.)

The function $d(Vb)/dV$ is plotted in Figure 5.1. For large values of V, it approaches unity for all modes. At cutoff, $d(Vb)/dV = 0$ for $l = 0$, 1, and equals $2(1 - 1/l)$ for $l \geq 2$. Figure 5.1 also shows that the mode of largest-order l has the largest group delay. The difference between this and the lowest mode is approximately $1 - 2/l$. For large V, the group spread is approximately equal to $(1 - 2/V)(n_0 - n_c)L/c$ for a multimode step-index fiber.

5.3 PULSE BROADENING

Next, we examine the influence of the source distribution on optical transmission properties of short pulses through a fiber. When a light pulse is coupled into a fiber, its power will be distributed among all modes. If there is no coupling among modes, the impulse response $P_{lm}(\lambda, z)$ of an lm mode for a spectral component λ at a position z is time independent and is determined primarily by the spectral distribution of the source $S(\lambda)$ and the process of excitation. It is therefore reasonable to assume the distribution function to be a product of two independent functions of the form

$$P_{lm}(\lambda, z) = S(\lambda)P_{lm}(z_0) \tag{5.13}$$

We shall analyze the effect of a finite spectral width on the pulse width of a short light pulse as it propagates a distance L in a fiber. The root-mean-square (rms) pulse width $\sigma(z)$ in terms of the group delay is defined by the expression

$$\sigma(z) = [\langle \tau_{lm}^2 \rangle - \langle \tau_{lm} \rangle^2]^{1/2} \tag{5.14}$$

where the notation $\langle \cdot \rangle$ represents the average value of the quantity in question. To perform the average, we shall make use of the time-independent impulse response function and integrate over the spectral width. Using Equation (5.13), we write

$$\langle \tau_{lm}^2 \rangle = \int \sum_{lm} P_{lm}(z_0) S(\lambda) \tau_{lm}^2 \, d\lambda \tag{5.15}$$

and

$$\langle \tau_{lm} \rangle = \int \sum_{lm} P_{lm}(z_0) S(\lambda) \tau_{lm} \, d\lambda \tag{5.16}$$

where the summation in Equations (5.15) and (5.16) is extended over all guided modes. For the steady-state situation, τ_{lm} is independent of z and is proportional only to L. Using Equation (5.14), we can determine the information-carrying capacity B in bits per second as defined by

$$B = \frac{1}{4\sigma} \tag{5.17}$$

To simplify the numerical calculation of Equation (5.14), it is instructive to expand the group delay τ_{lm} in a Taylor series about λ_0. We write

$$\tau_{lm}(\lambda) = \tau_{lm}(\lambda_0) + \tau'_{lm}(\lambda - \lambda_0) + \tfrac{1}{2} \tau''_{lm}(\lambda - \lambda_0)^2 + \cdots \tag{5.18}$$

where primes denote the derivatives with respect to λ and have been evaluated at λ_0.

Substituting Equation (5.18) into (5.14) and neglecting the higher-order terms, we obtain for the first term on the right-hand side of Equation (5.14),

$$\langle \tau_{lm}^2 \rangle = \sum_{lm} P_{lm}(z_0) \left\{ \tau_{lm}^2(\lambda_0) + \frac{\sigma_s^2}{2\lambda_0^2} [2\tau_{lm}(\lambda_0)\lambda_0^2 \tau''_{lm} + 2\lambda_0^2 \tau'^2_{lm}] \right\} \tag{5.19}$$

where σ_s is the rms spectral width of the source and is defined as

$$\sigma_s = \int_0^\infty (\lambda - \lambda_0)^2 S(\lambda) \, d\lambda \tag{5.20}$$

and the mean value λ_0 is given by

$$\lambda_0 = \int_0^\infty \lambda S(\lambda) \, d\lambda \tag{5.21}$$

In these definitions, normalized spectral distribution [e.g., $\int S(\lambda)\, d\lambda = 1$] has been assumed.

The second term on the right-hand side of Equation (5.14) can be written as

$$\langle \tau_{lm} \rangle^2 = [\Sigma P_{lm} \tau_{lm}(\lambda_0)]^2 + \left(\frac{\sigma_s}{\lambda_0}\right)^2 [\Sigma P_{lm} \tau_{lm}(\lambda_0) \Sigma P_{lm} \lambda_0^2 \tau_{lm}'']$$

$$+ \left(\frac{\sigma_s}{2\lambda_0}\right)^2 [\Sigma P_{lm} \lambda_0^2 \tau_{lm}'']^2 \tag{5.22}$$

By introducing the notation for a weighted-average value, for example,

$$\langle \tau(\lambda_0) \rangle \equiv \Sigma P_{lm} \tau_{lm}(\lambda_0)$$

into Equations (5.19) and (5.22), we can express the rms pulse width $\sigma(z)$ by rearranging the terms into two distinct groups, commonly referred to as the intermodal and the intramodel dispersions:

$$\sigma(z) = (\sigma_{\text{intermodal}}^2 + \sigma_{\text{intramodal}}^2)^{1/2} \tag{5.23}$$

where

$$\sigma_{\text{intermodal}}^2 = \langle \tau^2(\lambda_0) \rangle - \langle \tau(\lambda_0) \rangle^2$$

$$+ \left(\frac{\sigma_s}{\lambda_0}\right)^2 [\langle \lambda_0^2 \tau''(\lambda_0) \tau(\lambda_0) \rangle - \langle \lambda_0^2 \tau''(\lambda_0) \rangle \langle \tau(\lambda_0) \rangle] \tag{5.24}$$

and

$$\sigma_{\text{intramodal}}^2 = \left(\frac{\sigma_s}{\lambda_0}\right)^2 \langle \lambda_0^2 \tau'(\lambda_0)^2 \rangle \tag{5.25}$$

To calculate the pulse width, both the spectral function of the source and the impulse response function of the modes must be known. Usually, functional forms can be assumed to approximate these distributions. Before carrying this analysis further, we shall make some observations concerning the origins of various terms in Equations (5.24) and (5.25) and discuss the order of importance.

The intermodal broadening is a result of the delay difference among the modes. The leading term that is independent of the source spectrum is the dominating term, but its magnitude can be reduced significantly if a graded-index fiber with $\alpha = 2 - 2\Delta$ is chosen. The remaining term in Equation (5.24), which contains the zeroth and second order of derivatives, is only a small correction which is proportional to the square of the relative spectral width σ_s/λ_0. Typically, σ_s is about 10^{-3} μm for an injection laser source and about 2×10^{-2} μm for an incoherent LED source, all of which operate at ≈ 0.9 μm. The intramodal broadening represents an average spreading of each mode with a value also proportional to the square of σ_s/λ_0. It contains only the first derivative of the group delay.

5.4 MATERIAL DISPERSION

Intramodal broadening arises from two distinct effects: One is caused by the material dispersion and the other is contributed by the waveguide structure. The separation of these two effects has been made by the expression of the group delay as given by Equation (5.10). We can choose a fiber with a properly graded index profile such that the second term in Equation (5.10) is zero, so that the waveguide dispersions can be eliminated completely. After again differentiating Equation (5.10) with respect to λ and substituting into Equation (5.25), we obtain an expression for the pulse-broadening effect, which can only be attributed to the material dispersion:

$$\sigma_{\text{material}} = \frac{L}{c}\, \sigma_s \lambda \, \frac{d^2 n}{d\lambda^2} \qquad (5.26)$$

This expression represents the ultimate information-carrying capability for a graded-index fiber without modal and waveguide dispersion.

It is possible to eliminate the broadening effect caused by the material dispersion by choosing a proper wavelength such that $d^2 n/d\lambda^2 = 0$. Figure 5.2

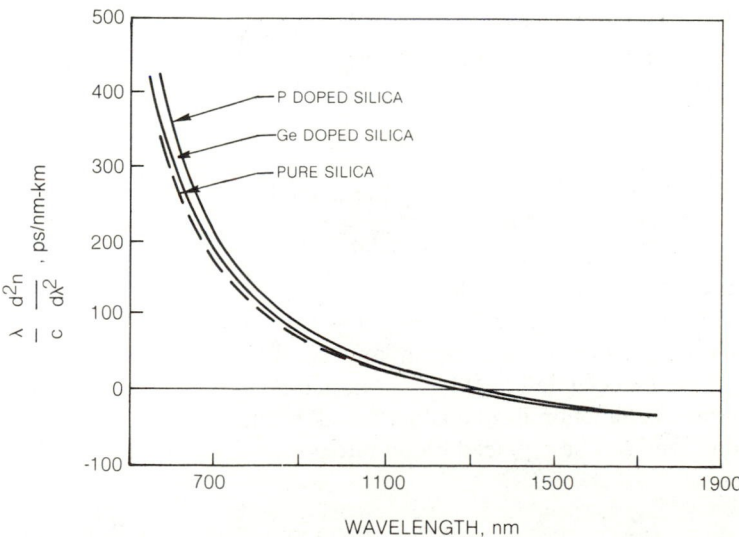

Figure 5.2 Material dispersion measurements for pure and doped silica glasses. [After D. N. Payne and A. H. Hartog, *Electron. Lett., 13,* 627 (1977).]

shows the effect of dispersion of most silicate fibers as a function of wavelength and also shows that the inflection point of the $n(\lambda)$ curve occurs at a λ value lying between 1.2 and 1.3 μm. For this reason and because the minimum attenuation loss also occurs in this wavelength region (see Figure 1.4), much research and development effort has gone into improving the performance of the light source and photodetection in this wavelength region, which lies beyond the emitting range of GaAs. It must be realized that the effect of material dispersion can be very serious only in fibers having very low modal dispersion.

5.5 INTERMODAL DISPERSION

The effect of intermodal dispersion on the pulse width can be analyzed by calculating the group delay time among the propagating modes and using Equation (5.24) to calculate the rms value of the pulse width σ. This has been done to some extent by treating a step-index fiber in Section 5.2. We consider here a graded-index fiber with an α profile. From Equation (4.127) we write

$$\beta_m = k_0 n_0 \left[1 - 2\Delta \left(\frac{m}{M} \right)^{2\alpha/(\alpha+2)} \right]^{1/2}$$

where $M(\alpha) = [\alpha/(\alpha + 2)]a^2 k_0^2 n_0^2 \Delta$. Substituting β_m into Equation (5.6) and neglecting the Δ^2 and higher-order terms, we obtain

$$\tau_m = \frac{L}{c} (n - \lambda n') \left[1 + \Delta \frac{\alpha - 2 - \varepsilon}{\alpha + 2} \left(\frac{m}{M} \right)^{2\alpha/(\alpha+2)} \right] + O(\Delta^2) \quad (5.27)$$

where

$$\varepsilon = - \frac{2n\lambda}{n - \lambda n'} \frac{\Delta'}{\Delta} \quad (5.28)$$

To calculate the rms value of the pulse width σ, we must perform the weighted averaging of τ_m^2 and τ_m by using Equations (5.19) and (5.22). It is necessary to acquire an expression for the impulse response function P_{lm}. Olshansky and Keck (Ref. 5.2) have obtained an expression for σ by assuming that the impulse function is a constant for all modes and that the summation in these equations can be replaced by an integral over all m modes. The result is given by

$$\sigma = \frac{L\Delta}{2c} (n - \lambda n') \frac{\alpha}{\alpha + 1} \left(\frac{\alpha + 2}{3\alpha + 2} \right)^{1/2} \left[\left(\frac{\alpha - 2 - \varepsilon}{\alpha + 2} \right)^2 \right.$$

$$+ \frac{2(\alpha - 2 - \varepsilon)(3\alpha - 2 - 2\varepsilon)(\alpha + 1)}{(2\alpha + 1)(\alpha + 2)^2} \, \Delta \qquad (5.29)$$

$$+ \left. \frac{2(3\alpha - 2 - 2\varepsilon)^2(2\alpha + 2)^2\Delta^2}{(5\alpha + 2)(3\alpha + 2)(\alpha + 2)^2} \right]^{1/2}$$

Using Equation (5.29), we can obtain an optimum α value such that σ is a minimum by letting $d\sigma/d\alpha = 0$. After considerable algebraic manipulation, we have

$$\alpha_{\text{optimum}} = 2 + \varepsilon - \frac{(3 + \varepsilon)(4 + \varepsilon)}{5 + 2\varepsilon} \Delta \qquad (5.30)$$

If we ignore the effect of spectral width (e.g., $\varepsilon = 0$) we get, from Equation (5.30), $\alpha_{\text{opt}} = 2 - (12/5)\Delta$. This result is in a good agreement with that calculated by the ray analysis given by Equation (4.110). When $\varepsilon \neq 0$ the result of Equation (5.30) deviates from that predicted by Equation (4.110). This difference is caused by introducing a finite spectral width for the source. Since the value of ε depends on the fiber material, the α value can vary over a rather wide range of wavelength, as shown in Figure 5.3.

Equation (5.29) has been plotted in Figure 5.4 as a function of α for three light sources with different spectral widths. The three curves represent an LED,

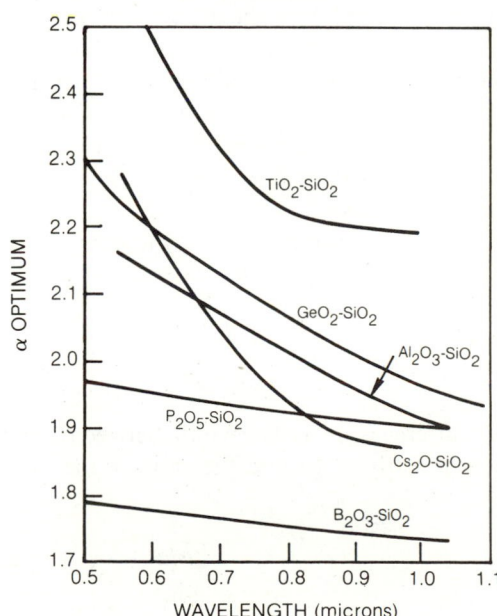

Figure 5.3 Optimum value of α as a function of λ for several types of composite glass. [After H. M. Presby and I. P. Kaminow, *Appl. Optics, 15,* 3029 (1976).]

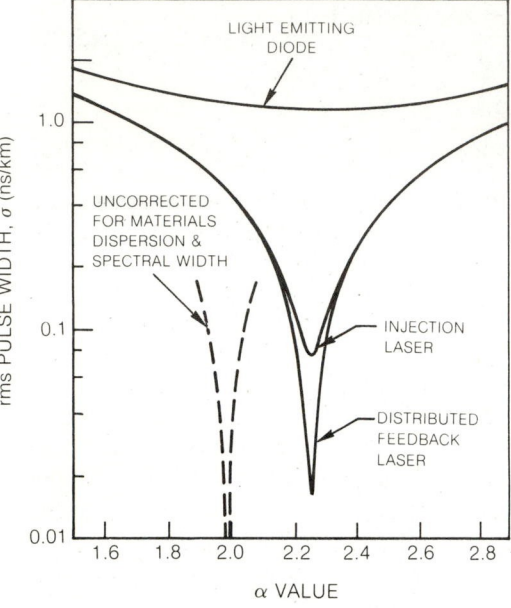

Figure 5.4 Calculated rms pulse spreading in a graded-index fiber as a function of the parameter α assuming a spectral width of 15, 1, and 0.2 nm for a LED, an injection laser, and a DFB laser, respectively. The uncorrected curve assumes no material dispersion and negligible spectral width. (From Ref. 5.2.)

an injection laser, and a distributed feedback (DFB) laser, having typical σ_s values of 150, 10, and 2 Å, respectively. The fiber parameters used in this calculation are $n_0 = 1.460$, $n_c = 1.452$, $\lambda n' = -0.014$, $\lambda^2 n'' = 0.02$, and $\lambda \Delta' = -0.0008$. For the LED source, a pulse broadening of less than 1.5 ns/km can be achieved if α is within 25% of the optimum value. For the injection laser, an α value within 5% of its optimum value will yield a pulse width of less than 0.2 ns/km. For a DFB laser, a width of less than 0.05 ns/km is predicted for fibers with the optimum α value. The dashed curve represents the rms width when both the material dispersion and the spectral width of the source are ignored.

For a step-index fiber, the rms width can be obtained directly from Equation (5.29) by letting $\alpha = \infty$:

$$\sigma_{\text{step}} = \frac{L\Delta}{2\sqrt{3}c} (n - \lambda n')(1 + 6\Delta + \frac{48}{5}\Delta^2) \qquad (5.31)$$

The difference in pulse width between the step-index and the graded-index fibers is fairly large and the ratio of the two σ values is approximately equal to

$$\frac{\sigma_{\text{graded}}}{\sigma_{\text{step}}} \simeq \Delta \times (1 - 3\Delta) \qquad (5.32)$$

The result of Equation (5.29) is obtained with the assumption that all modes of the fiber are excited equally. This is just a simplification to obtain an estimation of the pulse width. In reality, the mode distribution and impulse response can vary substantially and depend strongly on source distribution and coupling schemes. For example, it is possible to obtain a narrower pulse width by varying the incident beam waist or the launching position of a focused source than by using an unfocused incoherent source. On the other hand, an incoherent source can provide a shorter pulse width than can an unfocused laser source. However, these variations are usually very small and lead to only small corrections in the results obtained from the simple model above.

5.6 MODE COUPLING IN A MULTIMODE FIBER

Random coupling of modes in a multimode fiber can have a profound effect on pulse shape as the pulse propagates in a multimode fiber. It can be shown that for a Gaussian input pulse, the shape of the output pulse will remain approximately Gaussian; however, its width increases proportionally to the square root of the fiber length. This result is in a sharp contrast with the usual result that $\sigma \propto L$ for the case when mode coupling is ignored. In other words, a reduction in pulse dispersion can be achieved by inducing mode coupling in a multimode fiber. In practice, this must be compensated for by a certain loss penalty. Since the couplings between guided modes not only transfer energy among themselves but also transfer a certain amount of energy to the radiation modes, radiation losses are unavoidable.

Figure 5.5 shows the mode spacing and the regions of various modes. It is apparent that the spacing between guided-wave modes in β space decreases with decreasing mode number, and the coupling is strongest among its nearest neighbors. For the same reason, the radiation loss probably occurs through the coupling of higher-order modes. The reduction in pulse width can be viewed as a redistribution of power as a result of random mode coupling, during which the power contained in lower-order modes is transferred to higher-order modes, and vice versa. In the meantime, the power carried by higher-order modes is also transferred to the radiation modes. This constitutes a loss in the system. As a result of this loss mechanism, a center of gravity of the power distribution is

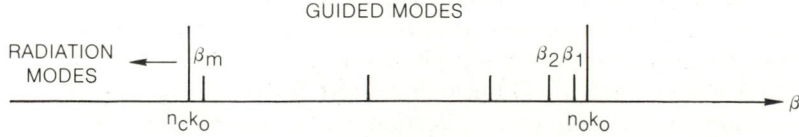

Figure 5.5   Regions defining guided-wave modes and radiation modes.

established around an average group velocity v_g, which more closely represents lower order modes with a negligible power transfer loss. This is possible because the power outflow from these modes has been compensated for by the power inflow from lower-order modes.

In a steady state situation, the coupled power equations assume the form

$$\frac{dP_m}{dz} = -(a_m + b_m)P_m + \sum_n c_{mn}P_n \tag{5.33}$$

where a_m is the power-loss coefficient of the mth mode that transfers directly to the radiation modes, and

$$b_m = \sum_n c_{mn} \tag{5.34}$$

and c_{mn} are the power coupling coefficients among various modes and depend on the imperfections of the waveguide. In the time-dependent case, we transform the derivative dP_m/dz from a stationary frame to a moving coordinate system traveling at velocity v_m. We then write

$$\frac{dP_m}{dz} = \frac{\partial P_m}{\partial z} + \frac{1}{v_m}\frac{\partial P_m}{\partial t} \tag{5.35}$$

Consequently, the time-dependent coupled power equations can be written as

$$\frac{\partial P_m}{\partial z} + \frac{1}{v_m}\frac{\partial P_m}{\partial t} = -(a_m + b_m)P_m + \sum_n c_{mn}P_n \tag{5.36}$$

The first term on the right-hand side of Equation (5.36) represents the power loss of the mth mode and the second term represents the power gain through random coupling. Exact solutions of this set of coupled equations are very complex. The model that will be followed here is that given by Gloge (Ref. 5.3), who treats this problem by assuming a modal continuum rather than thousands of individual modes. In this model the mode coupling problem can be described by a diffusion process.

We first consider fibers in which the steady-state mode distribution does not include modes close to cutoff. The output power from a fiber can be expressed as a function of time and output angle, which is related directly to the mode number for the case of a short input pulse. Of most interest is the impulse response that can be obtained by integrating over all angles at the output. Because Δ is very small for all fibers of interest, the maximum angle limited by the condition for critical internal reflection can be approximated by

$$\theta_{\max} \simeq \sqrt{2\Delta} \tag{5.37}$$

In a multimode fiber, the modes are so densely spaced that their distribution can be considered as continuous. The state of the fiber at a point z and time t can then be described by a distribution $P(\theta, z, t)$, where θ is a continuous variable.

Using the model above, Equation (5.36) can be modified by replacing the two terms on the right-hand side with the following:

1. The loss term is expressed by $-A\theta^2 P$, where A is the loss coefficient measured in m^{-1} rad^{-2}, and the attenuation effect is assumed to be proportional to the square of the characteristic angle. This assumption is reasonable, because the loss is expected to be greater for higher-order modes.

2. The mode coupling is found to occur essentially between closely adjacent modes and for this reason, takes the form of a diffusion process. The increase in power as a result of diffusion can be expressed by $(1/\theta)\partial/\partial\theta(\theta D\, \partial P/\partial\theta)$, a term typical for radial diffusion in the cylindrical configuration. D is a coupling coefficient and is assumed to be independent of θ.

Therefore, we rewrite Equation (5.36) as follows:

$$\frac{\partial P}{\partial z} + \frac{1}{v_g}\frac{\partial P}{\partial t} = -A\theta^2 P + \frac{1}{\theta}\frac{\partial}{\partial\theta}\left(\theta D\frac{\partial P}{\partial\theta}\right) \tag{5.38}$$

where the group velocity can be estimated from a simple ray picture that

$$v_g = \frac{c\cos\theta}{n} \simeq \frac{c}{n(1 + \theta^2/2)} \tag{5.39}$$

Substituting Equation (5.39) into (5.3), we have

$$\frac{\partial P}{\partial z} = -A\theta^2 P - \frac{n}{2c}\theta^2\frac{\partial P}{\partial t} + \frac{1}{\theta}\frac{\partial}{\partial\theta}\left(\theta D\frac{\partial P}{\partial\theta}\right) \tag{5.40}$$

where we have ignored the delay that is common to all modes. We now multiply Equation (5.40) by e^{-st} and integrate over t from $t = 0$ to $t = \infty$. With the help of the Laplace transform, e.g.,

$$p(\theta, z, s) = \int_0^\infty e^{-st} P(\theta, z, t)\, dt$$

we obtain

$$\frac{\partial p}{\partial z} = -A\, b^2\theta^2 p + \frac{1}{\theta}\frac{\partial}{\partial\theta}\left(\theta D\frac{\partial p}{\partial\theta}\right) \tag{5.41}$$

where

$$b = \left(1 + \frac{ns}{2cA} \right)^{1/2} \tag{5.42}$$

We see that the result of Equation (5.41) is identical to the steady-state case as described by Equation (5.33), with the exception that the loss coefficient contains a product of Ab^2. This indicates that in the time-dependent case for an impulse response, $s \neq 0$, hence $b \neq 1$. Physically, this implies that the spatial spread of energy during the pulse propagation is equivalent to the time spread of energy, which can lead to pulse narrowing in the time-dependent solution.

To obtain a closed-form solution of Equation (5.41), we assume a solution of the form

$$p(\theta, z, s) = f(z, s) \exp \left(\frac{-\theta^2}{\Theta^2} \right) \tag{5.43}$$

where Θ represents the angular width of the pulse, which varies from Θ_0 to Θ_z as z increases. Substituting Equation (5.43) into (5.41), we obtain

$$\frac{\partial f(z, s)}{\partial z} + f(z, s) \frac{2\theta^2}{\Theta^3} \frac{\partial \Theta}{\partial z} = -Ab^2 \theta^2 f(z, s) - Df(z, s) \left(\frac{4}{\Theta^2} - \frac{4\theta^2}{\Theta^4} \right)$$

By rearranging terms and separating the variables, we write

$$\frac{1}{f} \frac{\partial f}{\partial z} + \frac{4D}{\Theta^2} = -\frac{2\theta^2}{\Theta^3} \left(\frac{\partial \Theta}{\partial z} + \frac{Ab^2 \Theta^3}{2} - \frac{2D}{\Theta} \right) \tag{5.44}$$

Since this equation must hold for all values of θ, we obtain two separate equations:

$$\frac{\partial f(z, s)}{\partial z} = -\frac{4D}{\Theta^2} f(z, s) \tag{5.45}$$

and

$$\frac{\partial \Theta}{\partial z} = -\frac{A}{2} b^2 \Theta^3 + \frac{2D}{\Theta} \tag{5.46}$$

For very large z, the quantity $\partial \Theta / \partial z$ in Equation (5.46) must vanish, so that we obtain

$$\Theta_\infty = \frac{(4D/A)^{1/4}}{\sqrt{b}} \tag{5.47}$$

Substituting Equation (5.47) into Equation (5.45) we obtain

$$f(z, s) = f(0, s) \exp(-\sqrt{b} \, \gamma_\infty z) \tag{5.48}$$

where γ_∞ is the steady-state attenuation coefficient for a very long fiber $(z \to \infty)$.

The expression for γ_∞ in Equation (5.48) is given by

$$\gamma_\infty = \frac{4D}{\Theta_\infty^2 \sqrt{b}} = 2\sqrt{ADb} \tag{5.49}$$

For completeness we give the solutions of Equations (5.45) and (5.46), without going through the details, in the following:

$$\Theta^2(z, s) = \frac{\Theta_\infty^2}{b} \frac{b\Theta_0^2 + \Theta_\infty^2 \tanh(b\gamma_\infty z)}{\Theta_\infty^2 + b\Theta_0^2 \tanh(b\gamma_\infty z)} \tag{5.50}$$

and

$$f(z, s) = \frac{f(0, s)b\Theta_0^2}{\Theta_\infty^2 \sinh(b\gamma_\infty z) + b\Theta_0^2 \cosh(b\gamma_\infty z)} \tag{5.51}$$

where Θ_0 is the initial angular width at $z = 0$.

To find the time-dependent solutions, we must take the inverse Laplace transform of the results above. The closed-form Laplace transformation of Equation (5.43) exists only for the special cases when $z \ll 1/\gamma_\infty$ and $z \gg 1/\gamma_\infty$. In the former case of a short fiber, we replace $\sinh b\gamma_\infty z$ and $\tanh b\gamma_\infty z$ by their argument $b\gamma_\infty z$ and set $\cosh b\gamma_\infty z = 1$. We can rewrite Equation (5.43) with the help of Equations (5.42), and (5.47) to (5.51) as follows:

$$p(\theta, z, s) = \frac{f(0, s)}{1 + \gamma_\infty z} \exp\left[-\theta^2 \left(\frac{1}{\Theta_0^2} + \frac{nz}{2c} s \right) \right] \tag{5.52}$$

Equation (5.52) has the inverse Laplace transform

$$P(\theta, z, t) = F\left(0, t - \frac{n\theta^2 z}{2c}\right)(1 + \gamma_\infty z)^{-1} \exp\left(-\frac{\theta^2}{\Theta_0^2}\right) \tag{5.53}$$

The factor $(1 + \gamma_\infty z)^{-1}$ represents the loss in the short length of fiber; the expression $\exp(-\theta^2/\Theta_0^2)$, which describes the angular power distribution, is conserved under the transformation. The coefficient $F(0, t - n\theta^2 z/2c)$ shows that the portion of the input pulse propagating at an angle θ is delayed by $n\theta^2 z/2c$. The result of this calculation clearly shows that mode coupling has not affected the pulse shape after propagating a very short distance.

By integrating Equation (5.52) over all angles we obtain the total output power as

$$P(z, s) = 2\pi \int_0^\infty p(\theta, z, s)\theta \, d\Theta = \frac{\pi f(0, s)\Theta_0^2}{(1 + \gamma_\infty z)(1 + n\Theta_0^2 z s/2c)} \tag{5.54}$$

We now set $f(0, s) = 1$, which corresponds to an infinitesimally short input pulse of energy. The Laplace transformation of Equation (5.54) yields an impulse response of the fiber:

$$P(z, t) = \frac{2\pi c}{nz(1 + \gamma_\infty z)} \exp\left(-\frac{2ct}{n\Theta_0^2 z}\right) \tag{5.55}$$

Equation (5.55) can be derived without the use of the power-flow equation, because in a very short fiber, mode coupling is not expected to play an important role. The fact that coupled-mode theory leads to these results indicates that the diffusion model is valid for predicting the impulse response.

We shall now investigate the impulse response after a short pulse has propagated over a very long distance (e.g., $z \gg 1/\gamma_\infty$). In this case we let tanh $b\gamma_\infty z = 1$ and sinh $b\gamma_\infty z = \cosh b\gamma_\infty z = 1/2$ $\exp(b\gamma_\infty z)$. Substituting these approximations into Equation (5.43) and assuming that $\Theta_0 \simeq \Theta_\infty$, we obtain

$$p(\theta, z, s) = \frac{2b}{1 + b} \exp\left[-b\left(\frac{\theta^2}{\Theta_0^2} + \gamma_\infty z\right)\right] \tag{5.56}$$

Upon integrating Equation (5.56) over all angles, we get

$$p(z, s) = \frac{2\pi\Theta_0^2}{1 + b} \exp(-b\gamma_\infty z) \tag{5.57}$$

Substituting Equation (5.42) for b and taking the Laplace transform of $p(z, s)$, we get

$$P(z, t) = \Theta_0^2 \sqrt{\frac{\pi}{Tt}} \left(\frac{1}{2} + \frac{t}{\gamma_\infty z T}\right)^{-1} \exp\left(-\frac{t}{T} - \frac{\gamma_\infty^2 z^2 T}{4t}\right) \tag{5.58}$$

where

$$T = \frac{n}{2cA} = \frac{n\Theta_0^2}{2c\gamma_\infty} \tag{5.59}$$

The results of Equation (5.58) for pulse shapes are plotted in Figure 5.6 as a function of the delay time t/T for various normalized lengths $\gamma_\infty z$. These impulse responses are normalized for equal peak value. Expression (5.58) is very different from expression (5.55) for the uncoupled modes. From Equation (5.55) we see that when $t = T$, the pulse amplitude is reduced by a factor of e^{-1} if $z = 1/\gamma_\infty$. Therefore, the normalized length $1/\gamma_\infty$ is defined as the distance within which a 1-neper loss is incurred. This is equivalent to a total loss in decibels by the amount

$$\alpha(\text{dB}) = 4.35\gamma_\infty z \tag{5.60}$$

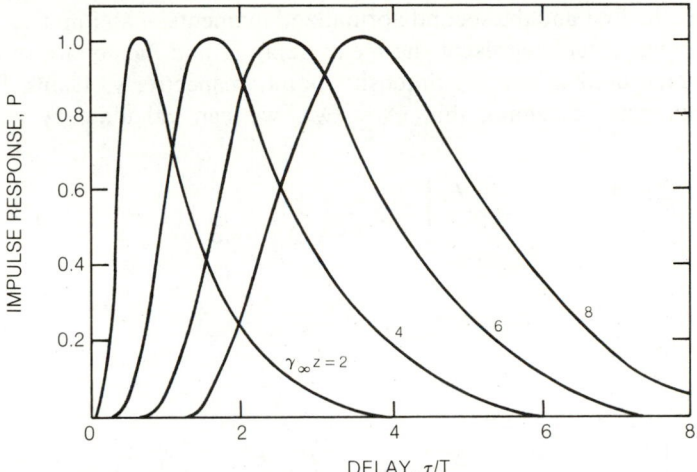

Figure 5.6   Normalized impulse response as a function of normalized time as a short pulse propagates in a multimode fiber over a normalized distance $\gamma_\infty z$. [From *The Bell System Technical Journal* (Ref. 5.3). Reprinted with permission of American Telephone and Telegraph Co., © 1973.]

5.7 PULSE DISTORTION

For a Gaussian input pulse, the output remains Gaussian, but its width is subject to change and dependent on the fiber length. In previous sections we have discussed phenomena involving pulse delay and pulse broadening in the absence of mode coupling. In this section we derive expressions for the mean pulse delay and the mean pulse width as a function of propagation distance z in a multimode fiber, in which mode coupling predominates. In the application of the diffusion model, we define the nth moments of an impulse response as

$$M_n = \int_0^\infty t^n P(z, t)\ dt \qquad (5.61)$$

Because of the general relation between $P(z, t)$ and its Laplace transform $p(z, s)$, we can express M_n in terms of the nth derivatives of p as

$$M_n = (-1)^n \frac{\partial^n p}{\partial s^n}\bigg|_{s=0} \qquad (5.62)$$

The normalized moments, m_n, can be simply expressed in terms of the nth derivatives of $\ln p$ as given by

$$m_n = \frac{M_n}{\displaystyle\int_0^\infty P\ dt} = (-1)^n \frac{1}{p}\frac{\partial^n p}{\partial s^n} = (-1)^n \frac{\partial^n \ln p}{\partial s^n} \qquad (5.63)$$

Physically, the first and the second normalized moments of an impulse response about its mean values represent the mean delay τ_θ and the square of the half-width σ_θ^2 measured at the $1/e$ intensity point, respectively. Using Equation (5.43) and again assuming that $\Theta_0 \simeq \Theta_\infty$, we can calculate τ_θ and σ_θ^2 as follows:

$$\tau_\theta = -\left.\frac{\partial \ln p}{\partial s}\right|_{s=0} = \frac{T}{2}\left[\gamma_\infty z + \left(\frac{\theta^2}{\Theta_0^2} - \frac{1}{2}\right)(1 - e^{-2\gamma_\infty z})\right] \quad (5.64)$$

and

$$\sigma_\theta^2 = \left.\frac{\partial^2 \ln p}{\partial s^2}\right|_{s=0}$$

$$= \frac{T^2}{4}\left[\gamma_\infty z + \left(\frac{\theta^2}{\Theta_0^2} - \frac{5}{4}\right) - 2\gamma_\infty z\left(2\frac{\theta^2}{\Theta_0^2} - 1\right)e^{-2\gamma_\infty z} \right. \quad (5.65)$$

$$\left. + e^{-2\gamma_\infty z} + \left(\frac{\theta^2}{\Theta_0^2} - \frac{1}{4}\right)e^{-4\gamma_\infty z}\right]$$

where T^2 is defined by Equation (5.59).

Figure 5.7 shows the variation of τ_θ and σ_θ as a function of $\gamma_\infty z$ for $\theta = 0$ and $\theta = \Theta_\infty$. For a very short length, the pulse propagates without broadening

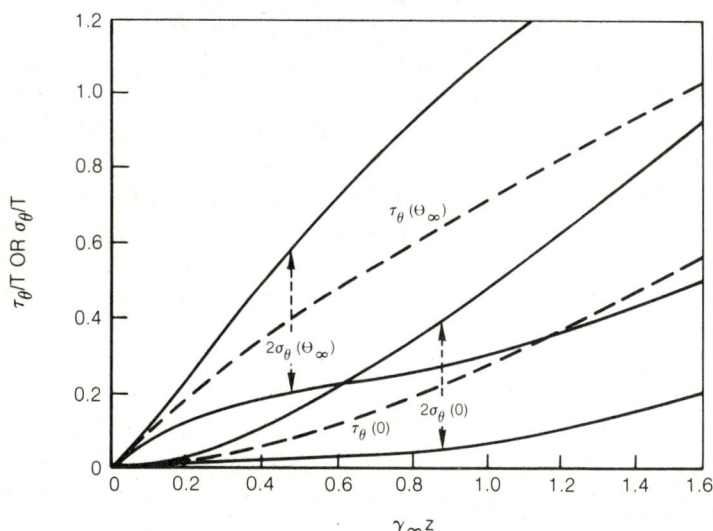

Figure 5.7 Time delay and pulse spreading as a function of fiber length. [From *The Bell System Technical Journal* (Ref. 5.3). Reprinted with permission of American Telephone and Telegraph Co., © 1973.]

and merely suffers a mode-dependent delay $n\theta^2/2c$, as expected. However, before the normalized length $1/\gamma_\infty$ is reached, the pulse in all modes begins to widen. Once $1/\gamma_\infty$ is passed, the pulse width in all modes increases essentially as $T(\gamma_\infty z)^{1/2}$. This result differs significantly from that when mode coupling is ignored. For $\gamma_\infty z \gg 1$, we can rewrite Equations (5.64) and (5.65) as

$$\tau_\theta = \frac{T}{2}\left(\gamma_\infty z - \frac{1}{2} + \frac{\theta^2}{\Theta_0^2}\right) \tag{5.66}$$

and

$$\sigma_\theta = \frac{T}{2}\left(\gamma_\infty z - \frac{5}{4} + \frac{\theta^2}{\Theta_0^2}\right)^{1/2} \tag{5.67}$$

To calculate the θ-independent delay and width, we must apply Equation (5.54) and perform the first and the second derivatives of $p(z, s)$. Without going through the details, the results are given in the following:

$$\tau = \frac{T}{2}\left[\gamma_\infty z + \frac{1}{2}(1 - e^{-2\gamma_\infty z})\right] \tag{5.68}$$

and

$$\sigma = \frac{T}{2}\left[\gamma_\infty z(1 - 2e^{-\gamma_\infty z}) + \frac{3}{4} - e^{-2\gamma_\infty z} + \frac{1}{4}e^{-4\gamma_\infty z}\right]^{1/2} \tag{5.69}$$

The ratio σ/T is shown in Figure 5.8 as a function of the normalized length $\gamma_\infty z$. For $z \ll 1/\gamma_\infty$, the width σ approaches $T\gamma_\infty z$, as expected. At $z = 1/4\gamma_\infty$, σ begins to follow a new asymptote,

$$\sigma = \frac{T}{2}\sqrt{\gamma_\infty z} \tag{5.70}$$

The derivation from the $T\gamma_\infty z$ curve as shown in Figure 5.8 is an indication of mode coupling. The effect of mode coupling reduces the width by a factor $\sqrt{4\gamma_\infty z}$ in exchange for an increase in the overall attenuation by $4.35\gamma_\infty$ dB/km. The physical contents of these results can best be summarized by defining a coupling length $L = 1/4\gamma_\infty$, at which the width of the impulse response changes from a linear to a square-root dependence of the length. Furthermore, if we introduce $\tau = \tau' + T\gamma_\infty z/2$ into Equation (5.58) with $\tau' \ll T\gamma_\infty z/2$, we can write the impulse response, for large z, as

$$P(z, t') = \Theta_0^2 \sqrt{\frac{2\pi}{\gamma_\infty z}} \exp\left(-\gamma_\infty z - \frac{2\tau'^2}{T^2\gamma_\infty z}\right) \tag{5.71}$$

Equation (5.71) indicates that the pulse shape changes from exponential to Gaussian in time with the variance $4\sigma^2$, as given by Equation (5.70). Beyond the coupling length L, the width of the impulse response increases only as the

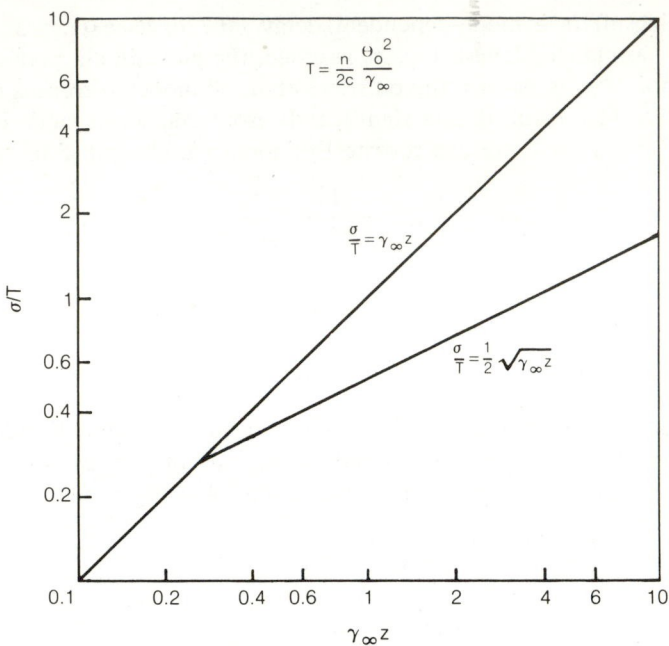

Figure 5.8   Asymptotic behavior of the width of an impulse response as a function of normalized length. [From *The Bell System Technical Journal* (Ref. 5.3). Reprinted with permission of American Telephone and Telegraph Co., © 1973.]

square root of the fiber length. Physically, the power carried by lower-order modes or small values of θ travels faster and tends to lead the pulse, but in the meantime, feeds back continuously into higher-order modes by diffusion. The power carried by these higher-order modes tends to fall behind the main pulse but tends to catch up with the main body of the pulse by diffusion. As a result, both the leading and trailing edges of the pulse are clipped to an extent that the characteristic width increases only as $\sqrt{L}$. Initially, a linear growth of the pulse width with length is only temporary while waiting for a redistribution of power as it approaches a steady-state condition.

5.8 SCATTERING AND ABSORPTION LOSSES

Because of the severity of optical power losses in fibers due to scattering and absorption, for a long time glass fibers were not seriously considered as a candidate for use as an optical transmission medium. Only within very recent years, since techniques for glass material preparation and fiber manufacturing have been improved to minimize these losses, has the field of fiber optics grown tremendously through a concerted research and development effort that has increased the level of understanding of fiber optical transmission by several orders of magnitude. We discuss these loss mechanisms here and leave a discussion of fiber fabrication techniques to Chapter 6.

As evident from the discussion of material dispersion, a longer-wavelength transmitter in the region 1.2 to 1.6 μm for a fiber optical system is a good choice. Another reason for choosing a longer wavelength is to minimize the scattering loss. This is a phenomenon inherent in all glasses, because all optical materials contain defects that scatter light as it propagates over a long distance. The amount of power scattered by those defects is dependent on the defect density and the scattering cross section, C. The value of C is a measure of the scattered power P_s in a single scattering event for an incident light intensity I_0 (watts/cm^2). By definition, we write

$$P_s = CI_0 \qquad (5.72)$$

In Equation (5.72), C has the dimensions of an area and is related to the geometrical cross-sectional area of the scatterer and the strength of the scattering interaction S as

$$C = \pi r^2 S \qquad (5.73)$$

where r is the average radius of the scatterer and S is the dimensionless factor, which depends on the ratio r/λ.

Consider that when a plane wave is incident on a spherical particle, the scattered wave in the far field and the field inside the sphere can be expanded in terms of spherical coordinates. The expansion coefficients can be determined by matching these fields at the boundary of the scatterer and also by satisfying the conditions at infinity. These coefficients are related to the scattering amplitudes, which, in general, are very complex functions and yield information concerning both the amplitudes and phase of the scattered field.

In all types of glass, the scatterers are primarily impurities such as oxides and transition metal ions, with sizes typically much smaller than the wavelength (e.g., $r/\lambda \ll 1$). In this limit commonly known as Rayleigh scattering, the expressions for the scattering amplitudes are considerably simplified. Assuming that the incident light is linearly polarized and that secondary waves are irradiated from the microscopic scatterers in the form of an induced electric dipole, with its moment $|\mu|$ parallel to the polarization, the scattering cross section in the far field can be expressed simply by

$$C = \frac{8\pi}{3\lambda^4} |\mu|^2 f(\theta) \qquad (5.74)$$

where $f(\theta)$ is an angular factor describing the radiation pattern of a dipole. It should be noted that the most outstanding character in this expression is the $1/\lambda^4$ dependence, which has already been illustrated in Figure 1.4. For the case of a simple dipole, the perpendicular polarization scatters isotropically and the parallel polarization scatters as $\cos^2 \theta$, which yields an equal distribution in both the forward and backward directions, and zero at the right angle.

The total scattered power per unit length is a product of P_s and the number density of the scatterers in the fiber. Since the scatterers distribute randomly in the glass, there exists a density fluctuation over the length of the fiber. This can be traced back to thermal fluctuations that cause the impurities to diffuse in the form of Brownian motion in the molten state before solidification. The magnitude of this density fluctuation is proportional to the product of the softening point of the glass, T_s, and the isothermal compressibility, κ. In terms of these parameters, one obtains a scattering cross section given by (Ref. 5.4)

$$C = \frac{8\pi^3}{3\lambda^4} (n^2 - 1)kT_s\kappa \tag{5.75}$$

where k is the Boltzmann constant. For fused silica with a softening temperature near $1500°C$, we obtain from Equation (5.75) a loss of 1.7 dB/km at 0.85 μm, which is consistent with the experimental results. According to Equation (5.75), a lower softening point may lead to a lower scattering loss. In reality, this situation is complicated by the fact that in materials with a lower softening temperature, a variety of impurities can easily be introduced. This creates a composition fluctuation which can affect the refractive index much more severely than can the density fluctuation. Consequently, the composition fluctuation in high-index glass can introduce higher losses than those in lower-index glass. In both cases the $1/\lambda^4$ dependence is the dominating factor. There are other scattering processes for which the scattering cross section is relatively independent of λ. This occurs when r/λ approaches unity at which point Mie scattering begins to dominate. In this case, the fibers are usually made with very poor quality because they contain very large scattering centers with scale sizes greater than λ. Consequently, loss mechanism other than scattering can play an important role.

In addition to scattering losses, there are absorption losses in glass which arise from both the intrinsic structure of the material and impurity absorption. The intrinsic absorption originates from a charge transfer between various energy bands with characteristic spectra lying primarily in the ultraviolet region. However, these bands are sufficiently wide, with their spectral wings well extended into the near-infrared region; therefore, they could cause some absorption loss. However, recent measurements indicate that for wavelengths beyond 0.8 μm, band-edge absorption is almost certainly less than 1 dB/km.

The impurity absorption, on the other hand, is caused by metal ions such as Fe, Cu, V, and Cr. At an impurity level of 10 ppb, the loss figure could run up to 20 dB/km. Highly purified silica glass is now routinely made without discerning the loss component due to impurities. However, in highly doped glass, losses due to metal ions are troublesome. Special care must be taken to reduce the impurity level in this glass.

The transition metal ions have incompletely filled inner electron shells which give rise to their characteristic absorptions by inducing transitions between those levels. Unlike metals, the formation of oxidation in glass leaves the transition ions with unfilled levels. Even though transitions between different oxidation states are forbidden, the perturbation introduces a splitting in these levels that is responsible for the spectra observed. The coloration observed in heavily doped glass can be used to characterize different impurities under different conditions of oxidation. Proper balance between oxidation and reduction through control of the partial pressure of the oxygen in the melt can reduce the absorption loss to a minimum. In general, when the glass is overly reduced, the absorption tends to increase at longer wavelengths. If the opposite is true, the absorption loss tends to increase at shorter wavelengths. The process of the reduction is to convert ion species from one type to another (e.g., from Fe^{3+} to Fe^{2+}, etc.). To accomplish such a chemical reaction, the partial pressure of oxygen may be varied by many orders of magnitude. There are many ways to oxidize the glass in the melt: for example, by adding oxides of arsenic and antimony, or by using bubbling gas such as CO and CO_2 to stir the melt. At present the production of high-purity glass is possible only by chemical vapor disposition technique for the growth of glass preforms that is discussed in Chapter 6.

Another mechanism responsible for absorption loss involves vibrational energy associated with some of the common bonds present in glass. In most cases the vibrational spectra of glass lie in the infrared region 2 to 10 μm, in which the overtones of the fundamental stretching vibration of the hydroxyl ion (OH) play an important role. The fundamental is centered around 2.8 μm, with its first three overtones at 1.4, 0.97, and 0.75 μm, respectively. Attempts have been made by the Corning Glass Works and others to identify the OH overtones by matching individual absorption lines. However, the measured line shape deviates significantly from the expected Lorentzian profile. This type of measurement is extremely difficult unless the glass under investigation can be made without any loss other than that due to the OH stretching mode. To reduce OH content, a technique has been developed by heating the oxide powders of glass in an oven at 250°C for a few days. The glass is otherwise prepared in the usual way but under a controlled-humidity atmosphere, and the melt is bubbled with gases of varying dew points. From the measurement of absorption loss in this glass, which is about only 1 dB/km at 0.9 μm, the estimated OH content corresponds to only 1 ppm weight of water.

5.9 MICROBENDING LOSSES

Optical losses associated with the cabling process can be introduced as a result of microbending of the fiber. This problem becomes serious when the radius of curvature $R(z)$ of the bend is small but large compared with the radius of the fiber. Because of the stiffness of the fiber, $R(z)$ normally changes slowly and continuously with fiber length. This type of loss is found most often in single-mode or quasi-single-mode fiber systems, where very large bit rates are desired. In these systems, only the fundamental mode is excited. Due to the bending, the fundamental mode power is eventually lost through coupling to higher-order and/or radiation modes.

Problems of this type are usually solved by using coupled-mode theory, which involves a system of coupled equations that are usually solved by perturbation methods. Several methods have been introduced, with varying degrees of accuracy. Unfortunately, all these methods are rather tedious and involved. One must first deduce the eigenvalue equations by using proper boundary conditions involving microbending before one can calculate the coupling coefficient between the HE_{11} mode and the leaky mode. In the end, it is very difficult to compare the theoretical results accurately with measurements because they depend strongly on the shapes of the correlation and statistical distribution functions of these micro defects. A common technique often used in these analyses is to introduce an effective refractive index n_e such that the fields in the bend can be approximated by those in the straight section. This involves a transformation of the refractive index of the bending fiber in such a way that the equation reduces to that of an equivalent straight fiber. Coupling between the normal modes of a fiber, due to microbending, can be treated by introducing a perturbation term involving the effective refractive index. In this approximation, it is necessary to introduce an additional loss term, because as the mode propagates from a straight section into a curved section, the field distribution could change significantly. Therefore, in addition to a pure bending loss a_b as in the case of an abrupt change of curvature R, a transition loss a_T may also occur as a result of mode conversion in the transition region where the modal energy is redistributed.

At small values of R, the pure bending loss is predominant. As R increases, it has been shown (Ref. 5.5) that the transition loss can exceed the bending loss for the same fiber. The calculated values of the loss coefficients a_b and a_T for the bending and transition cases are listed in Table 5.1. These results are obtained specifically for a single-mode fiber with $V = 2.4$ and $\Delta = 0.002$ at $\lambda = 1 \ \mu m$. It is clear from these results that these two loss coefficients are strong functions of σ^{-1}, which is the rms deviation of $R(z)$. Furthermore, these loss coefficients depend strongly on Δ values and correlation length in the case of a random deformation. In general, microbending losses decrease rapidly with increasing Δ; therefore, it is very important to use a single-mode step-index fiber with a numerical aperture as high as possible for long-distance tele-communication systems.

TABLE 5.1   Calculated Microbending Loss a and Transition Loss a_T
as a Function of the RMS Deviation of $1/R(z)$

σ^{-1} (cm)	a_b (dB/m)	a_T (dB/m)
6	2.6	0
12	0.015	0
40	0	5.5
400	0	0.055

PROBLEMS

5.1. Calculate the group delay between the fastest and the slowest mode in a 1-km-long step-index fiber with $n_0 = 1.5$ and $\Delta = 0.003$, using a light source at 0.9-μm wavelength.

5.2. Derive Equation (5.26).

5.3. For a typical LED source emitting in the wavelength region 0.8 to 0.9 μm, the spectral width is about 20 nm. Calculate the material dispersion of a silicate fiber with a LED source using the results of Figure 5.2, over a fiber length of 1 km.

5.4. Using the result of Equation (5.29), show that for $\alpha = \infty$,

$$\sigma_{\text{step}} = \frac{1}{2\sqrt{3}\,c}\,L(n - \lambda n')\Delta \times \left(1 + 6\Delta + \frac{48}{5}\,\Delta^2\right)$$

5.5. Using Equation (5.29), show that for $\alpha = 2$, and ε is equal to zero

$$\sigma_{\text{graded}} = \frac{1}{2\sqrt{3}\,c}\,L(n - \lambda n')\Delta^2$$

5.6. Assuming an effective numerical aperture NA of a fiber to be $n\theta_\infty$, derive an expression for the width of the impulse response for the case $z \gg 1/\gamma_\infty$ as a function of NA.

5.7. For a fixed fiber length L which is much larger than $1/\gamma_\infty$, calculate the loss penalty for an increase in bandwidth $B = 1/4\sigma$ by a factor of 2.

REFERENCES

5.1. D. Gloge, *Appl. Opt.*, *10*, 2252 (1971).

5.2. R. Olshansky and D. B. Keck, *Appl. Opt.*, *15*, 483 (1976).

5.3. D. Gloge, *Bell Syst. Tech. J.*, *6*, 801 (1973).

5.4. R. D. Maurer, *J. Chem. Phys.*, *25*, 1206 (1956).

5.5. W. A. Gambling, H. Matsumura, and C. M. Ragdale, *Opt. Quantum Electron.*, *11*, 43 (1979).

Glass Materials,
Fiber Fabrication, and
Characterization Techniques

6.1 INTRODUCTION

Low-loss fibers with $a < 5$ dB/km can now be routinely manufactured by using ultrapure glass materials and advanced fiber drawing techniques. To make good optical fibers, many requirements for the growth of glass materials with varying indices while keeping the fiber defect-free at a long length must be met simultaneously. We first discuss the structure of glass and its physical properties. Methods by which different glass with desired properties can be formed into fibers are then introduced. The chapter includes a section on various measurement techniques for the characterization of fiber index profile, dispersion, and losses.

6.2 GLASS MATERIALS

It is interesting to note that the earth's crust is composed of approximately 62% oxygen and 21% silicon. These two elements are the major constituents of glass, with some minorities such as metal ions either substituting for silicon in the tetrahedral structure or coordinating themselves so as to form voids between the silicon tetrahedra. The silicon tetrahedra are arranged such that the oxygen atoms actually form the most closed packing. The structure of a typical silicate glass is shown in Figure 6.1. The exact composition of glass can vary tremendously, but it must contain predominately oxygen, silicon, boron, sodium, and aluminum, the network-forming atoms, listed in Table 6.1. There

(a)

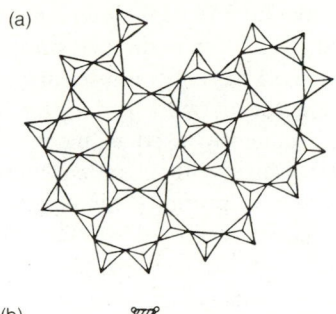

(b)

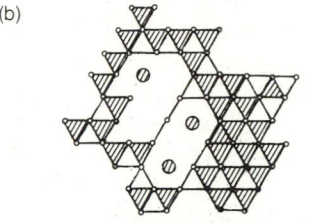

Figure 6.1   (a) Lattice arrangement of SiO$_4$ tetrahedra in a glass; (b) two-dimensional network of Si–O group, including the effect of additional network modifiers.

- ○ OXYGEN SITE
- ▽ SILICON IS LOCATED IN ANOTHER PLANE CONTAINING THE VERTICES OF THE TETRAHEDRA
- ⊘ NETWORK MODIFIER

are other oxides (see Table 6.1), called network modifiers, which serve to change or modify the basic properties of glass, such as index of refraction, thermal expansion, absorption coefficient, and melting point.

The strongest bond in glass is the Si—O bond in the silicon tetrahedra. The absence of symmetry in this structure allows the strength of the bond to vary from one tetrahedron to the next. The interatomic distances in each tetrahedron are virtually the same, 1.62 Å. Each silicon atom is tetrahedrally coordinated by four oxygen atoms, and each oxygen atom is bonded to two silicon atoms. If other oxides are added to the silica glass, the total number of oxygen atoms present in the glass is increased. Consequently, some of the oxygen atoms that are bonded to only one silicon atom must pick up additional bonds with other atoms present. For example, addition of sodium tends to break up the Si—O network to form a sodium silicate glass. As a result, the structure

TABLE 6.1   Some of the Most Common Types of Glass and Their Composition

Network Former	Network Modifier
SiO$_2$	K$_2$O
B$_2$O$_3$	MgO
Al$_2$O$_3$	CaO
Na$_2$O$_3$	PbO

is less tightly bonded and its melting temperature is lowered. On the other hand, metal ions can diffuse into the glass and be distributed randomly among the voids in the network, as shown in Figure 6.1. There are other types of glass: for example, where the network is formed by B_2O_2, Na_2O_3, and so on. These materials are frequently used in industrial products but are not suitable for optical fibers because of their optical quality.

The distinctive property of all glass is that it undergoes a continuous decrease in viscosity when heated. Therefore, glass softens gradually instead of going through an abrupt melting stage as encountered in crystals. This is, of course, a unique property of an amorphous solid. If a glass has been stressed, some type of preferred orientation effect takes place in the immediate vicinity of the induced strains and local anisotropy results. Therefore, the index of refraction depends on the previous thermal history of the glass. The refractive index and the thermal expansion coefficient can also vary with the composition of the glass. This is the most common way to control the difference in refractive indices between the core and the cladding. By varying the concentration of the modifiers in a silicate network, one can obtain the desired Δ along an isothermal expansion curve. This is usually done with two additional oxides. For example, the sodium calcium silicate group is one such system. Another interesting system is the sodium borosilicate group, because it not only has the freedom to modify the index while holding the expansion coefficient constant, but also has a relatively low softening point, a matter of great importance for fiber manufacturing. However, great care must be taken in selecting the composition for this system because there exists a relatively wide region in the phase diagram in which the glass product is found to be unstable and has a tendency to separate into two different glass groups, one with a silica network and the other with a boric oxide network. For a very large refractive index difference, the lead silicate group is the system to use. However, losses in lead silicate fibers have been found to be much greater than those in other silicate glass.

Of course, there are many possible combinations that form glass. The groups mentioned above can be formed by the conventional melting technique of oxide powders. In this way, they can be manufactured in large quantities at relatively low cost. For high-purity, low-loss glass, the material preparation has to be modified from that of the melting method. The most successful technique has been chemical vapor deposition (CVD), which is discussed in the following section.

6.3 PREFORM PRODUCTION

There are many techniques for making glass fibers. For example, glass fibers can be drawn directly from the melt of the oxide powders in a crucible. This is a relatively simple and economical method for making glass fibers, and in this case it is not necessary to prepare glass preforms. However, to produce fibers with extremely low loss, a two-step process is often used: first, the production of an ultrapure glass preform, and second, drawing fibers from this preform. Preforms consist of glass rods with the desired index profile. Many techniques have been developed to make preforms. The most attractive ones are those that provide for rapid growth of solid layers of glass with a high degree of purity. The difficulty still exists of making a homogeneous preform of long length with a desired graded-index profile. The fundamental limitation is the finite length of the preform, which limits the total length of the fiber that can be drawn.

We begin with the simple technique used by the Corning Glass Works to produce the first remarkably low-loss ($\sim$20 dB/km) fibers in the late 1960s. This technique, commonly known as the "soot" process, involves hydrolyzing a mixture of $SiCl_4$ and O_2 with an additive of either $TiCl_4$ or $GeCl_4$ vapors to produce a soot of either Ti-doped or Ge-doped SiO_2 material, deposited on pure SiO_2 glass. Since the refractive index of these doped materials is higher than that of pure silica, this deposition technique has also been utilized to make the preform material for drawing optical fibers. The process involves proper injection of a stream of doped SiO_2 particles deposited on a pure silica tube. A layer of soot is formed on the inner surface of the tube. After accumulating a sufficient thickness, the tube is heated and collapsed to form a preform rod that has a core with a higher refractive index and a cladding with a lower index.

Since this early experiment by the Corning Glass Works, many modifications and improvements have been introduced by using a variety of dopants, forming the soots on the outside surface of a removable mandrel, and so on. It has been shown that using this soot process, the preforms can be made at very rapid deposition rates and at a reasonably high purity level. This process can also be used to produce graded-index fibers with well-controlled profiles. In fact, all CVD processes offer this capability. CVD is a process in which chemical reactions take place at a relatively lower temperature and deposition is initiated from a heated surface. It is a very reliable process because pressure, flow rate, and temperature can all be controlled very accurately. A typical CVD system is shown in Figure 6.2. A fused quartz tube is placed in an oven with differential temperature zoning. As the gas mixture flows into the heated region, reaction with the heated surface occurs and a glass layer is deposited at a rate governed by these parameters. A simpler system, which is frequently used, consists of a rotating tube heated by a multiple-burner torch. If the temperature is very high, rotating the tube is often needed to prevent the tube from sagging. At high temperatures, the distinction between the soot and CVD processes becomes less clear.

The products of chemical reactions for these mixtures are given in Table 6.2, together with a comparison of refractive indices. The exact value for the refractive index depends on the thermal history of the glass, because it undergoes an anomalous change in refractive index as a result of quenching. Because borosilicate glass has the lowest refractive index, it is often used as the cladding material. Table 6.3 shows a group of doped silica core fibers with borosilicate glass as the cladding material. Also shown in Table 6.3 are some

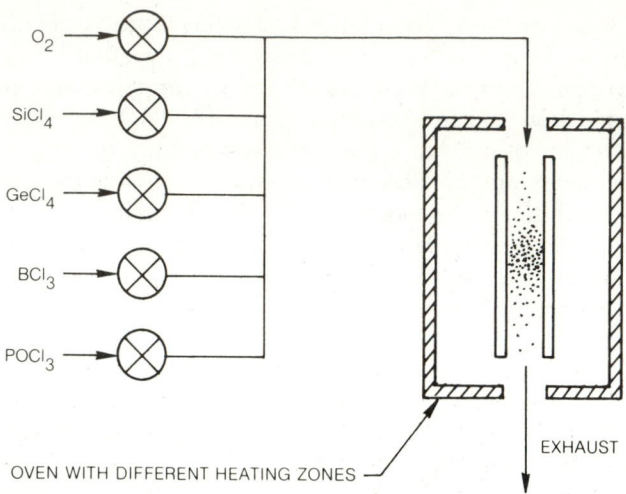

Figure 6.2   Schematic diagram of a CVD system for the growth of glass preforms.

typical values of Δn between the core and the cladding. The exact Δn values for these fibers depends not only on the thermal history but also on the mole concentration of the doping.

The transmission loss of fibers in the near infrared is due primarily to the OH absorption. The most significant reduction of OH ions can be accomplished by preconsolidation treatment of the CVD-produced preform, using, for example, $SOCl_2$ at temperatures up to 1450°C. The dehydration effect of heat treatment of the preform in $SOCl_2$ vapor over a long period of time ($\sim$5 h) reduces the OH ion content from a level of 30 ppm to below 0.1 ppm. Figure 6.3 shows a typical reduction of OH content as a function of temperature using $SOCl_2$ treatment. The OH residual content decreases rapidly to a level of 0.3 ppm in the vicinity of 700°C. Further reduction in the OH content occurs only very gradually with increasing temperature. A minimum level of OH residual of about 7 parts per billion has been obtained. This level of OH residual corresponds to an absorption loss of 0.45 dB/km at 1.39 μm.

To produce a graded-index preform with a smooth profile, it is necessary to deposit several hundred layers of doped silicate glass with a slightly different

TABLE 6.2   Glass Networks and Their Mixtures

Mixture	Network	Refractive Index
$SiCl_4$, O_2	SiO_2	n_0
$GeCl_4$, O_2	GeO_2	$n > n_0$
$POCl_3$, O_2	P_2O_5	$n > n_0$
BCl_3, O_2	B_2O_3	$n < n_0$

TABLE 6.3   Doped Silica Glass Fibers Using Borosilicate Glass as Cladding

Core		Cladding		
Dopant	Network	Dopant	Network	Δn (%)
P_2O_5	SiO_2	B_2O_3	SiO_2	0.8
GeO_2	SiO_2	B_2O_3	SiO_2	1.2
GeO_2, B_2O_3	SiO_2	B_2O_3	SiO_2	1.3

mole concentration. This requirement makes the CVD a rather tedious process to follow. Several methods have been introduced to assist the CVD process for rapid growth of a large number of layers. We shall mention only one of them, which makes use of plasma-augmented CVD to increase the reaction rate in the hot zone. In a plasma-augmented CVD, an inert gas (Ar) is excited by an inductive radio-frequency (RF) circuit. The discharge must be sustained under the deposition environment. The introduction of endothermic materials such as oxygen and other oxides tends to quench the plasma and therefore requires higher sustaining RF power. Plasma-augmented CVD can not only provide a very rapid deposition rate, but can also initiate the growth at relatively low temperature that eliminates the possibility of tube deformation.

To produce very long fibers, a method called vapor-phase axial deposition (VAD) has recently been introduced and has gained considerable popularity. This method has been shown to offer an important advantage by avoiding the formation of cracks due to thermal mismatch between the core and the cladding, a problem that often occurs in the conventional CVD process. Figure 6.4 illustrates schematically the essential features of the VAD method. Gaseous mixtures such as $SiCl_4$, $GeCl_4$, $POCl_3$, and O_2 are fed into an oxygen–hydrogen burner, which produces a stream of glass soot resulting from the flame

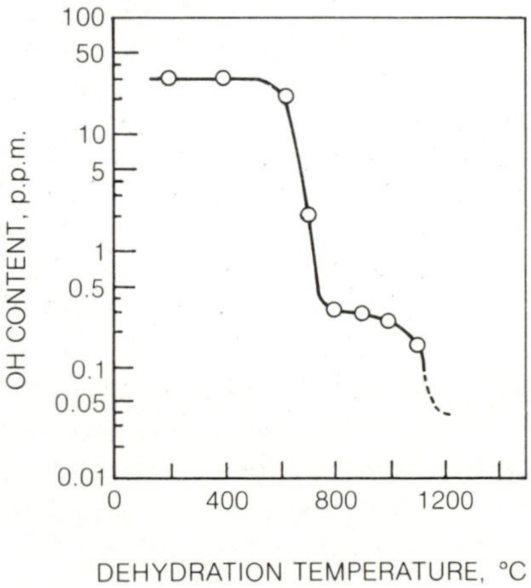

Figure 6.3   OH ion content as a function of dehydration temperature. [After T. Edahiro, M. Kawachi, S. Sudo, and H. Takata, *Electron. Lett., 15,* 482 (1979).]

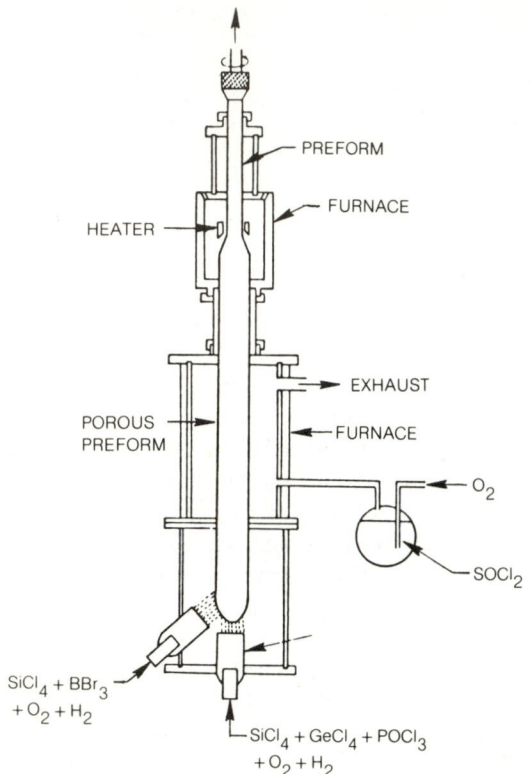

Figure 6.4 Apparatus for fabrication of low-OH-content optical fibers by VAD method. [After S. Sudo, M. Kawachi, T. Edahiro, T. Izawa, T. Shioda, and H. Gotok, *Electron. Lett., 14,* 534 (1978).]

hydrolysis. This stream of fine glass particles is directed toward one end of the starting rod, at which a porous glass rod is grown in the axial direction. The starting rod is rotated about its axis and moved upward at a speed consistent with the growth rate of the preform. Glass particles of lower refractive index are deposited on the porous glass rod from another oxygen–hydrogen burner to form the cladding. The porous glass rod is then vitrified to a bubble-free transparent fiber preform in an electric resistance furnace at temperatures of approximately 1650°C. By varying the mixtures and adjusting the flow rates, it is possible to create a graded-index profile; however, the reproducibility and control of the index profile is more difficult to obtain by this method than by plasma-augmented CVD.

The reduction of OH ions is accomplished by using $SOCl_2$ gas to create a dehydration reaction with the OH ions and H_2O molecules contained in the porous glass rod. The dehydration process takes place in an electric furnace at a temperature of about 800°C. Under these conditions, OH ions and H_2O molecules will diffuse to the surface and react with $SOCl_2$ to form HCl and SO_2 gases (dehydration process). The dehydrated preform produced by this method has been used to produce 20-km-long fibers with a measured loss coefficient of

~1 dB/km at 1.2 μm. This fiber has a 60-μm core diameter, a 150-μm cladding diameter, and a Δ value of 0.0014.

6.4 FIBER FABRICATION

Fibers can be drawn either from a preform or directly from melts of oxide powders. From the manufacturing-cost point of view, it is more economical to produce high-purity preforms than to purify powders and to maintain cleanliness. But at large production rates, manufacturers face the problem of frequent replacement of preforms, which can lead to a certain production waste and high manufacturing cost resulting from repeated shutdown and startup cycles. If the specification as to optical loss can be relaxed, a continuous process that offers very high output can easily be carried out using the double-crucible technique. A double-crucible configuration for fiber pulling from melts is shown in Figure 6.5. Two concentric crucibles, each with a specially designed nozzle, are configured with their axes in the vertical direction. The inner crucible is filled with a composite core material and the outer crucible is filled with cladding material. As the molten glass flows through these nozzles under Poiseuille flow conditions, the ratio of core radius to cladding radius is determined by the simple expression

$$\frac{a_{core}}{a_{clad}} = \sqrt{\frac{Q_{core}}{Q_{clad}}} \tag{6.1}$$

The quantity Q in Equation (6.1) represents the volumetric flow and is given by

$$Q = \frac{\pi P r^4}{8 \eta l} \tag{6.2}$$

where P is the pressure difference across the nozzle, η the viscosity, and r and l the radius and the length of the nozzle, respectively.

With a double-crucible apparatus, fiber drawing can be very simple if powder materials can be fed directly into the crucibles. However, because the production of crude glass from powders usually requires several stages of processing, including melting, mixing, oxidation, or reduction, it is very difficult to obtain homogeneous and high-purity glass fibers directly from powders. Instead, premelted glass has to be poured into the crucibles. An alternative method is to fill the crucible with small pieces of glass and subsequently to melt the glass slowly, allowing plenty of time for gas bubbles to escape.

This double-crucible method has also been used to produce graded-index fibers. It is accomplished by selecting a glass pair that allow interdiffusion to occur. Thallium, in particular, is a suitable dopant for the core because it can greatly increase the refractive index and is easily diffusible. Diffusion occurs as

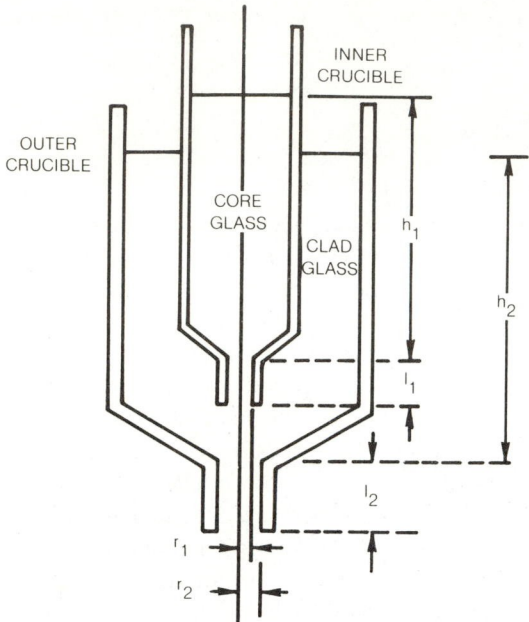

OUTER
CRUCIBLE

INNER
CRUCIBLE

CORE
GLASS

CLAD
GLASS

h_1

h_2

l_1

l_2

r_1

r_2

r — RADIUS OF NOZZLE
l — LENGTH OF NOZZLE
h — HEIGHT OF GLASS COLUMN

Figure 6.5 Cross-sectional view of a double crucible showing parameters important for controlling the dimension of the core and the cladding of a fiber.

soon as the core glass enters the molten cladding glass in the vicinity of the nozzles, where both the core and the cladding flow together during the production. Using the simple diffusion equation with radial symmetry, the concentration of the diffusible species can be written as

$$N\left(\frac{r}{a}\right) = N_0 \int_0^\infty \exp\left(-\frac{Dtu^2}{a^2}\right) J_0\left(\frac{u}{a}r\right) J_1(ur) \; du \qquad (6.3)$$

where N_0 is the initial concentration, D the diffusion coefficient, t the transit time through the nozzle region, and a the radius of the core. Figure 6.6 is a plot of the normalized concentration as a function of r/a for different values of the diffusion parameter Dt/a^2. The normalized concentration is directly related to the graded-index profile. The diffusion parameter is related to the volumetric flow rate Q_{core} by the expression

$$\frac{Dt}{a^2} = \frac{D\pi l}{Q_{\text{core}}} \qquad (6.4)$$

The model above is an oversimplification of the actual situation. Fortunately, the measured profile for a diffused double-crucible fiber matches very closely the α profile for an α value of approximately 2.4.

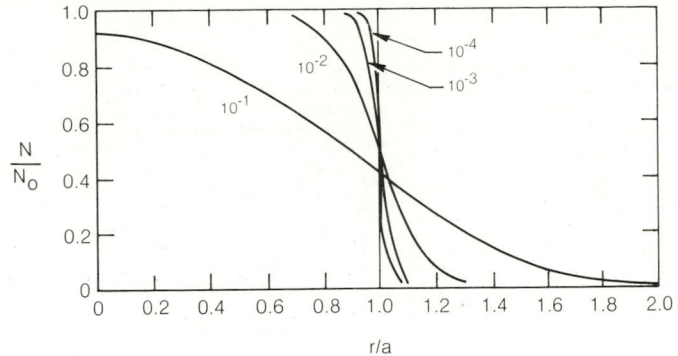

Figure 6.6 Index profiles calculated by a diffusion model for various values of Dt/a^2. [After K. B. Chan, P. J. B. Clarricoats, R. B. Dyott, G. R. News, and M. A. Sarva, *Electron. Lett, 6*, 748 (1970).]

Fiber drawing directly from a preform is a simple and straightforward process. The index profile is dictated primarily by that of the preform, as discussed in Section 6.3. To obtain a uniform fiber, one must control both the pulling speed and the feeding speed of the preform. The equation of continuity requires that

$$A_f V_f = A_p V_p \qquad (6.5)$$

where A_f, A_p and V_f, V_p are the area and velocity of the fiber and the preform, respectively. A schematic for drawing fibers from a preform is shown in Figure 6.7. A detailed balance of Equation (6.5) can be accomplished by careful inspection of the fiber diameter with a sensing device as shown in Figure 6.7. A change in fiber diameter can be corrected by changing either the winding or the feeding speed, or both. It is assumed that a stable condition is maintained during the drawing process.

6.5 FIBER OPTICAL COUPLING

Before discussing various measurement techniques, it is essential to gain a good understanding of the power-transfer mechanism between a source and a fiber or between two fibers. The parameters involved in fiber coupling are the surface area, the emitting angle, the field of view, the numerical aperture of the emitter and the receiver. If a lens is used, the parameters of the lens, such as the *f*-number and the magnification factor, must also be taken into account. For an emitting surface A_s, the intensity distribution is defined as

$$I = \int_{A_s} B \, \cos \theta \, dS \qquad (6.6)$$

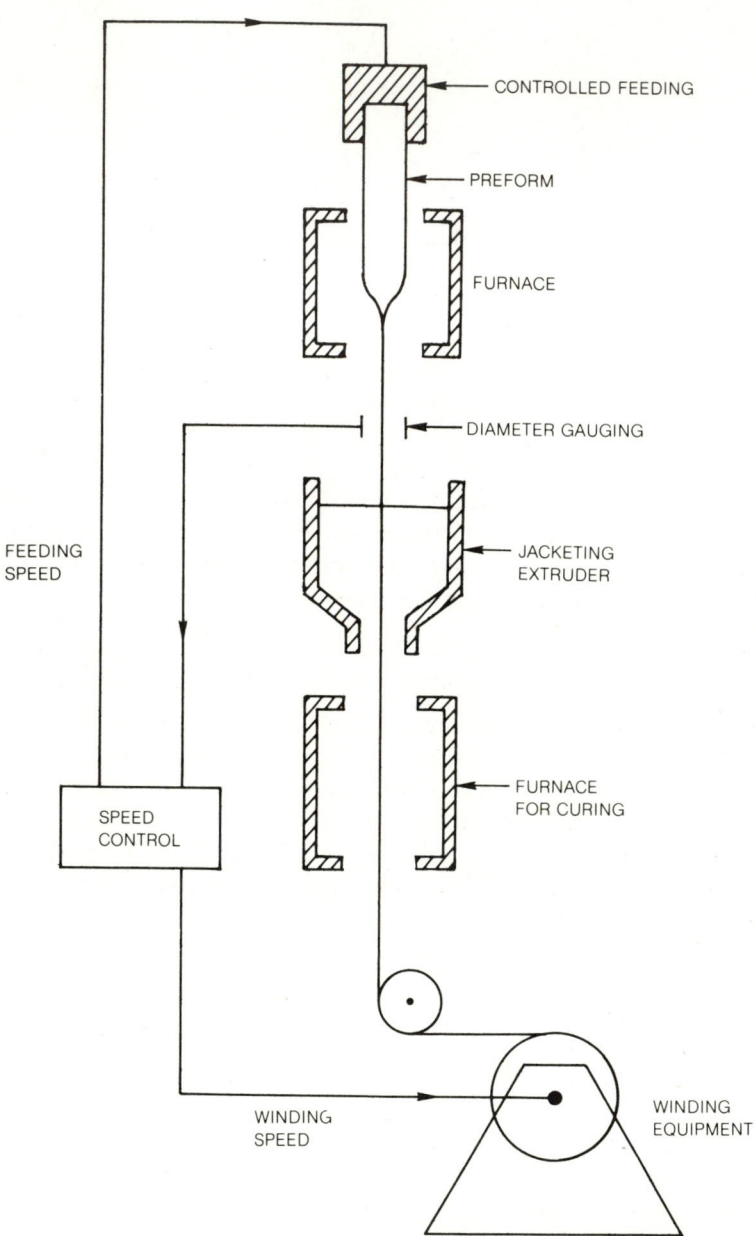

Figure 6.7   Apparatus for pulling optical fibers.

where B is the radiance of the source and θ is the angle measured from the normal of the element dS. For most sources, B can be approximated by the expression

$$B(\theta) = B_0 \cos^n \theta \tag{6.7}$$

where $n \geq 1$. In the case of a light-emitting diode (LED), $n \simeq 1$ and the distribution is commonly called Lambertian. A direct power transfer from A_s to a receiver surface A_r, as shown in Figure 6.8, can be computed by carrying out a double surface integral, as given by

$$P = \frac{\int_{A_r} \int_{A_s} B \; dS \cos \theta_r dS_r}{R^2} \tag{6.8}$$

where B is the radiance of the source in units of watts per cm^2-sterad. For a Lambertian source such as LED, $B = B_0 \cos \theta$. Since

$$\cos \theta_r \, dS_r \simeq 2\pi R^2 \sin \theta_s \, d\theta_s$$

Equation (6.8) can be expressed as

$$P = 2\pi A_s \int B_0 \cos \theta \sin \theta \, d\theta \tag{6.9}$$

The power received by a multimode fiber through direct coupling can be calculated using Equation (6.9) by integrating θ from 0 to θ_{NA}, where θ_{NA} is the angle that corresponds to the numerical aperture of the fiber. For a small numerical aperture, it can be shown that the fractional power coupled into the fiber is approximately equal to θ_{NA}^2.

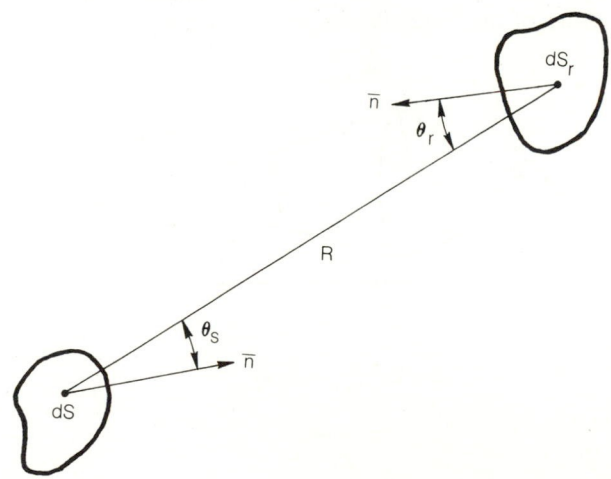

Figure 6.8   Geometric relationships between the element of an emitting surface and the element of a receiving surface.

When a lens is used, the total power collected by a lens within the solid angle subtended by the lens (see Figure 6.9) can also be obtained from Equation (6.9). For a source having a constant radiance B_0, we obtain

$$P_{\text{lens}} = 2\pi B_0 A_s (1 - \cos \theta_l) \tag{6.10}$$

where

$$\theta_l = \tan^{-1} \left[\frac{M}{2(M+1)f} \left(1 + \frac{d_s}{d_l} \right) \right] \tag{6.11}$$

In Equation (6.11), the quantities M, f, d_s, and d_l are defined as

$$M = \text{magnification power}$$
$$d_l = \text{diameter of the lens}$$
$$d_s = \text{diameter of the source}$$
$$f = f\text{-number of the lens}$$

The power emitted by the source into the full hemisphere is $P_T = 2\pi A_s B_0$. The power emitted by the source into a cone defined by $2\theta_s$ (see Figure 6.9) is

$$P_s = 2\pi A_s B_0 (1 - \cos \theta_s) \tag{6.12}$$

Substituting Equation (6.12) into (6.10), we can express the power collected by a lens in terms of the source power as

$$P_{\text{lens}} = P_{\text{source}} \frac{1 - \cos \theta_l}{1 - \cos \theta_s} \tag{6.13}$$

The total power received by the fiber through a lens can also be obtained by taking into account the mismatch between the angles of focusing θ_f and the numerical aperture of the fiber. It can be shown that

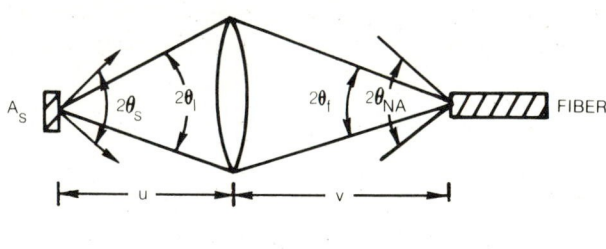

$$\tan \theta_1 = \frac{1}{M} \tan \theta_\ell$$

$$M = \frac{v}{u}$$

Figure 6.9 Geometric relationships for coupling between an emitter and a fiber by using a thin lens.

$$P_{\text{fiber}} = P_{\text{source}} \frac{1 - \cos \theta_l}{1 - \cos \theta_s} \frac{1 - \cos \theta_{\text{NA}}}{1 - \cos \theta_f} \tag{6.14}$$

where θ_f is also a property of the lens, and can also be expressed in terms of M, f, d_s, and d_l by the same geometric relation as that obtained for θ_l:

$$\theta_f = \tan^{-1} \frac{1 + d_s/d_l}{2(M + 1)f} \tag{6.15}$$

In the small-angle approximation, Equation (6.14) becomes

$$P_{\text{fiber}} \simeq P_{\text{source}} \left(\frac{\theta_{\text{NA}}}{\theta_s}\right)^2 \left(\frac{d_f}{d_s}\right)^2 \tag{6.16}$$

From Equation (6.16) we see that optimum coupling occurs when $\theta_s = M\theta_{\text{NA}}$ and $d_f = Md_s$. The only advantage of using a lens is to provide proper matching of the angles and the apertures. In most cases $\theta_{\text{NA}} < \theta_s$; therefore, an emitter must be selected that has as small an aperture as possible in order to satisfy simultaneously all the requirements noted above.

6.6 INDEX PROFILE MEASUREMENTS

The exact profile of either a step-index or a graded-index fiber is a very important parameter for determining the optical transmission properties of the fiber. There are several methods for measuring index profile by either direct or indirect means. Direct methods involve measurements of either the reflected or the transmitted optical power in the near and far fields. A more precise but rather tedious method is to measure fringes generated by the phase difference between the incident and refracted light from a fiber sample having a polished planoparallel facet with a phase-contrast microscope. The difficulty with the latter technique is that very thin slices of fiber samples with planoparallel surfaces must be prepared at a thickness no more than 50 μm. Attempts have been made to fabricate these samples by cladding the fiber in a glass tube filled with cement. A thin slice is cut from this rod and polished to optically flat surfaces. Another method for measuring the index profile involves dispersion measurements of time-resolved impulse response through a long fiber. This type of measurement not only gives information about the index profile, but also, and more important, the ultimate information-carrying capability or the bandwidth of the fiber. We shall present a detailed discussion of only one of the simplest index-measuring techniques and leave the pulse dispersion measurements to the next section.

The apparatus and experimental arrangement for measurements of the near-field power profile are shown in Figure 6.10. A light-emitting diode (LED) can be used as the source, which provides uniform illumination across the face

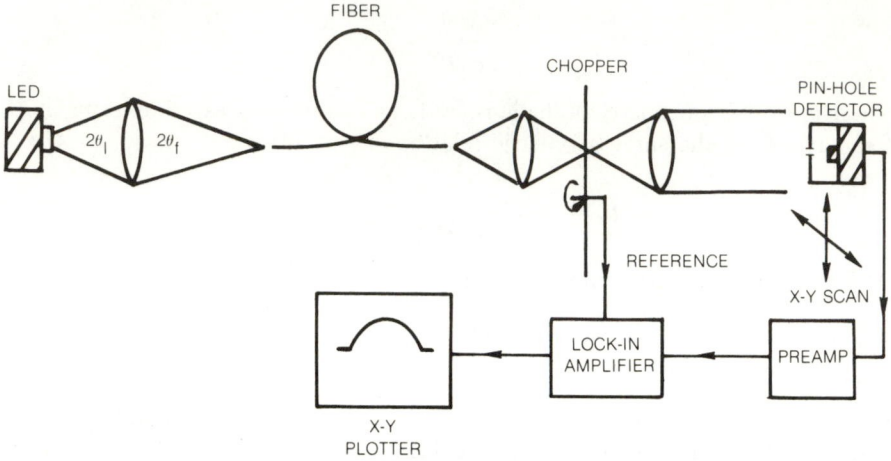

Figure 6.10   Experimental arrangement for measurements of the near-field output power as a function of normalized core radius.

of the fiber using a microscope objective lens. The fiber sample is prepared with a fixed length of approximately 1 m. Two ends of the fiber must be optically polished. The detector with a pinhole is mounted on an X-Y scanner and used to trace out the near-field output profile.

The relation between the index profile of a graded-index fiber and the power distribution in the near field can easily be established if the effects of nonuniform excitation, loss differences, and coupling among modes are ignored. If the fiber is uniformly illuminated, the power per unit solid angle at any point in the cross section is constant. For a step-index fiber, the numerical aperture is given by $(n_0^2 - n_c^2)^{1/2}$. For a graded-index fiber, we shall define a local numerical aperture $A(r)$ at a point r in the cross section as

$$A(r) = [n^2(r) - n_c^2]^{1/2} \qquad (6.17)$$

The ratio of the power at a point r to the average power P_0 received by a fiber having a constant numerical aperture A_0 can be expressed as

$$\frac{P(r)}{P_0} = \frac{A^2(r)}{A_0^2} = \frac{n^2(r) - n_c^2}{n_0^2 - n_c^2} \qquad (6.18)$$

If the fiber length is very short, it is reasonable to assume that all modes are attenuated equally and propagating through the fiber without coupling. Therefore, the same power distribution should hold for the transmitted power. Substituting Equation (1.2) into (6.18) for $n(r)$, we obtain

$$P(r) = P_0 \left[1 - \left(\frac{r}{a}\right)^{\alpha} \right] \qquad (6.19)$$

The power profile described by Equation (6.19) resembles the index profile and is exactly the same for small Δ values. Figure 6.11 is a plot of the near-field power distribution for various values of α.

In practice, it is necessary to correct the near-field profile to account for the contribution from the leaky modes because they are usually present in graded-index fibers, and they can introduce errors in the profile measurement. The output power expression, therefore, must include a correction factor $C(r, z)$ due to leaky modes:

$$\frac{P(r)}{P_0} = \frac{n^2(r) - n_c^2}{n_0^2 - n_c^2} C(r, z) \tag{6.20}$$

where

$$C(r, z) = 1 + \frac{4}{\pi[n^2(r) - n_c^2]} \int_0^{\pi/2} d\phi \int_0^{\pi/2} \cos\theta \sin\theta \, d\theta \, \exp\left(-\frac{\alpha}{a} z\right) \tag{6.21}$$

The second term on the right-hand side of Equation (6.21) represents the contribution due to leaky modes. The factor $\cos\theta$ represents the angular distribution of a Lambertian source, and the factor $\exp(-\alpha z/a)$ represents a leakage of power from the end face of the fiber. The integration is taken over the emission angle θ and the projection angle ϕ.

Converting Equation (6.21) into mode description, and assuming that the index of the external medium is unity, we can make the following changes of variables. From Snell's law, we have

$$\sin\theta = n(r) \sin\theta_g$$

or

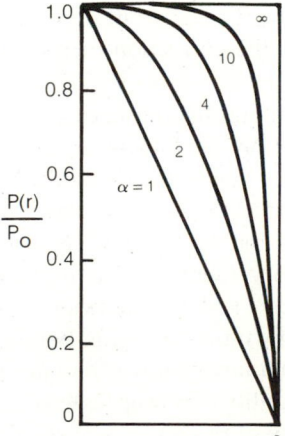

Figure 6.11   Plot of near-field output power profiles for various values of α.

$$\cos \theta \, d\theta = n(r) \cos \theta_g \, d\theta_g$$

where θ_g is the corresponding angle inside the guide. Now using the relations, e.g., Equation (4.120) among the components of the k vector, we have

$$\cos \theta_g = \frac{\beta}{k_0 n(r)}$$

and

$$\sin \phi = \frac{l/r}{\sqrt{k_0^2 n^2(r) - \beta^2}}$$

Substituting these changes of variables into Equation (6.21), we can write the correction factor in the form

$$C(r, z) = 1 + \frac{4}{\pi[n^2(r) - n_c^2]} \int_0^l \frac{dl}{a^2 k_0^3 r} \int_{u_l}^{u_u} \frac{\exp(-\alpha z/a) u \, du}{\sqrt{n^2(r) - n_c^2 + u^2/k_0^2 a^3 - l^3/k_0^2 r^2}} \tag{6.22}$$

where

$$u^2 = a^2[k_0^2 n^2(0) - \beta^2], \qquad V^2 = a^2 k_0^2[n^2(0) - n_c^2]$$

and the upper limit $u_u = \sqrt{V^2 + l^2}$ if the separation between guided modes and leaky modes occurs at $r = a$. The lower limit u_l, for a parabolic profile, is

$$u_l = \sqrt{V^2 \left(\frac{r}{a}\right)^2 + \left(\frac{a}{r}\right)^2 l^2} \qquad (r > 0) \tag{6.23}$$

where l values for leaky modes can vary from 0 up to V. The minimum value for u_l is obviously $V(l = 0)$. For small values of l, u_l values increase rapidly near the center of the core. To carry out the integral numerically, we must choose a value for r at which the correction factor is required, and compute u_l for each value of l.

Before performing the numerical integration, we must calculate the normalized attenuation coefficient α/a. We shall follow the approach of Adams et al. (Ref. 6.1) using the result of WKB approximation. This method is not only very accurate but offers some physical insight into the problem. Figure 6.12 defines the region that separates the oscillatory and evanescent fields. The oscillatory fields, which represent the guided modes, are bound within caustic points r_1 and r_2. The evanescent fields, which represent the leaky modes, are confined within the region bound by r_2 and r_3. Using the concept of quantum mechanical tunneling, we shall define a tunneling coefficient, T, which represents the probability of power leakage in the region bounded by r_2 and r_3. It is defined as the ratio of the squares of the field amplitudes in these regions. By

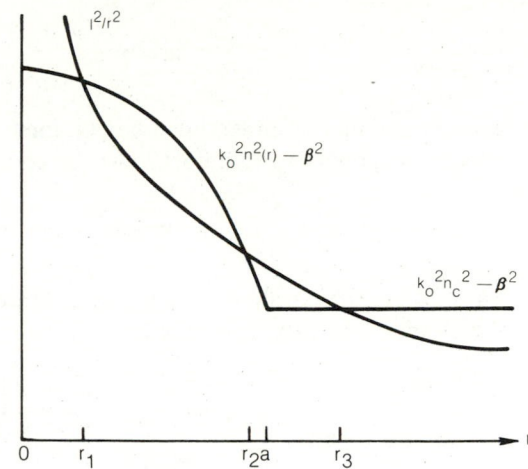

Figure 6.12 Graphical solutions of the points r_1, r_2, and r_3 that define the regions for guided and evanescent fields.

making use of the expression Equation (4.117)] for the phase factor of these fields, it can be shown that

$$T = \exp\left[-2\int_{r_2}^{r_3}\sqrt{\frac{l^2}{r^2} - k_0^2 n^2(r) + \beta^2}\ dr\right] \qquad (6.24)$$

The attenuation coefficient $\alpha\ (r, z)$, normalized to the core radius a, can be expressed in terms of T as

$$\alpha\ (r, z)\ \Delta z = aT \qquad (6.25)$$

where Δz is the distance between two internal reflections. Using the result of WKB approximation, we can write

$$\Delta z = 2\int_{r_1}^{r_2}\frac{\beta\ dr}{\sqrt{k_0^2 n^2(r) - \beta^2 - l^2/r^2}} \qquad (6.26)$$

Equations (6.24) and (6.26) can be integrated for a specific index profile. For simplification, we shall make the following assumptions:

1. The true profile will not deviate much from the perfect parabolic profile ($\alpha = 2$).
2. The integral of Equation (6.24) will be evaluated only from a to r_3, because the contribution from r_2 to a is negligible.

With these assumptions, we can obtain a closed-form solution for α. Substituting the parabolic index profile into Equation (6.26), we obtain an arcsine solution for Δz as given by

$$\Delta z = \frac{\beta a^2}{V^2} \arcsin \left[\frac{u^2 - 2V^2 r^2/a^2}{\sqrt{u^4 - 4V^2 l^2}} \right]_{r_1}^{r_2} \tag{6.27}$$

where the limits of integral can be obtained by solving the intersections of two curves as shown in Figure 6.12. At the caustic, we have

$$\beta^2 = k_0^2 n^2(r) - \frac{l^2}{r^2} \tag{6.28}$$

Substituting the parabolic index profile again into Equation (6.28) and solving the quadratic equation, we obtain

$$r_{1,2}^2 = \frac{u^2 \pm \sqrt{u^2 - 4V^2 l^2}}{2V^2} a^2 \tag{6.29}$$

Substituting the results of Equation (6.29) into (6.27) and letting $r_2 = a$, we obtain a simple result for Δz, given by

$$\Delta z = \frac{\beta a^2}{V} \pi \tag{6.30}$$

An analytic solution can also be obtained for T if a parabolic profile is introduced into Equation (6.24). In the case that $r_2 = a$, a simple expression for T is given by (Ref. 6.1)

$$T = \left[\frac{u^2 - V^2}{(l + \zeta)^2} \right]^l \exp(2\zeta) \tag{6.31}$$

where

$$\zeta = \sqrt{l^2 - u^2 + V^2} \tag{6.32}$$

With the results of Equations (6.30), (6.31), (6.25), and (6.22), it is possible to evaluate numerically the correction factor C as a function of fiber length z and the V parameter. As expected, C approaches unity as z approaches infinity. The expression given by Equation (6.22) is rather complex and can be simplified by introducing a normalization parameter $X = (1/V) \ln(z/a)$. As a result, a single set of curves $C(r, z)$ can be obtained to specify the near-field intensity profile completely for a length of a fully excited fiber having an arbitrary index profile. Most important is the fact that the same set of curves can be used to correct the measured intensity distribution to give the true refractive index profile of the fiber. Figure 6.13 gives a set of curves for $C(r, z)$ as a function of the normalized radius r/a for several values of X. These curves represent a good approximation not only for all X values normally encountered, but also for a wide range of near-parabolic and power-law variations.

   As an example, let us take a 1-m-long graded-index fiber having a core distribution of 80 μm and a numerical aperture of 0.18. Using $\lambda = 0.9 \mu$m, we

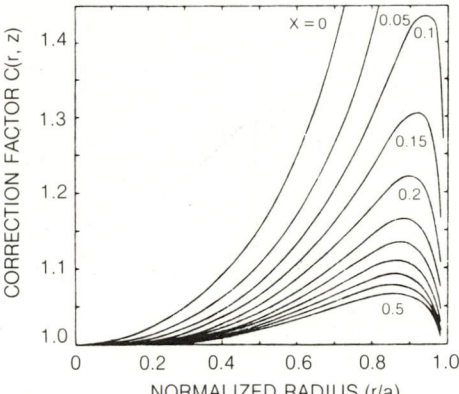

Figure 6.13 The correction factor for near-field output power measurements as a function of the radius of the core for various values of $X = \ln(z/a)/V$. (From Ref. 6.1. Reprinted with permission of North-Holland Publishing Company, Amsterdam.)

calculate the X value to be 0.2. From these curves we see that a correction to the index profile is only about 8% greater than the value measured in the near field for a normalized radius of 0.6 and rises to 20% for r/a to be 0.85. The result of this approximation is proven to be good to within 2% for a wide range of index profiles, provided that $z/a > 10^3$.

6.7 DISPERSION MEASUREMENTS

Modal and material dispersion of a multimode fiber can be determined by measuring either the impulse response in the time domain or the spectral response in the frequency domain of the output. In the former case, the pulse broadening can be measured either by sampling the waveform in real time with very high-resolution electronics or by analyzing its Fourier transform in slow time with a minicomputer. Frequency-response measurements, on the other hand, are relatively simple to perform; however, they require the use of a very broadband optical modulator. Since these techniques are very useful in modern research, we shall discuss these methods in some detail. Mechanisms responsible for dispersion in fiber have been discussed in Chapter 5. Since all fibers are dispersive, it is expected that a short pulse after propagating through a fiber will be broadened in width and distorted in shape to a certain extent, depending on the fiber parameters. For a step-index fiber, optical loss usually increases with increasing mode order. The impulse response, illustrated in Figure 6.14(a), can therefore be expressed as

$$P(t) = P_0 \sum_{k=1}^{N} \delta(t - \tau_k L) e^{-\alpha_k L} \tag{6.33}$$

where $\delta(t)$ is the Dirac delta function, P_0 a normalizing constant, and τ_k and α_k the group delay and attenuation coefficient of the kth mode in the step-index fiber.

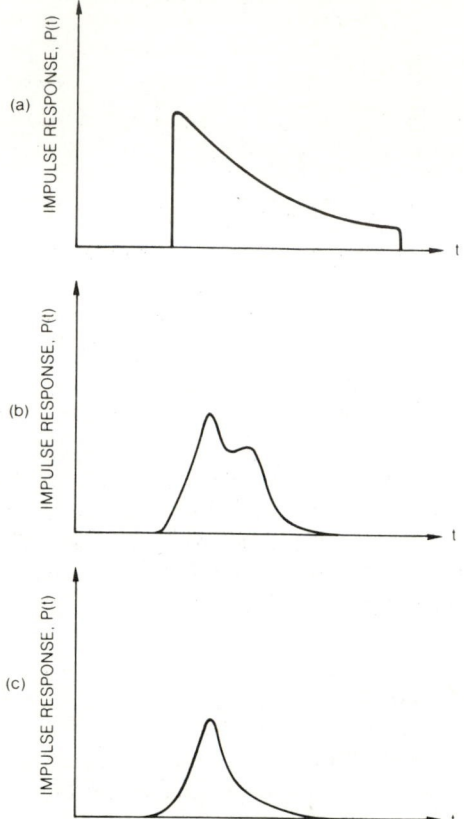

Figure 6.14 Waveforms of various possible impulse responses.

For a graded-index fiber, the impulse response is illustrated in Figure 6.14(b) for the case of no mode coupling. The dip in the response curve shown in Figure 6.14(b) often occurs in practice because of imperfections in the index grading. As a result, a short input pulse could excite two distinct groups of modes. Figure 6.14(c) shows the impulse response of a fiber involving strong mode coupling along its length ($\gamma_\infty z \gg 1$). In this case the expression for the impulse response is a Gaussian-like function [see Equation (5.71)] with width σ proportional to $\sqrt{L}$.

If we assume that the fiber is a linear filter, which is a valid assumption under normal circumstances when the optical power in a fiber is very low, we can write the input and output relationship in terms of a convolution integral given by

$$P_{out}(t) = \int h_f(t - \tau)P_{in}(\tau) \ d\tau \tag{6.34}$$

Equation (6.34) can sometimes be expressed by the symbol $h_f(t) * P_{in}(t)$. Physically, this integral represents the sum of impulse responses at the observation time t to impulses of strength $P_{in}(\tau) \ d\tau$ occurring at τ. The time delay $t - \tau$ at the instant of observation is a result of dispersion. It must be realized that the output waveform distortion is not only caused by the fiber but also by the measuring instrument with an impulse response function $h_i(t)$. Therefore, the output is a series of convolution integrals given by

$$P_{\text{out}}(t) = P_{\text{in}}(t)*h_f(t)*h_i(t) \tag{6.35}$$

To eliminate the effect of the instrument, we must replace the test fiber with a short "strap" fiber. The output through the strap is simply

$$P'_{\text{out}}(t) = P_{\text{in}}(t)*h_i(t) \tag{6.36}$$

We now take Fourier transforms of Equations (6.35) and (6.36) and divide the two results. The net result is the frequency response of the fiber alone, because a convolution in time domain transforms to a multiplication in the frequency domain.

The impulse function $P(t)$ can be expressed in terms of its power-transfer function $F(\omega)$ by the Fourier transform as

$$P(t) = \frac{1}{2\pi} \int_{-\infty}^{\infty} e^{-i\omega t} F(\omega) \; d\omega \tag{6.37}$$

We now examine the convolution integral

$$P_2(t) = \int d\tau \; P_1(t - \tau)P_0(\tau) \tag{6.38}$$

where P_1 and P_0 have the corresponding transfer functions $F_1(\omega)$ and $F_0(\omega)$ as indicated by Equation (6.37). In terms of F_1 and F_0, we write Equation (6.38) as

$$\begin{aligned}
P_2(t) &= \left(\frac{1}{2\pi}\right)^2 \int d\tau \int d\omega' \; e^{-i\omega'(t-\tau)} F_1(\omega') \int d\omega \; e^{-i\omega\tau} F_0(\omega) \\
&= \left(\frac{1}{2\pi}\right)^2 \int d\tau \; e^{i\tau(\omega'-\omega)} \int d\omega' \; e^{-i\omega't} F_1(\omega') \int d\omega \; F_0(\omega)
\end{aligned} \tag{6.39}$$

since

$$\int d\tau \; e^{i\tau(\omega'-\omega)} = 2\pi\delta(\omega' - \omega) \tag{6.40}$$

Substituting Equation (6.40) into (6.39), we have

$$\begin{aligned}
P_2(t) &= \frac{1}{2\pi} \int d\omega \int d\omega' \; \delta(\omega' - \omega) F_1(\omega') F_0(\omega) e^{-i\omega't} \\
&= \frac{1}{2\pi} \int d\omega \; e^{-i\omega t} F_1(\omega) F_0(\omega)
\end{aligned} \tag{6.41}$$

Equation (6.41) indicates that the transfer function of a convolution integral $P_2(t)$ is the product of the transfer functions $F_1(\omega)$ and $F_0(\omega)$. That is,

$$F_2(\omega) = F_1(\omega)F_0(\omega) \tag{6.42}$$

To obtain the impulse response for the fiber, we take the inverse Fourier transform of the frequency response. The experimental procedure in this case involves first taking measurements of $P_{out}(t)$ and $P'_{out}(t)$ and processing these pulse shape data with a minicomputer using a fast Fourier transform routine (FFT). The division yields the frequency response of the fiber. If necessary, a deconvolution process can be executed to obtain the impulse response.

It is very desirable to observe the pulse shape in real time; however, in a time scale on the order of a few picoseconds, most optical detectors are inadequate to resolve the waveform. It is necessary to use an ultrafast optical shutter such as a Kerr cell, or alternatively, a streak camera to reach the necessary time resolution. Figure 6.15 shows an experimental arrangement for real-time measurements of impulse response. In this setup, a 0.53-μm mode-locked Nd-doped glass laser with a pulse width of about 7 ps at the second harmonics of the laser ($\lambda = 1.06$ μm) is used as the source. The key component in this measuring apparatus is the optical shutter, which is made of an optically

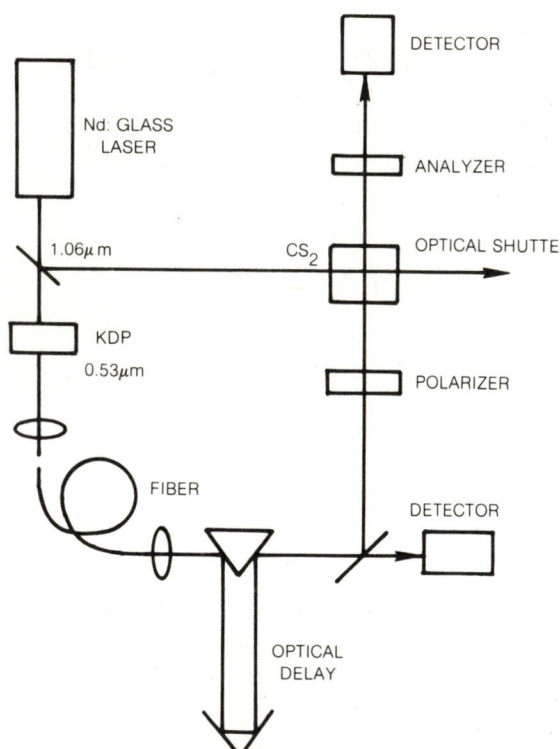

Figure 6.15 Apparatus for real-time measurements of ultra-short pulse waveform by using an optical shutter. (From Ref. 6.2. Reprinted with permission of the Institute of Electrical and Electronics Engineers, Inc., © 1972.)

active carbon disulfide CS_2 gas cell. When the laser pulse at its fundamental wavelength with a very high peak power is passing through the cell, the optical field induces an instantaneous birefringence in the CS_2 gas for a duration comparable to the pulse width. This phenomenon is commonly known as the Kerr effect. The Kerr cell is placed between two crossed polarizers. In the absence of a 1.06-μm laser pulse, the signal pulse at 0.53 μm is totally extinguished by the crossed polarizers. During the brief opening period of the optical shutter, the signal pulse is allowed to transmit and to be recorded by a sensitive photomultiplier. The signal pulse at 0.53 μm is generated in a nonlinear KDP crystal which is responsible for the frequency doubling. After propagating through the fiber, this pulse is delayed by a variable optical delay line, which consists of a moving prism retroreflector. By means of this varying delay, the waveform of a distorted signal pulse can be sampled across its entire width with the optical shutter. A reference detector is placed in front of the shutter to provide a normalization factor for canceling any fluctuation in the sampled signals through the cell. This optical shutter is considered to be one of the fastest signal-processing techniques available at present.

Figure 6.16 shows a log plot of the profile of an undistorted signal pulse obtained by this sampling procedure. The sampled profile exhibits a width of about 24 ps at two $1/e$ points, indicating that significant broadening has already taken place as a result of this sampling procedure. This is expected, because at $\lambda = 0.53$ μm, both the material dispersion and the scattering loss can be significant and can contribute to broadening and pulse-shape distortion in addition to the effect of modal dispersion.

The experimental arrangement for making a detailed analysis of the output waveform of an impulse response with a fast Fourier transform technique is

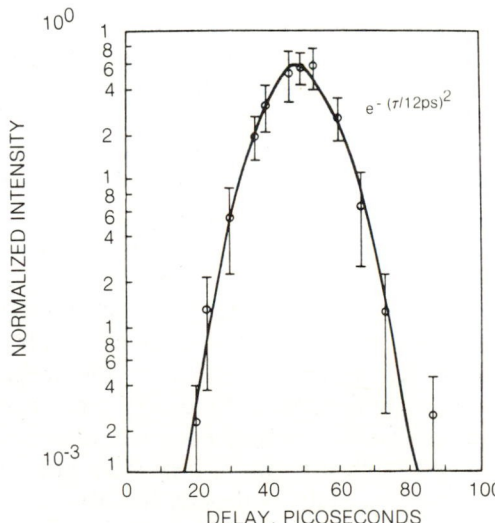

Figure 6.16 Time resolution of a CS_2 Kerr cell activated by a 1.06-μm Nd : glass laser with a pulse width of 10 ps. The input pulse width is 7 ps. (From Ref. 6.2. Reprinted with permission of the Institute of Electrical and Electronics Engineers, Inc., © 1972.)

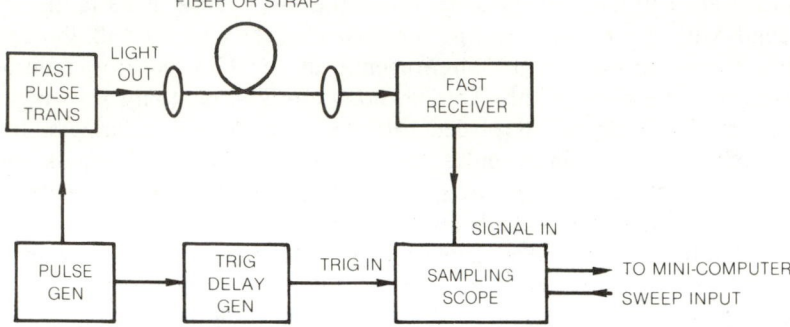

Figure 6.17 Apparatus for the measurement of the impulse response of a fiber by the Fourier transform technique.

shown in Figure 6.17. The output of a diode laser with a typical pulse width of ≥ 200 ps is coupled into the testing fiber using a lens. The output waveform is detected by a fast avalanche photodiode APD, and displayed on a sampling oscilloscope synchronized by delay trigger pulses. The sampled waveform is then digitized and processed using a minicomputer. The processed results of the averaged pulse response and its Fourier transform can be displayed with a printout. The transfer function of the fiber is obtained by dividing the Fourier transform of the averaged pulse response of the fiber by that of the strap.

By using a long-wavelength diode laser ($\lambda \geq 1.2 \, \mu$m), the material dispersion can be reduced. Major limitations on accuracy and reproducibility of the method above are detector nonlinearity, jitter in the delay trigger, laser instability, and spectral width. Factors such as nonlinearity and jitter can produce large errors in the deconvolution process. Using a normalization process with a strap, the jitter effect can be removed, and with sufficient optical attenuation, the system nonlinearity can be minimized.

As an alternative to time-domain measurements, one can measure transfer functions directly in the frequency domain. Figure 6.18 shows the experimental apparatus required for making frequency-response measurements of a fiber. In this case an incoherent light source can be used, but its spectral width must be filtered by a monochromator. Therefore, a monochromatic source is generated and passed through an electro-optic modulator. The output from the modulator is coupled into the fiber or the strap. The output from the fiber is captured by a broadband receiver, whose output is fed to a spectrum analyzer through a calibrated attenuator. As the modulation frequency is varied over a wide range, the spectrum analyzer records the receiver output as a function of the modulation frequency. By dividing the two sets of measurements obtained for the fiber and the strap, the transfer function of the fiber alone is obtained. This method is clearly simpler than the other two methods in the time domain. Furthermore, this method is insensitive to problems associated with system

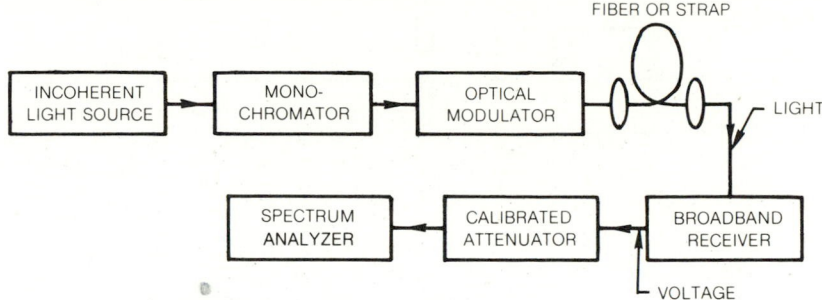

Figure 6.18   Apparatus for the measurement of frequency response of a fiber.

nonlinearity as for the case of the pulse sampling methods. On the other hand, the method in the frequency domain requires an external optical modulator and a monochromator, both of which are essential for this type of measurement. The spectral purity of the source must be kept at a very high level in order to maintain a phase synchronization of the incoming waves entering the modulator. Otherwise, a chirping effect occurs and can lead to large dispersion.

If a source has a very broad spectral width, material dispersion becomes significant. This effect must be separated from the broadening due to modal dispersion. A simple way to measure the material dispersion of a fiber is to observe the delay through the fiber using narrow pulses from lasers at two different wavelengths. In this way the material dispersion can be obtained by dividing the delay difference due to the wavelength difference of the lasers and normalizing to the fiber length.

6.8 FIBER CONNECTION AND SPLICING

One practical problem associated with applications of fiber optics is the difficulty of connecting and disconnecting fibers. This is due the fact that the small cross sections of the fibers require extremely high precision in alignment. A severe loss penalty can result if lateral displacement between two parallel end faces of the fibers exceeds a small fraction of their radius. More serious problems occur when the end faces are rough and not parallel. Clearly, this problem is much more severe for single-mode fibers than for multimode fibers because the diameter of a single-mode fiber is only a few micrometers. We shall treat the case of a single-mode fiber and introduce techniques which have been developed for splicing single-mode fibers at relatively low loss. These techniques can readily be extended to multimode fibers to yield negligible coupling losses.

Figure 6.19 illustrates two types of misalignment that can cause losses in fiber connections. Rigorous calculation of the joint loss in terms of the HE_{11}

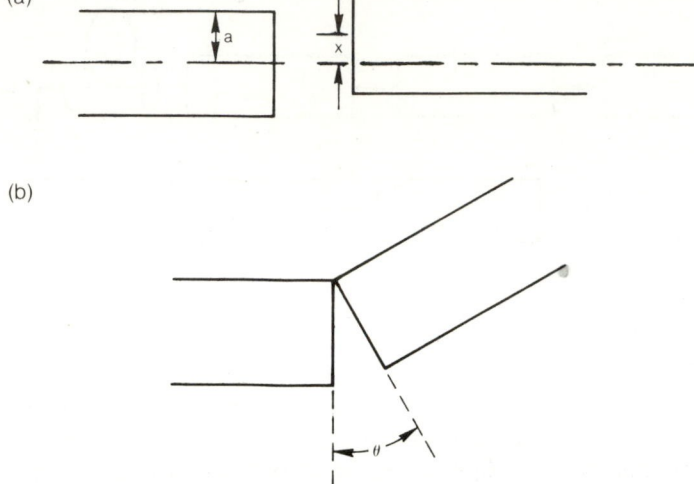

Figure 6.19   Two basic types of fiber coupling misalignment.

mode is difficult. If one replaces the HE_{11} field by a Gaussian function as an approximation, the calculation can be considerably simplified. When two identical single-mode fibers are displaced by a normalized distance $d = x/a$, where x is the offset from fiber axis and a is the fiber radius, as shown in Figure 6.19(a), the loss α_d in decibels is given with a reasonable degree of accuracy by the simple expression (Ref. 6.3)

$$\alpha_d = 2.17 \left(\frac{d}{\omega} \right)^2 \qquad \text{decibels} \qquad (6.43)$$

where ω, the spot size of the HE_{11} mode normalized to a, is given by (Ref. 6.4)

$$\omega = \frac{1}{\sqrt{2}} (0.65 + 1.62 V^{-1.5} + 2.88 V^{-6}) \qquad (6.44)$$

When the end faces of the two fibers are tilted by an angle θ, as shown in Figure 6.19(b), the loss α_θ in decibels is given by (Ref. 6.3)

$$\alpha_\theta = 2.17 \left(\frac{\theta \omega n V}{NA} \right)^2 \qquad \text{decibels} \qquad (6.45)$$

If the joint between two fibers is misaligned with both the lateral and angular displacements, the total loss α_T, to a good approximation, is given by the expression (Ref. 6.3)

$$\alpha_T = 3.6 d^2 + \left(\frac{n \theta}{NA} \right)^2 (7.6 + 7 d^2) \qquad (6.46)$$

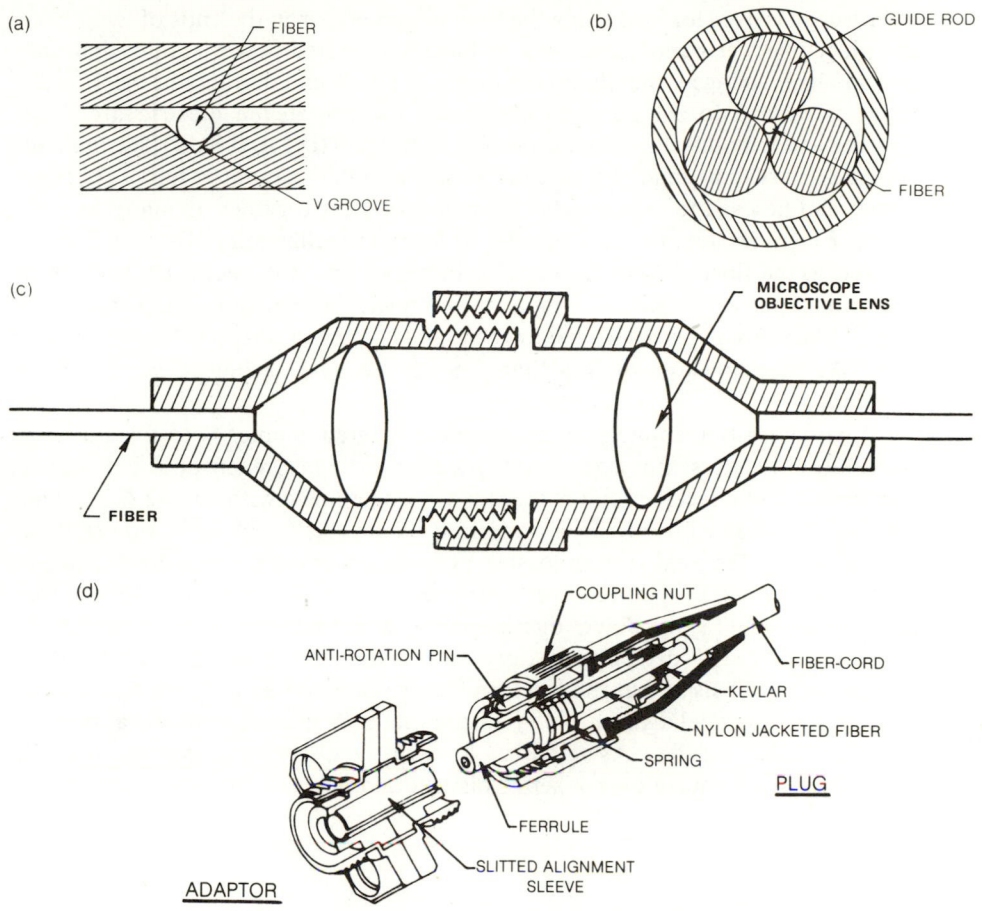

Figure 6.20   Examples of demountable coupling technique.

To maintain the joint within an acceptable tolerance, several techniques have been developed, as illustrated in Figure 6.20. These techniques are widely used and provide a very reproducible result, with typical joint losses of about 0.2 to 3 dB, depending on the precision in making these adaptors and preparing the fiber end faces. In all cases the primary goal is to assure that the lateral and angular misalignment of two fibers is minimized. The methods illustrated in Figure 6.20 are useful for making demountable connectors. In particular, the connector, as shown in Figure 6.20(d), has been used very successfully in the field (Ref. 6.5). The best method for making a permanent connection between two fibers is to fuse the two prealigned ends with a short burst of electric arc. Practical fusion-splicing equipment is now available for use with multimode fibers in the field, because most fiber optical systems in service today utilize multimode fibers. This equipment usually consists of a microscope and an electric arc. The microscope is used for inspecting the fiber alignment when the ends of two fibers are placed between two electrodes. A battery-operated power supply that can provide high voltage typically on the order of 10 kV and currents in the 10-mA range is needed to energize the electrodes to generate an electric arc. The splicing loss with this technique is usually very small (0.1 to 0.3 dB). Permanent splicing can also be made by thermal fusion to obtain similar results. Thermal fusion can be used to make a star coupler that joins together as many as 100 fibers. Fibers are first twisted together to form a ropelike joint. Tension is then applied to the fibers before fusion. For better results it is necessary to form a biconical taper at the joint. Using a star coupler, the optical power can be fed into N fibers from any one of the fiber bundles on the opposite side of the taper joint. An insertion loss of less than 0.6 dB has been achieved with such a coupler.

An interesting application of biconical tapered coupler is to form a data distribution network that contains N parallel terminals, each of which has a transmitter and a receiver and is arranged in accordance with Figure 6.21. The ratio of the received power P_R by one of the receivers to the input power P_I is determined by the total system losses. For this system the losses are (1) fiber-to-transmitter and fiber-to-receiver coupling losses, (2) the insertion loss of the star coupler, and (3) the power distribution loss through the star coupler, which is equal to $10 \log N$ (dB). The first two loss factors are independent of the total number of terminals N, but the last loss factor is a function of N. An interesting feature of the parallel distribution system is that as N increases, the total system loss saturates rapidly. As a rule, a parallel distribution system has a considerable advantage over a series distribution network for $N > 10$.

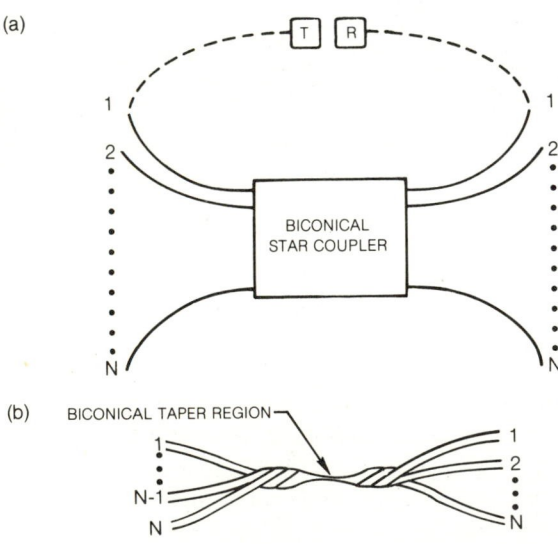

(a)

(b) BICONICAL TAPER REGION

Figure 6.21 (a) Data distribution network using a biconical star coupler; (b) schematic diagram showing a biconical taper joint.

One of the most commonly used techniques to obtain a reasonably good end face is to cleave the fiber under tension with a sharp diamond-tipped or tungsten carbide knife edge. A straight scratch on the surface produces a stress concentration and can lead to a very clean cleavage. If the initial scratch is not very sharp, a faulty cleavage often occurs at which one or more long tails are found to attach to the otherwise perfect cleavage plane. Another method that yields a consistently good cleavage plane is to cut the fiber with a focused carbon dioxide (CO_2) laser beam. This is not a very convenient method to use in field service, but it is certainly the most reliable method to use in a manufacturing facility.

The best method for measuring the joint loss is to measure the power transmitted through the joint between two short fibers normalized to the same length as that of a fiber without the joint. Other methods, such as the pulse echo and time-domain reflectometry (TDR) could be used to estimate splicing losses and to locate imperfections in a long length of fiber. Echo and reflection measurement techniques have obvious limitations, which can arise from many factors, such as end reflection, scattering, absorption, dispersion, and imperfections other than splicing. Therefore, results are often difficult to interpret. The distance between most repeater stations is in the 10km range, so the typical insertion loss of the fiber can easily exceed 50 dB. To detect a return signal by these methods, a very large dynamic range (>120 dB) for the detection sensitivity is required.

PROBLEMS

6.1. A LED source has an output of 1 mW and a maximum emitting angle of $90°$. The diameter of its emitting surface is 100 μm. Calculate its radiance in units of W cm^{-2} sr^{-1}.

6.2. Calculate the percent power that can be coupled into a fiber directly from a LED source that emits into a full hemisphere. The fiber has a numerical aperture $\theta_{NA} = 0.3$.

6.3. Derive Equation (6.11), assuming that the simple lens formula applies.

6.4. If a laser diode is used as the source and it has an emitting angle $\theta_s = 15°$, repeat Problem 6.2 and compute the improvement factor for using the laser source over the LED.

6.5. Calculate the near-field power profile emitting from a fiber with a parabolic index profile.

6.6. Calculate caustic points r_1 and r_2 for a fiber with a parabolic index profile, in terms of its parameters V, a, u, and l.

6.7. For a 1-m-long graded-index fiber having a value of 50 for the parameter V, determine the correction factor for the near-field power profile at three points, corresponding to $r/a = 0.2$, 0.5, and 0.8.

6.8. Calculate the joint loss of two single-mode fibers ($V = 2.4$) having NA $= 0.1$ and $n = 1.5$, if only the angle between the two fibers is misaligned by $2°$. What would be the joint loss if these two fibers are displaced by $\frac{1}{2}\omega$, where ω is the spot size of the HE_{11} mode?

REFERENCES

6.1. M. J. Adams, D. N. Payne, and F. M. E. Sladen, *Opt. Commu. 17,* 204 (1976).

6.2. D. Gloge, A. R. Tynes, M. A. Duguag, and J. W. Hanson, *IEEE J. Quantum Electron., QE-8,* 217 (1972).

6.3. W. A. Gambling, H. Matsumura, and C. M. Ragdale, *Electron. Lett., 14,* 618 (1978).

6.4. D. Marcuse, *Bell Syst. Tech. J., 56,* 703 (1977).

6.5. J. Minowa, M. Saruwatari, and N. Suzuki, *IEEE J. Quantum Electron., QE-18,* 705 (1982).

7

Light Emission Processes
in Semiconductors

7.1 INTRODUCTION

The process in which electrical energy is converted to light in solids is called electroluminescence. Light emission occurs during the recombination of an electron in the conduction band with a hole in the valence band or an electron at a donor with a hole at an acceptor. The radiative transitions for all light-emitting systems involves spontaneous or stimulated emission and absorption of photons. For semiconductor light sources the selection rules that determine the radiative transition probability can be relaxed significantly by the continuous nature of energy bands. As a result, the spectrum is considerably broadened. Semiconductors can be categorized into two groups, one of which is associated with direct bandgap materials and the other with indirect bandgap materials. The radiative transition probability is usually very high for direct bandgap materials. However, complications can arise when the material is heavily doped or is subjected to a high injection of current. Detailed theoretical treatments on these subjects is beyond the scope of this text. In this chapter we review briefly the fundamentals of the quantum theory of semiconductors without getting into too much mathematical detail. We discuss the mechanisms and circumstances by which various radiative transitions occur. In particular, we treat in some detail the probability of electron–hole transitions between the conduction and the valence band of a direct bandgap material. The result of this treatment will lead directly to calculation of the recombination rate between electrons and holes. From the results of this analysis it is possible to estimate the quantum efficiency, current density, and switching rate for semiconductor light-emitting devices.

7.2 QUANTUM MECHANICAL DESCRIPTION OF SEMICONDUCTORS

The simple model commonly used to describe the motion of an electron in a semiconductor is to consider an electron in a periodic potential with a period determined by the lattice of the semiconductor. This potential is caused by the periodic charge distribution associated with the ion cores situated on the lattice sites plus a constant term due to the contribution of all other free electrons in the crystal. The wave function for the single electron in this potential can be obtained by solving Schrödinger's equation. The solution for a single electron then provides a set of states that can be shared by all the electrons in the crystal subject to the limitations of the Pauli exclusion principle, which states that no two electrons may occupy the same quantum state specified by a given quantum number, unless the state is degenerate. In an infinite one-dimensional crystal, Schrödinger's equation is of the form

$$\frac{d^2\psi}{dx^2} + \frac{2m}{\hbar^2}[E - V(x)]\psi = 0 \tag{7.1}$$

where $\hbar(h/2\pi)$ is Planck's constant, m the mass of a free electron, E the energy associated with the electron, and $V(x)$ the periodic potential, which can be expressed in terms of a Fourier series as (Ref. 7.1)

$$V(x) = -\gamma \frac{\hbar^2}{2m} \sum_{n=-\infty}^{\infty} C_n e^{-i2\pi nx/a} = \sum_{\infty}^{\infty} V_n e^{-i2\pi nx/a} \tag{7.2}$$

In this form, $V(x)$ is valid for small values of γ.

According to the Bloch theorem, we can write the wave function in the form

$$\psi(x) = e^{ikx}u(x) \tag{7.3}$$

where $u(x)$ is a periodic function of period a, which is the lattice constant. Equation (7.3) indicates that $\psi(x)$ is of the form of plane waves with a propagation vector $\mathbf{k}$, modulated by a periodic function $u(x)$ with a period of the crystal lattice. For $\gamma = 0$, the wave function must be that for a free particle, and thus we write

$$\psi(x) = B_0 e^{ik_0 x} \tag{7.4}$$

and the energy of a free particle is simply

$$E = \frac{\hbar^2 k_0^2}{2m} \tag{7.5}$$

For $\gamma \neq 0$, we solve Equation (7.1) in the vicinity of band edges by using the free-particle approximation. We assume that

$$k = k_0 \qquad \text{if} \qquad k \neq \frac{n\pi}{a}$$

At lattice sites or band edges,

$$k_n = \pm \frac{n\pi}{a}, \qquad \text{where } n = 0, 1, 2, \ldots \qquad (7.6)$$

In the vicinity of a band edge, we shall assume a wave function of the form

$$\psi(x) = e^{ikx}(B_0 + \gamma B_n e^{-i2\pi nx/a})$$

$$= B_0 e^{ikx} + \gamma B_n e^{ik_n x} \qquad (7.7)$$

If $k \to n\pi/a$, $k_n \simeq -n\pi/a$. $\psi(x)$ is simply a superposition of two traveling waves in opposite directions. Physically, we can describe the electron as undergoing Bragg reflection at the band edge. Substituting Equations (7.7), (7.5), and (7.2) into (7.1), we get

$$B_0(k_0^2 - k^2)e^{ikx} + \gamma B_n(k_0^2 - k_n^2)e^{i(k - 2\pi n/a)x}$$

$$+ \gamma B_0 \sum_{n' \neq 0} C_{n'} e^{i(k - 2\pi n'/a)x} + \gamma^2 B_n \sum_{n \neq 0} C_{n'} e^{i(k - 2\pi n/a) - 2\pi n'/a} = 0 \quad (7.8)$$

Multiplying Equation (7.8) by e^{-ikx} and integration from $x = 0$ to $x = a$ we obtain by using the orthogonality relations, an equation, which came from the first and last terms of Equation (7.8) when $n' = -n$, of the form

$$(k_0^2 - k^2)B_0 + \gamma^2 C_n B_n = 0$$

Similarly, multiplying Equation (7.8) by $e^{-ik_n x}$ and integrating from $x = 0$ to $x = a$, we obtain another equation:

$$C_n B_0 + (k_0^2 - k_n^2)B_n = 0$$

The foregoing two homogeneous equations involving B_0 and B_n have nontrivial solutions only if

$$(k_0^2 - k^2)(k_0^2 - k_n^2) - \gamma^2 C_n^2 = 0$$

or

$$k_0^2 = \frac{1}{2}\left[(k^2 + k_n^2) \pm \sqrt{(k^2 - k_n^2) + 4\gamma^2 C_n^2}\right]$$

Therefore, we have an $E(k)$ relationship given as

$$E(k) = \frac{\hbar^2}{4m}\left[k^2 + \left(k - \frac{2\pi n}{a}\right)^2 \pm \sqrt{\left[k^2 - \left(k - \frac{2\pi n}{a}\right)^2\right]^2 + \left(\frac{4m\,|V_n|}{\hbar^2}\right)^2} \right]$$

(7.9)

From these results we see that at the band edge $k = \pm\, n\pi/a$, internal Bragg reflection takes place and is accomplished by a discontinuity at the energy gap in the E versus k curve (Figure 7.1). The width of the bandgap is given by $2\,|V_n|$, where V_n is the nth Fourier coefficient in the series expansion of the periodic lattice potential as given by Equation (7.2). Therefore, at the band edge, we write

$$E = \frac{\hbar^2}{2m}\left(\frac{n\pi}{a}\right)^2 \pm |V_n|$$

Since the number of atoms in the crystal is very large, the allowed values for E and k, although discrete, are very close together and, for practical purposes, can be regarded as quasi-continuous bands of allowed values. This description applies also to a three-dimensional crystal, except that in this case more than two values of k are allowed for each eigenvalue of E. The k values can be both real and imaginary.

Figure 7.1 is a plot of the $E(k)$ relation, which indicates the allowed and the forbidden regions of E. The dashed curve in Figure 7.1 represents the energy for a free particle, which is a parabolic function of k as given by Equation (7.5). As E increases, the allowed bands become very broad, whereas the forbidden bands become very narrow. These features are shown in Figure 7.2 and are quite general, in the sense that they are relatively

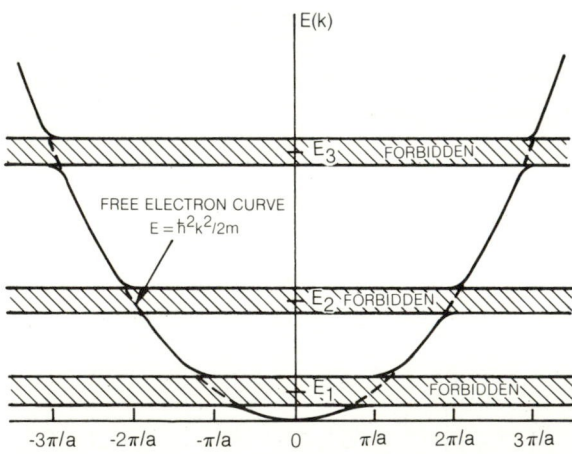

Figure 7.1   General features of the $E(k)$ curve for an electron in a semiconductor with a lattice constant a.

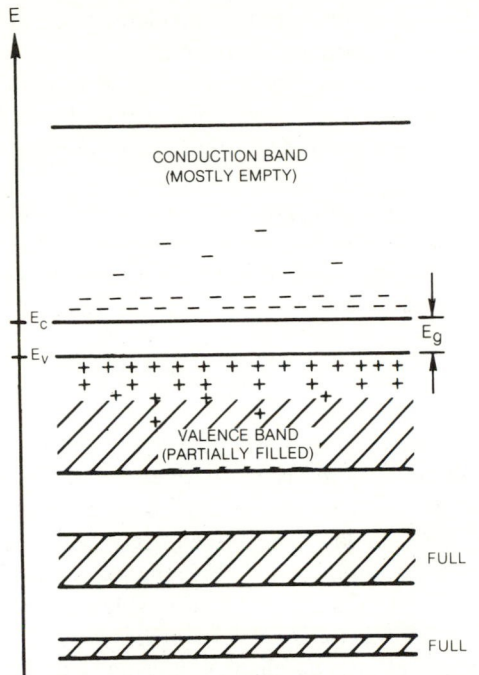

Figure 7.2 Energy-band diagram of a
semiconductor.

independent of the precise form of the potential function. It is to be noted that
the electrical conductivity can only arise from bands which are partially filled.
This occurs in the proximity of a *pn* junction in a semiconductor, which is
discussed in Chapter 8. If the bandgap energy E_g extending across the forbidden
zone is small, there will be an appreciable number of electrons excited thermally
from the top of the valence band across the gap to the bottom of the conduction
band.

Band theory also shows that the masses of electrons and holes in the bands
generally have values different from those of a free electron. This is a direct
consequence of the fact that the energy of these particles is a function of k and
we must therefore deal with the group velocities v_g of these particles, as defined
by

$$v_g = \frac{d\omega}{dk} = \frac{2\pi}{h}\frac{dE}{dk}$$

where we have introduced Planck's relation $E = \hbar\omega$. It is to be noted from
Figure 7.1 that $E(k)$ always has zero slope at the edges of the allowed bands,
$k = \pm n\pi/a$. This implies that at these points, v_g is zero, indicating that the
particle is at rest. The electron or the hole can be regarded as undergoing an
internal reflection by the lattice potential. In that instance, the effective masses

of these particles may be considered to be infinitely large. Therefore, in dealing with the dynamic behavior of electrons and holes in the bands, we introduce an effective mass m_e for electrons and m_h for holes, which can be significantly different from the mass m of a free particle. When the values of E become very large, the function $E(k)$ approaches the free energy of the particle as given by Equation (7.5).

Using DeBroglie's relation stating that the momentum of a particle p is equal to $\hbar k$, where k is the wave number associated with a wave packet, we can show that the electron's effective mass m_e associated with the wave packet can be expressed as

$$m_e = \frac{\hbar^2}{d^2E/dk^2} \tag{7.10}$$

Fortunately, the relation between E and k is almost always parabolic or nearly parabolic over the range of energies of interest. What is important is that the effective mass be essentially constant over a sufficient energy interval δE in which a collision process can occur within a time of the order of the radiative recombination lifetime. This amount of energy typically is of the order of a few kT ($kT \simeq 0.025$ eV at $300°K$), which is very small compared with the bandgap energy E_g (1 to 2 eV). In other words, in the interval between successive collision events, an electron is usually confined to a short segment of the $E(k)$ curve, which can be regarded to a good approximation as parabolic. For k values near the edge of the conduction band [e.g., $k = (n\pi/a) + k'$, where $k' \ll \pi/a$], $E(k)$ deviates from the free electron energy curve by a small amount, as shown in Figure 7.1.

In the neighborhood of the band edge, we express $E(k)$ in terms of k' as

$$E(k') = \frac{\hbar^2}{2m}\left[\left(\frac{n\pi}{a}\right)^2 + k'^2 + \frac{2m|V_n|}{\hbar^2}\sqrt{1 + 4k'^2\left(\frac{n\pi}{a}\right)^2\left(\frac{\hbar^2}{2m|V_n|}\right)^2}\right] \tag{7.11}$$

The free-particle energy and the bandgap energy are defined by the expressions

$$E_n = \frac{\hbar^2}{2m}\frac{n\pi}{a} \qquad \text{and} \qquad E_g = 2|V_n|$$

Substituting E_n and E_g into Equation (7.11) and expanding the radical, we obtain an approximation for $E(k')$ as

$$E(k') = E_n + \frac{E_g}{2} + \frac{\hbar^2 k'^2}{2m}\left(1 + \frac{4E_n}{E_g}\right) \tag{7.12}$$

Differentiating Equation (7.12) twice with respect to k', we obtain

TABLE 7.1   Constants for Selected III–V Binary Compounds

Compound	E_g (eV)	χ (eV)	m_e/m	m_h/m	ε	n	σ (W/cm-deg)
AlAs	2.163		0.15	0.79	10.1	3.178	0.91
AlSb	1.58	3.64	0.12	0.98	14.4	3.4	0.57
GaP	2.261	4.0	0.82	0.60	11.1	3.452	0.77
GaAs	1.424	4.05	0.067	0.48	13.1	3.655	0.44
GaSb	0.726	4.03	0.042	0.44	15.7	3.82	0.33
InP	1.351	4.4	0.077	0.64	12.4	3.450	0.68
InAs	0.360	4.45	0.023	0.40	14.6	3.52	0.27
InSb	0.172	4.59	0.0145	0.40	17.7	4.0	0.17

$$m_e = \frac{m}{(1 + 4E_n/E_g)} \tag{7.13}$$

For GaAs, $E_g = 1.42$ eV and $E_n = \chi + E_g/2 = 4.75$ eV; therefore, a value for the effective mass of an electron was found to be approximately $0.067m$. In all direct bandgap materials, the effective mass of a hole in the valence band is approximately one order of magnitude larger than that of an electron. For GaAs, $m_h \simeq 0.48m$. Table 7.1 gives the effective mass of electrons and holes for a number of binary compounds that are commonly used in making heterostructure light-emitting devices. Also listed in this table are values for the electron affinity χ, which is the energy required to liberate an electron from the conduction band edge to free space, and is useful for the calculation of the energy difference of the conduction bands ΔE_c between two dissimilar materials. The values for the dielectric constant ε, the refractive index n, the thermal conductivity σ, and the bandgap energy at $300°K$ are also included.

7.3 CARRIER DISTRIBUTION AND CONCENTRATION

To extend the analysis for a single electron or a single hole to a distribution of electrons in the conduction band or holes in the valence band, we introduce a term $\rho(E)$, which represents the density of states at any particular energy E, and we multiply it by the fractional occupation of states. This description is analogous to that used for describing the mode distribution in a fiber. In the case of a semiconductor, the energy states are modes of the electronic wave functions, which are analogous to the optical modes. The major distinction between the two cases is that in a quantum mechanical system the electron distribution is determined by Fermi–Dirac statistics, which are a direct consequence of the Pauli exclusion principle that prevents more than one particle from occupying an identical state. Fermi–Dirac statistics allow us to describe the statistical behavior of a system of particles such as conduction electrons in semiconductors.

The probability that an electronic state at a particular energy E is occupied by an electron is given by the Fermi–Dirac distribution:

$$f(E) = \frac{1}{\exp(E - F)/kT + 1} \tag{7.14}$$

where F is the Fermi energy level, k the Boltzmann constant, and T the absolute temperature. Figure 7.3 is a plot of this distribution as a function of E for several values of temperature. At absolute zero, this distribution function is simply a step function for which

$$f(E) = \begin{cases} 1 & \text{for } E < F \\ 0 & \text{for } E > F \end{cases}$$

As the temperature increases, the edges of the step start to round off and the distribution function loses its steplike character and varies much more slowly with energy. It can be shown that in the high-temperature limit the Fermi–Dirac distribution approaches the classical Maxwell–Boltzmann distribution. From Equation (7.14), we have for $E = F$

$$f(F) = \frac{1}{2}$$

This implies that a quantum state at the Fermi level has a probability of occupation of ½.

We now introduce the concept of the density of states per unit energy for electrons and holes in a semiconductor. It can be visualized by taking a cube of unit volume a^3 isolated within the crystal. The components of **k** in each state must satisfy the boundary conditions

$$k_x = \frac{m\pi}{a}, \qquad k_y = \frac{n\pi}{a}, \qquad k_z = \frac{l\pi}{a}$$

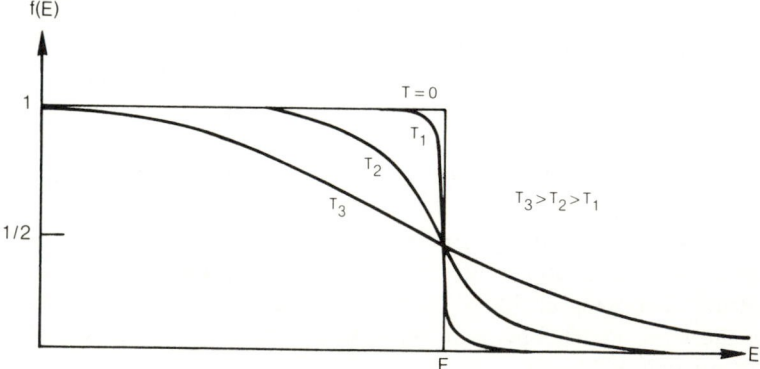

Figure 7.3   Temperature dependence of Fermi–Dirac distribution function $f(E)$.

Each state occupies the same volume in **k** space: π^3/a^3. The number of states per unit volume $\rho(k)$ in an interval dk can be determined by considering the volume of a spherical shell of thickness dk in one quadrant, which has a value given by $\frac{1}{2}\pi k^2\,dk$. Therefore, the number of states that occupy the volume above for a given k value is

$$\rho(k)\,dk = \frac{k^2}{\pi^2}\,dk \tag{7.15}$$

where the volume of a unit cell is assumed to be unity and a factor of 2 has been introduced into Equation (7.15) to allow for two possible spin directions.

Using the expression for a free particle, e.g., $E - E_c = \hbar^2 k^2/2m_e$, we can express the density of states in terms of E for the conduction band as

$$\rho_c(E)\,dE = \frac{8\sqrt{2}}{h^3}\,\pi m_e^{3/2}\,\sqrt{E - E_c}\ dE \tag{7.16}$$

where E_c and m_e are the band edge and the effective mass of the electron in the conduction band, respectively. Similarly, the density of states in the valence band is given by

$$\rho_v(E)\,dE = \frac{8\sqrt{2}}{h^3}\,\pi m_h^{3/2}\,\sqrt{E_v - E}\ dE \tag{7.17}$$

where E_v and m_h are the band edge and the effective mass of the hole in the valence band. Equations (7.16) and (7.17) indicate that the density of states is proportional to $m_{\text{eff}}^{3/2}$. From the difference in effective mass for electrons and holes, we see that the density of states in the valence band is almost 25 times greater than that in the conduction band.

When the system is in thermal equilibrium, one can define a single Fermi energy level that uniquely determines the distributions of electrons and holes in these bands over the entire range of energy. As excess electrons are injected into one or both bands, the equilibrium condition is disturbed and it is necessary to establish two distinct distributions, each of which corresponds to a quasi-Fermi level. If we introduce F_c and F_v as the quasi-Fermi levels of the conduction and valence bands, respectively, we can write the occupation factors f_c and f_v for the two bands as

$$f_c = \frac{1}{\exp(E - F_c)/kT + 1} \tag{7.18}$$

and

$$f_v = \frac{1}{\exp(E - F_v)/kT + 1} \tag{7.19}$$

Using Equations (7.16) to (7.19), we can calculate the total concentration of electrons and holes by integrating the product of the density of states and the occupational factor over the entire energy range:

$$n = \int \rho_c(E - E_c)f_c \, dE \tag{7.20}$$

and

$$p = \int \rho_v(E_v - E)(1 - f_v) \, dE \tag{7.21}$$

The carrier concentration in the conduction band as given by Equation (7.20) can be written explicitly as

$$n = \frac{4\pi}{h^3} (2m_e)^{3/2} \int_{E_c}^{\infty} \frac{(E - E_c)^{1/2} \, dE}{\exp[(E - F_c)/kT] + 1} \tag{7.22}$$

Similarly, we can write the expression for the hole concentration as

$$p = \frac{4\pi}{h^3} (2m_h)^{3/2} \int_{-\infty}^{E_v} \frac{(E_v - E)^{1/2} \, dE}{\exp[-(E - F_v)/kT] + 1} \tag{7.23}$$

To carry out these integrals, it is necessary to determine F_c and F_v. For very lightly doped material, the Fermi level lies outside the band. If F_c is at least $3kT$ below the conduction band edge in n-type material, and F_v is at least $3kT$ above the valence-band edge in p-type material, the exponential term in the denominators of Equations (7.22) and (7.23) is large compared to unity. We can obtain an approximate value for n. By letting $x = (E - F_c)/kT$ and changing the limits of integration, a simple expression for the electron density is obtained from Equation (7.22):

$$n = \frac{4\pi}{h^3} (2m_e)^{3/2} \exp\left(\frac{F_c - E_c}{kT}\right) \int_0^{\infty} x^{1/2} \exp(-x) \, dx$$

Since the value of the integral is $\sqrt{\pi}/2$, we obtain the result

$$n = N_c \exp\left(\frac{F_c - E_c}{kT}\right) \tag{7.24}$$

where

$$N_c = 2 \left(\frac{2\pi m_e kT}{h^2}\right)^{3/2} \tag{7.25}$$

Similarly, we have

$$p = N_v \exp\left(\frac{E_v - F_v}{kT}\right) \tag{7.26}$$

where

$$N_v = 2 \left(\frac{2\pi m_h kT}{h^2} \right)^{3/2} \tag{7.27}$$

At 300°K, the N_c and N_v values in GaAs are 4.35×10^{17} cm^{-3} and 8.87×10^{18} cm^{-3}, respectively.

Equations (7.20) and (7.21) are useful only if the impurity concentration is very low. However, most semiconductor light-emitting devices are either made of heavily doped materials or are operated under high injection currents. Under these conditions, it is necesary to modify the density of states by adding various types of tails at the edges of the conduction and valence bands. These smearing effects on the band structure caused by high density of impurities or electrons are difficult to treat theoretically. A usual approach is to modify the normal density-of-state functions with an exponential tail state function with an empirical parameter that governs the effective depth of the tail. The parameter must be determined experimentally from the measurements of the emission and absorption spectra. More details can be found in the book by Thompson (Ref. 7.2).

7.4 EFFECTS OF DOPING

In an intrinsic material, the densities of electrons and holes are nearly equal. This condition occurs when $F_c = F_v$, as shown in Figure 7.4(a). Substituting $F = F_c = F_v$ into Equations (7.24) and (7.26), we have

$$n = N_c \exp \left(\frac{F - E_c}{kT} \right)$$

$$= N_v \exp \left(\frac{E_v - F}{kT} \right) = p$$

From the product of $np = n_i^2$, we obtain an expression for intrinsic carrier concentrations,

$$n_i = \sqrt{N_c N_v} \exp \left(- \frac{E_g}{2kT} \right) \tag{7.28}$$

which can be described simply by Boltzmann statistics. Figure 7.4(b) shows an n-type material in which the Fermi level is close to the conduction band. This shift is primarily a result of the ionization of the donors by an amount

$$\delta_c = E_c - F_c \tag{7.29}$$

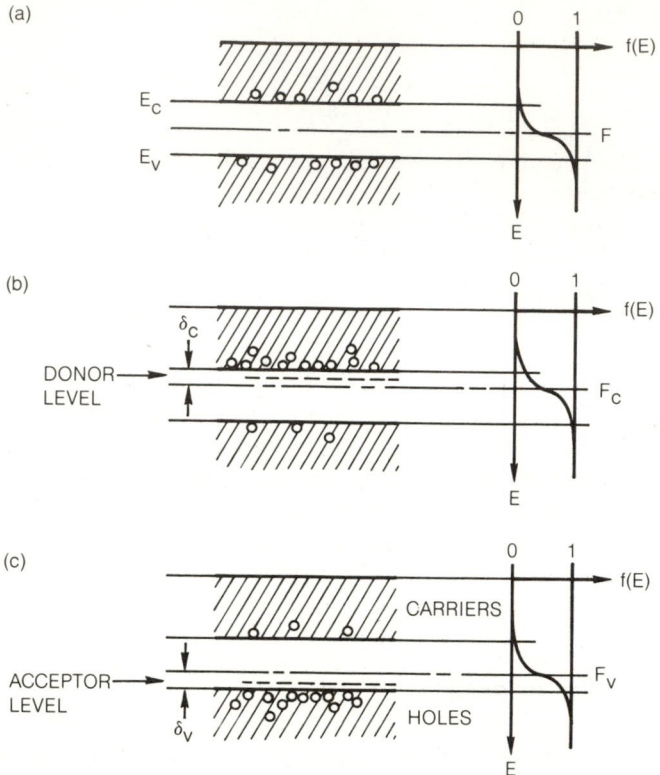

Figure 7.4 Energy levels and charge distribution of semiconductors: (a) intrinsic, (b) *n*-type; and (c) *p*-type materials.

Figure 7.4(c) shows *p*-type material in which the Fermi level is close to the valence band. This is a result of the capture of electrons by the acceptors, thus creating holes in the valence band. Physically, we must consider the bonding structure within a solid. In the case of silicon and germanium, each atom has four valence electrons, and each electron is shared with a nearest neighbor. Ideally, there are no free electrons at 0°K. As the temperature increases, electrons are excited into the conduction band with a concentration proportional to Boltzmann's factor, and, in the meantime, holes are created in the valence band. Under an applied electric field the electrons and the holes are moving in opposite directions at different velocities depending on their effective masses. In GaAs the nearest-neighbor atoms have unequal numbers of valence electrons. Five valence electrons are associated with the As atom and three are associated with the Ga atom. But the total number remains eight. Thus there is a net charge difference between the neighboring atoms.

The introduction of dopants into a semiconductor causes changes in the carrier density and in ionization energy E_i. Consequently, the intensity and the wavelength of the light emitted from the material will be affected. In Zn-doped GaAs, where the Ga atom is replaced by a Zn atom, the impurity acts as an acceptor. In Te-doped GaAs, the As atom is replaced by a Te atom and forms a donor. There are various dopants in GaAs that fall into three major categories: (1) simple donors, (2) simple acceptors, and (3) complex centers. They can be introduced into GaAs by various techniques involving epitaxial growth, which are discussed in Chapter 8. Table 7.2 lists various types of dopants and their ionization energies in GaAs.

From band theory, the effect of doping is described by forming the bandtail states as shown in Figure 7.5. These states become significant when the impurity concentration approaches N_c and N_v. The net effect of bandtail states is to reduce the separation between the valence and the conduction band edges. Therefore, optical transitions involving emission and absorption occur at

TABLE 7.2   Ionization Energies of Dopants in GaAs

Type	Element	E_i (eV)
Simple donors	S	0.006
	Se	0.006
	Te	0.006
	Sn	0.006
	C	0.006
	Ge	0.006
	Si	0.006
Simple acceptors	Cd	0.03
	Zu	0.03
	Mg	0.03
	Be	0.03
	C	0.03
	Su	0.2
	Pb	0.12
	Ge	0.038
	Si	0.035
Comples center	Ge (acceptor)	0.08
	Si (acceptor)	0.1
Transition metals	Cr	0.8
	Mn	0.1
	Fe	0.2–0.5
	Co	0.1–0.5
	Cu	0.15
	Ag	0.24

Source: Data compiled by H. Kressel and J. K. Butler, *Semiconductor Lasers and Heterojunction LEDs,* Academic Press, Inc., New York, 1977.

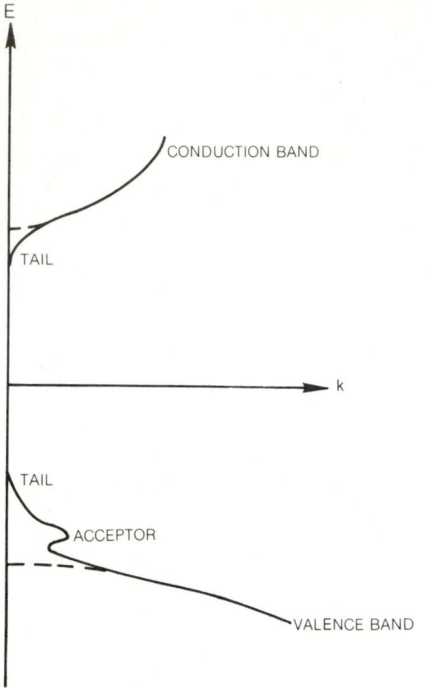

Figure 7.5 Effect of p-type impurities on density-of-state distribution for conduction and valence bands by introducing the bandtails to $E(k)$ curves.

energies less than the bandgap energy of the undoped material. Another consequence of doping is that the radiative efficiency will increase with increasing electron or hole concentration in n- or p-type material, provided that the nonradiative lifetime remains constant.

7.5 RADIATIVE TRANSITIONS AND RECOMBINATION RATES

In thermal equilibrium the conduction band usually contains only a few filled states and the valence band has only a few empty states. Therefore, the radiative transitions through spontaneous emission in this case are insignificant. One of the most efficient ways to produce electroluminescence is to inject currents through a pn junction by applying forward bias to the junction. During the recombination of injected electrons with holes, light is emitted from the junction. The efficiency and the rate of recombination are very different in direct versus indirect bandgap materials. For direct bandgap semiconductors such as GaAs, InP, GaSb, and so on, a band-to-band radiative recombination has the highest probability, because this process involves the most direct interaction between an electron, a hole, and a photon. As shown in Figure 7.6(a), the conduction band edge which is the minimum occurs at the same k value as for the maximum (band edge) of the valence band. In this case, the electrons and

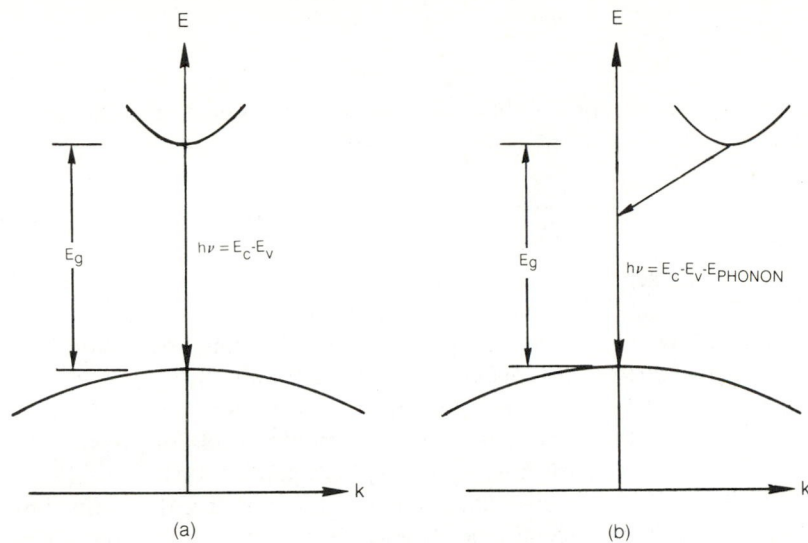

Figure 7.6 Processes occurring during a radiative transition in (a) direct bandgap and (b) indirect bandgap materials.

holes have a large overlapping range of k values to satisfy the momentum conservation requirement for the recombination process. This requirement is commonly referred to as the k selection rule. In these materials, the radiation emitted can also be absorbed by exciting an electron in the valence band back to the conduction band. For this case, the density of the emitted photons is an important factor and will be treated in the next section.

For indirect bandgap materials such as AlAs, GaP, Si, and so on, light emission involves a multistep process in which recombination of an electron and a hole can occur only through collision with lattice vibrations or phonons as illustrated in Figure 7.6(b). Therefore, the efficiency and rate of recombination are considerably reduced for these materials. A direct bandgap material can be converted to indirect by an alloy processing through which a binary can be made into a ternary or a quaternary. For example, a certain amount of aluminum can be introduced into GaAs to form the $Al_x Ga_{1-x}As$ compound, which becomes an indirect material when the composition factor x reaches 0.45. This process, which has been used widely in making heterostructures, is discussed in detail in Chapter 8. For heavily doped indirect materials, however, radiative transitions between deep impurity levels can be very efficient, because wave functions of spatially confined impurity states can extend quite far in momentum space, and consequently, relax the k selection rule.

The probability for radiative recombination, P_r, which is directly proportional to the recombination rate R, is related to the radiative lifetime τ_r as

$$P_r = \frac{1}{\tau_r} \tag{7.30}$$

If we let P_{nr} be the probability for nonradiative transitions, the quantum efficiency for luminescence can be expressed as

$$\eta = \frac{P_r}{P_r + P_{nr}} = \frac{1}{1 + \tau_r/\tau_{nr}} \tag{7.31}$$

Equation (7.31) represents the ultimate limiting value, because for actual devices the quantum efficiency is usually reduced by other losses, such as absorption and reflection. The efficiency for a band-to-band transition in direct bandgap semiconductors such as GaAs, InP, and GaSb, as well as the ternary and quaternary alloys of these compounds, is very high and can approach 100%. Therefore, these materials are commonly used for making highly efficient LEDs and LDs for fiber optical system applications.

Both the quantum efficiency and the transition rate are determined by the radiative lifetime. This value also determines the switching rate of the source. In this section we introduce various possible radiative transitions and derive expressions for the rates of these transitions. For simplicity, we deal first with transitions between two discrete levels and then extend the treatment by introducing the density of states and proper selection rules to make it suitable for an energy band.

Let E_1 and E_2 be the energy of a state within the valence and conduction bands, respectively. There are, in general, three types of radiative transitions: spontaneous emission, stimulated emission, and absorption. Each transition is associated with a probability coefficient: A_{21}, B_{21}, and B_{12}, respectively. The order of appearance for the subscripts indicates the direction in which these processes occur. For example, emission processes occur only from state 2 to state 1 and absorption from 1 to 2. The rate of a transition is related to the product of the occupation factor of the initial state and the nonoccupation factor of the terminal state with the appropriate probability coefficient. We then write the spontaneous emission rate R_{sp} as

$$R_{sp} = A_{21} f_2 (1 - f_1) \tag{7.32}$$

where f_1 and f_2 can be obtained by using Equations (7.18) and (7.19), which are given as

$$f_1 = \frac{1}{\exp(E_1 - F_v)/kT + 1} \tag{7.33}$$

$$f_2 = \frac{1}{\exp(E_2 - F_c)/kT + 1} \tag{7.34}$$

The transition rates for stimulated emission and absorption, on the other hand, depend not only on these occupation factors but also on the density of photons $\rho_p(\varepsilon)$, where $\varepsilon = h\nu$. Therefore, the absorption transition rate can be written as

$$R_{ab} = B_{12}f_1(1 - f_2)\rho_p(\varepsilon) \qquad (7.35)$$

and the downward stimulated emission rate can be written as

$$R_{st} = B_{21}f_2(1 - f_1)\rho_p(\varepsilon) \qquad (7.36)$$

The spectral density of photons $\rho_p(\varepsilon)$ can be determined in a manner similar to the concentration of electrons and holes. However, the distribution law that governs photons is different from that for charged particles, because photons are not subject to the Pauli exclusion principle, and they are regarded as a system of indistinguishable particles having integer spin. The Bose–Einstein distribution law applies to such a system. The average number of photons in a single quantum state is given by the expression

$$\langle N \rangle = \left[\exp\left(\frac{\varepsilon}{kT}\right) - 1 \right]^{-1} \qquad (7.37)$$

Therefore, the photon density distribution per unit volume and unity energy ε can be obtained by multiplying $\langle N \rangle$ by the density of states, $\rho(\varepsilon)$. From Equation (7.15), we can express the density of states in terms of ε. Noting that $k = 2\pi n_0\varepsilon/hc$ and $dk = (2\pi/hc)(n_0 + \varepsilon\, dn_0/d\varepsilon)\, d\varepsilon$, we have

$$\rho_p(\varepsilon)d\varepsilon = \frac{k^2}{\pi^2}\, dk = \frac{8\pi n_0^2\varepsilon^2}{h^3c^3}\left(n_0 + \varepsilon\,\frac{dn_0}{d\varepsilon}\right) d\varepsilon \qquad (7.38)$$

where n_0 is the refractive index of the material. Using Equations (7.37) and (7.38) and ignoring the material dispersion, we obtain

$$\rho_p(\varepsilon) = \langle N \rangle \rho_p(\varepsilon) = \frac{8\pi n_0^3\varepsilon^2}{h^3c^3}\left[\exp\left(\frac{\varepsilon}{kT}\right) - 1\right]^{-1} \qquad (7.39)$$

We can now establish the relationships between the three probability coefficients by using detailed balancing between the upward and downward transitions. At thermal equilibrium, we obtain the following relations:

$$R_{ab} = R_{st} + R_{sp} \qquad (7.40)$$

and by definition,

$$F_c = F_v \qquad (7.41)$$

Substituting Equations (7.32), (7.35), and (7.36) into (7.40), we obtain

$$\rho_p(\varepsilon) = \frac{A_{21} f_2 (1 - f_1)}{B_{12} f_1 (1 - f_2) - B_{21} f_2 (1 - f_1)} \qquad (7.42)$$

Using the results of Equations (7.39) and (7.41), we can separate Equation (7.42) into temperature-dependent and temperature-independent terms as follows:

$$\exp\left(\frac{\varepsilon}{kT}\right)\left(\frac{8\pi n_0^3 \varepsilon^2}{h^3 c^3} B_{12} - A_{21}\right) = \left(\frac{8\pi n_0^3 \varepsilon^2}{h^3 c^3} B_{21} - A_{21}\right) \qquad (7.43)$$

For all values of T, we see that the quantities inside the parentheses on both sides of Equation (7.43) must vanish. Hence we obtain the following relations:

$$A_{21} = \frac{8\pi n_0^3 \varepsilon^2}{h^3 c^3} B_{21} \qquad (7.44)$$

and

$$B_{21} = B_{12} \qquad (7.45)$$

Even though the analysis above is carried out for a system in thermal equilibrium, the probability coefficients are independent of the nature of the system, and can be used for systems that deviate from thermal equilibrium. In this case we assume that the field intensity changes only slowly with frequency and is essentially constant over the entire range of the emission spectrum.

In thermal equilibrium, as shown by Figure 7.7(a), a photon with energy $\varepsilon = h\nu \geq E_g$ will be absorbed by the system and excite a valence electron into the conduction band. This additional electron will again return by recombination to the valence band after an average time τ_{sp} that defines the spontaneous emission lifetime. Under nonequilibrium conditions, as shown by Figure 7.7(b), the photon can induce a downward event, which is called stimulated emission. In this case population inversion is established by injecting electrons over potential barriers at heterojunctions, as discussed in Chapter 8. Since an equal number of holes is generated in the valence band to maintain charge neutrality, the states in the valence band up to F_v are empty. Because the conduction band is filled up to F_c, photon energies greater than E_g but less than $F_c - F_v$ cannot be absorbed, but these photons can induce downward electronic transitions by recombining electrons from the conduction band with holes in the empty valence band. Therefore, the requirement for stimulated emission is

$$F_c - F_v > h\nu$$

In other words, to sustain stimulated emission, it is necessary to satisfy the condition

$$R_{st} > R_{ab} \qquad (7.46)$$

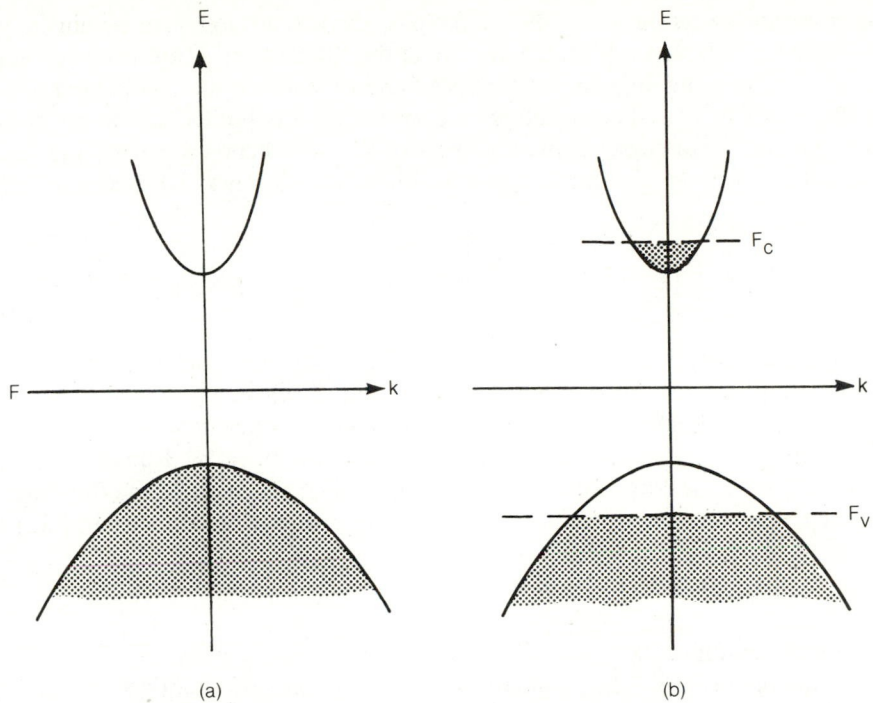

Figure 7.7 Electron energy diagram as a function of density of states for systems (a) in thermal equilibrium and (b) in nonequilibrium.

Substituting Equations (7.35) and (7.36) into (7.46), we obtain

$$f_2(1 - f_1) > f_1(1 - f_2)$$

This condition is equivalent to

$$\exp[(F_c - F_v)kT] > \exp\left(\frac{E_2 - E_1}{kT}\right)$$

which implies that $F_c - F_v > E_2 - E_1 = h\nu$. The net rate, defined by the difference between the downward and the upward transitions, can be reduced to a simple form, given by

$$R_{net} = R_{st} - R_{ab} = \frac{A_{21}(f_2 - f_1)}{\exp(\varepsilon/kT) - 1} \qquad (7.47)$$

To extend the foregoing analysis for transition rates between two levels E_1 and E_2 to the case involving a band-to-band transition, it is necessary to take into account the density of states as given by Equations (7.16) and (7.17) and to integrate the rate expressions over the entire energy range allowed by appropriate selection rules. For example, the absorption rate as given by Equation (7.35) depends on the density of the filled states in the valence band, $\rho_v(E_v - E_1)f_1$, and also on the density of empty states in the conduction band, $\rho_c(E_2 - E_c)(1 - f_2)$. The selection rule in this case is limited to the condition that the amount of energy separation between the two bands is exactly the same as that given by $h\nu$. With these modifications we can rewrite Equation (7.35) as

$$R_{ab} = \int_{-\infty}^{\infty} B_{12}\rho_v(E_v - E_1)\rho_c(E_2 - E_c)f_1(1 - f_2)\rho_p(\varepsilon)\,\delta(E_2 - E_1 - \varepsilon)\,dE \tag{7.48}$$

Equation (7.48) implies that the transition from the valence band to the conduction band is allowed only between states where ρ_v and ρ_c separated exactly by $h\nu$. Under this circumstance, k selection rules can be ignored.

Similarly for spontaneous emission, the rate given by Equation (7.32) depends on the density of the filled conduction band states, $\rho_c f_2$, and the density of the empty valence band states, $\rho_v(1 - f_1)$. We can rewrite Equation (7.32) as

$$R_{sp} = \int_{-\infty}^{\infty} A_{21}\rho_c\rho_v f_2(1 - f_1)\delta(E_2 - E_1 - \varepsilon)\,dE \tag{7.49}$$

The same modification must be made to the stimulated emission rate as given by Equation (7.36). To evaluate these rates, one of the two probability coefficients A_{21} or B_{12} must be determined. The calculation of these coefficients requires extensive quantum mechanical knowledge which involves the use of time-dependent perturbation theory, as described in most texts on quantum mechanics. Basically, one must calculate the matrix element for the interaction Hamiltonian, H, involving an electron and the radiation field between the two states in the semiconductor. This treatment is omitted here because of the difficulty in obtaining the appropriate wave functions associated with a highly doped semiconductor, whose band structure deviates significantly from the simple parabolic model. Several models have been introduced to modify the smearing effect of band edges by adding bandtails in the structure, as illustrated in Figure 7.5. However, verification of these models is still a subject of current research. The density of states deviates significantly near the band edges from the usual parabolic dependence for a semiconductor with an intermediate doping level. The hole impurity band structure, which lies fairly deep in the valence band, is characteristic of p-type doping at levels around 10^{18} cm^{-3}. In n-type materials the impurity bands are much less conspicuous than those shown in Figure 7.5, because the donor band is relatively shallower and merges with the conduction band quickly at levels greater than 10^{16} cm^{-3}.

It is possible to relate these probability coefficients to the absorption coefficient $\alpha(\varepsilon)$, which is a readily measurable quantity. Experimentally, one obtains the value of $\alpha(\varepsilon)$ by measuring the incident and the transmitted power, P_0 and P_t, through a given material thickness t as given by

$$\alpha(\varepsilon) = \frac{1}{t} \ln \frac{P_0}{P_t} \tag{7.50}$$

The product of the absorption coefficient, $\alpha(\varepsilon)$, and the incident photon flux, $v_g \rho_p(\varepsilon)$, defines the net absorption rate, which yields the difference in the number of photons between upward and downward transitions. Using Equations (7.35) and (7.36), we can write for the case of two discrete levels that

$$\alpha\, v_g \rho_p(\varepsilon) = [B_{12} f_1(1 - f_2) - B_{21} f_2(1 - f_1)] \rho_p(\varepsilon)$$

The equation above can readily be reduced to the following simple form:

$$\alpha(\varepsilon) \frac{c}{n} = B_{12}(f_1 - f_2) \tag{7.51}$$

where the material dispersion has been ignored. Substituting Equation (7.51) into (7.32) and using the relation between A_{21} and B_{21} given by Equation (7.44), we can write the spontaneous emission rate as

$$R_{sp} = -\frac{8\pi n_0^2 \varepsilon^2}{h^3 c^2} \alpha(\varepsilon) \gamma \tag{7.52}$$

where the factor $\gamma = f_2(1 - f_1)/(f_2 - f_1)$ can be further reduced to

$$\frac{1}{\gamma} = 1 - \exp\left[\frac{\varepsilon - (F_c - F_v)}{kT}\right] \tag{7.53}$$

Similarly, we obtain the relation between the net stimulated emission rate and the absorption coefficient as given by

$$R_{net} = -\frac{8\pi n_0^2 \varepsilon^2}{h^3 c^2} \alpha(\varepsilon) \tag{7.54}$$

It is interesting to note that the rate expression above is independent of occupation factors, and the difference between R_{sp} and R_{net} is given only by a Boltzmann factor, $\exp(\varepsilon - \Delta F)kT$, with $\Delta F = F_c - F_v$ being the difference in the quasi Fermi levels. If $\Delta F \gg \varepsilon$, $R_{net} \simeq R_{sp}$. This means that the spontaneous emission rate approaches that of the stimulated emission under very high injection.

7.6 CARRIER LIFETIME

From the previous analysis of spontaneous and net stimulated transitions, it is possible to evaluate the radiative lifetime as limited by a band-to-band recombination. As defined by Equation (7.30), the recombination rate resulting from a specific transition is related to the carrier lifetime as

$$\tau_r = \frac{n}{R} \tag{7.55}$$

where n is the carrier concentration given by Equation (7.20) or (7.21). R is the corresponding transition rate. From measured values of α at various injection levels, one obtains the corresponding τ_r values. We shall first consider the case involving either spontaneous emission or low injection level. The spontaneous recombination rate R_{sp} can be expressed by Equation (7.49) for the spontaneous emission rate. In terms of both the electron and hole concentrations and by making use of Equation (7.20) and (7.21), we write

$$R_{sp} = \frac{8\pi n_0^3 \varepsilon^2}{h^3 c^3} \int_0^{n} \int_0^{p} B(n, \, p) \, dn \, dp \tag{7.56}$$

where B is the rate coefficient for spontaneous emission. This is a useful form, especially in the case of very low injection, which results in a relatively undisturbed distribution of electrons and holes in the noninverted condition. To a good approximation, it is possible to take $B(n, \, p)$ outside the integral, and we can write a simple expression for the rate as

$$R_{sp} = B' \, np \tag{7.57}$$

where B' is related to the rate constant and is equal to $(8\pi n_0^3 \varepsilon^2/h^3 c^3)B$. For the case of a very high injection level, the k value of at least one of the carrier types becomes very large, so that the k selection rule is relaxed to the extent that B is a constant over most of the integral. Therefore, we can again write the recombination rate in the simple form given by Equation (7.57). This simple relationship is no longer true when the Fermi levels move from within the bandgap into the bands. In this case, the value of B or the matrix element is strongly dependent on the energy of the initial and final states.

Under conditions of thermal equilibrium, the product np is equal to n^2. Hence we define the noninverted recombination rate $R_{sp}^\circ = B'n^2$. Under nonequilibrium conditions, additional carriers $\Delta N = \Delta P$ are injected into the material. Therefore, the total recombination rate must be modified by the expression

$$R_{sp} = R_{sp}^\circ + R_{sp}^{ex} \tag{7.58}$$

where the recombination rate of the injected excess carrier is given by

$$R_{sp}^{ex} = R_{sp} - R_{sp}^\circ = B'(n + \Delta N)(P + \Delta P) - B'np$$
$$= B' \Delta N(n + p + \Delta N) \tag{7.59}$$

The radiative lifetime for the excess carriers is defined by

$$\tau_r = \frac{\Delta N}{R_{sp}^{ex}} = \frac{1}{B'(n + p + \Delta N)} \tag{7.60}$$

TABLE 7.3   Radiative Recombination Rates, Lifetimes, and Other Parameters
for p-type GaAs at 297°K

p_0 (cm^{-3})	E_g (eV)	μ (cm^2/V-s)	B' (10^{-10} cm^3/s)		τ_r (ns)	
			Exptl.	Theor.[a]	Exptl.	Theor.[a]
1.2×10^{18}	1.408	162	3.2	1.9	2.6	4.4
2.4×10^{18}	1.403	102	2.8	1.7	1.5	2.4
1.6×10^{19}	1.381	67	1.7	0.9	0.37	0.7

[a]From R. E. Fern and A. Onton, *J. Appl. Phys.*, *42*, 3499 (1971).

At high currents (e.g., $\Delta N > n + p$) the radiative lifetime can be approximated by the expression

$$\tau_r \simeq (B'\Delta N)^{-1} \qquad (7.61)$$

Table 7.3 gives the values of B' and τ_r for three cases of p-type GaAs. The values are obtained from the absorption measurements under low injection levels. From Equation (7.57) we can express the radiative lifetime τ_r for p-type material in terms of $B'p_0$ or

$$\tau_r = \frac{1}{B'p_0} \qquad (7.62)$$

Also included in Table 7.3 are the calculated values of B' and τ_r based on the bandtails model. The calculated and the measured values differ by only a factor of about 2. Figure 7.8 shows the radiative lifetime as a function of the hole concentration p_0 at 300°F. The data points represented by the squares are obtained by measuring the diffusion length in GaAs. From simple diffusion theory, one obtains for the diffusion length Λ

$$\Lambda = (D\tau_r)^{1/2} \qquad (7.63)$$

where D is the diffusion coefficient and is related to the mobility μ as

$$D = \frac{\mu kT}{e} \qquad (7.64)$$

At hole concentration levels greater than 10^{18} cm^{-3}, good agreement exists among these lifetime data. Below this concentration level, the lifetime is dominated by nonradiative recombination, which sharply degrades the quantum efficiency. At levels above 10^{18} cm^{-3} the quantum efficiency approaches 100% for most direct bandgap materials, and a recombination lifetime shorter than 1 ns is expected.

Besides the spontaneous lifetime, there are two other lifetimes relevant to the operation of a semiconductor laser: (1) the stimulated lifetime and (2) the

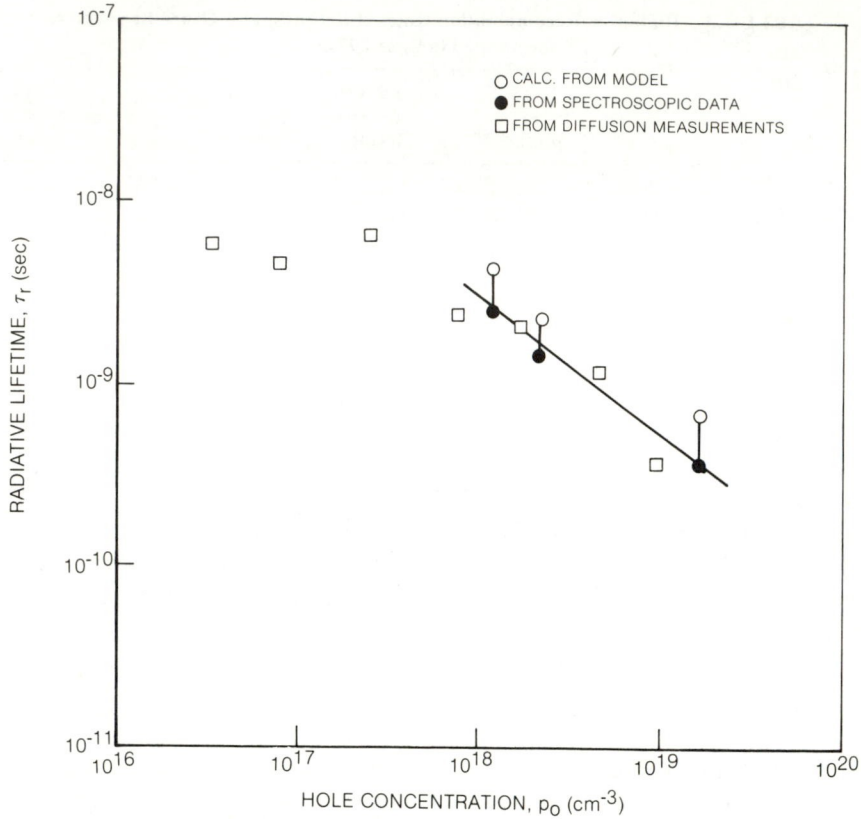

Figure 7.8 Radiative lifetime t_r as a function of hole concentration at 300°K. (From Ref. 7.3. Reprinted with permission of Academic Press, Inc., New York.)

photon lifetime. The values for the former are of the order of 10^{-11} s and decrease with increasing light intensity. The latter has values on the order of 10^{-12} s for GaAs diode lasers and decreases with increasing losses of the cavity.

7.7 LIGHT AND CURRENT RELATIONS

The current density J in an active layer with a thickness d is defined as

$$J = \frac{edR}{\eta} \tag{7.65}$$

where R is the recombination rate per unit volume and η is the quantum efficiency associated with the transition. Near threshold or at low injection levels, we assume that the stimulated recombination process is negligibly small and R_{sp} can account for all the luminescent process. In this case we can relate R in Equation (7.65) to the rate of spontaneous emission R_{sp} as given by Equation (7.52) and write

$$J(A/cm^2) = -\frac{ed\,8\pi n_0^2}{\eta h^3 c^2} \int \varepsilon^2 \gamma(\varepsilon)\alpha(\varepsilon)\,d\varepsilon \tag{7.66}$$

In practice the solution of this integral is not straightforward because of the difficulty in obtaining satisfactory relationships for the density-of-state functions and the precise dependence of α on ε. However, it is possible to obtain an estimate of the optical gain or loss versus injection current density by examining Equation (7.65) in a semiqualitative way. We rewrite Equation (7.65) as follows:

$$J = \frac{ed}{\eta} \int R_{sp}(\varepsilon)\,d\varepsilon \tag{7.67}$$

where the integral is carried out over all energy values to include the states in the conduction and valence bands relevant to the luminescent process. We now evaluate R_{sp} by using Equation (7.52) at an energy value ε_m at which a maximum transition rate occurs. Dividing Equation (7.67) by $R_{sp}(\varepsilon_m)\,\Delta\varepsilon$ and assuming that the process is above threshold [e.g., $-\alpha(\varepsilon) = g(\varepsilon)$], where g is the gain coefficient, we can express Equation (7.67) by the approximation (Ref. 3.)

$$J(A/cm^2) = \frac{8\pi edn_0^2\varepsilon_m^2 g(\varepsilon_m)\,\Delta\varepsilon\,\gamma}{\eta c^2 h^3} \int_0^\infty \frac{R_{sp}(\varepsilon)\,d\varepsilon}{R_{sp}(\varepsilon_m)\,\Delta\varepsilon} \tag{7.68}$$

where the integral in Equation (7.68) defines the line-shape factor and is approximately of the order of unity. Therefore, the current density can be approximated by the expression

$$J(A/cm^2) \simeq \frac{8\pi edn_0^2\varepsilon_m^2\gamma\,\Delta\varepsilon}{\eta c^2 h^3} g(\varepsilon_m) \tag{7.69}$$

Equation (7.69) indicates that, above threshold, the gain is proportional to the current density and is independent of material properties. The major factors that control the gain are the line width $\Delta\varepsilon$, the photon energy for maximum gain, and the γ factor.

Actually, one can obtain a family of curves for $g(\varepsilon)$ from Equation (7.69) at various injection levels. For a given injection current, there exists a corresponding quasi-Fermi level separation ΔF. The gain coefficient is expected to reach a peak value $g(\varepsilon_m)$ at ε_m, which lies between E_g and ΔF. It is intuitive that g versus J must have a zero value at threshold. Figure 7.9 shows the calculated peak gain values $g(\varepsilon_m)$ as a function of injection current density near liquid-nitrogen ($80°K$) and room ($300°K$) temperatures for both undoped

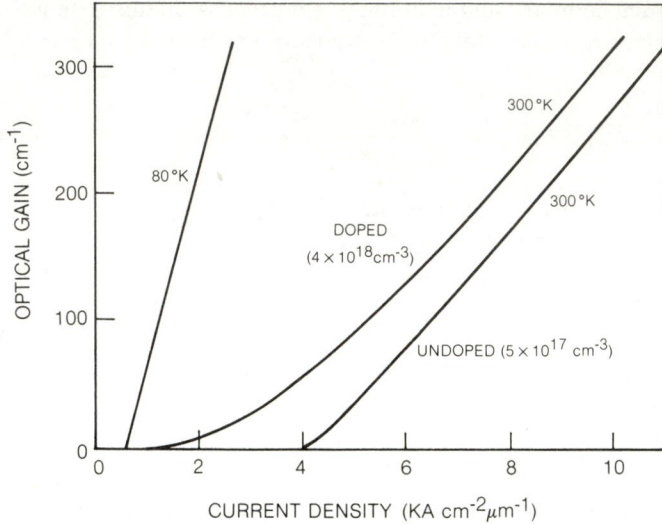

Figure 7.9 Calculated gain values as a function of injection current density at 80°K and 300°K. [After F. Stern, *IEEE J. Quant. Elect., QE-9*, 290 (1973). Reprinted with permission of the Institute of Electrical and Electronics Engineers, Inc., © 1973.]

and heavily doped GaAs. In this case the injection current is confined to an active layer having a thickness of 1 μm. For undoped materials, a sharp cutoff point is expected. As injection level increases, gain coefficient increases, and also the energy ε_m at which the gain is maximum. This upward shift in energy or downshift in wavelength is a direct consequence of an increase in the separation energy between the quasi-Fermi levels. For highly doped materials, cutoff is more or less gradual because bandtail states are present with significant impurity levels. In these calculations, a number of approximations has to be introduced to allow for the effect of the bandtails on the density of states and on the transition probability. Therefore, the results shown in Figure 7.9 should be used only as an estimate, not for precise comparisons. At low currents, higher doping increases the optical gain because, the deeper the bandtail, the lower the Fermi energy required to produce inversion. However, the number of states occupied by the injected electrons is much less than that in an undoped material having the usual parabolic band. At low temperatures, there is very little difference in gain between doped and undoped materials.

At high injection levels, the line width decreases considerably; however, it is compensated by a change in γ. Over a large range of currents, the product of $\gamma \Delta \varepsilon$ is relatively constant. Therefore, over more than a 20-dB dynamic range of gain, a linear relationship between gain and current density can be assumed. More conveniently, we can write that

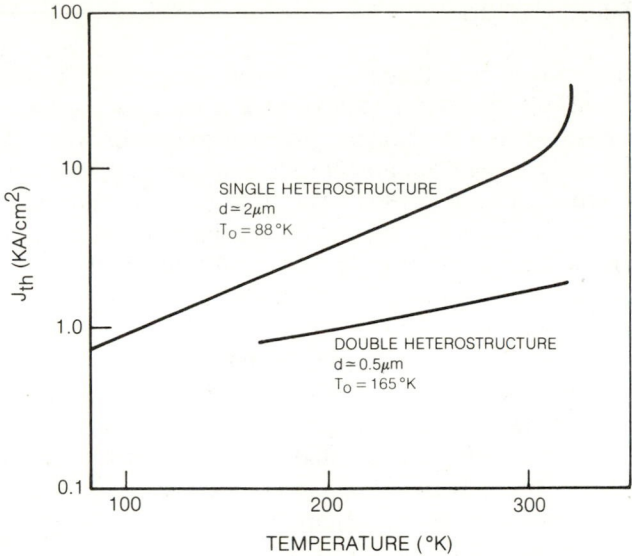

Figure 7.10   Temperature dependence of threshold current density for single- and double-heterostructure lasers. [After I. Hayaski, M. B. Danish, and F. K. Reinhart, *J. Appl. Phys.*, *42*, 1927 (1971).]

$$g = C(J - J_{\text{th}}) \tag{7.70}$$

where C is inversely proportional to T and J_{th} varies as $T^{3/2}$. At room temperature, C for GaAs is of the order of $0.045/\text{A-cm}^{-1}\text{-}\mu\text{m}$ and J_{th} is of the order of 4000 A/cm^{-2}. The temperature dependence of J_{th} is shown in Figure 7.10, indicating that the threshold current density increases with temperature in all types of semiconductor lasers. Because many factors are involved, no simple expression can rigorously describe the behavior over a wide temperature range. An empirical law often used to approximate the temperature effect is given in the form

$$J_{\text{th}} \propto \exp\left(\frac{T}{T_0}\right) \tag{7.71}$$

where T_0 is determined experimentally and found to be a constant over a certain operating range. As temperature increases slightly above the room temperature, a superlinear increase in threshold current occurs and causes a complete cessation of laser action. For this reason, most earlier studies were done at liquid-nitrogen temperature (77°K). At 300°K the threshold current for a single-heterostructure laser is about 10,000 A/cm^2. It can be reduced to the 1000-A/cm^2 level if a multiple-layer structure is employed. The details of heterostructure lasers are given in Chapter 8.

7.8 LASER OSCILLATION

As injection current is increased above J_{th}, the gain as given by Equation (7.70) becomes positive. To sustain laser oscillation, it is necessary to form an optical feedback circuit known as a laser cavity. For semiconductor lasers, the cavity is commonly formed by using two parallel reflecting surfaces which are cleavage planes of the semiconductor crystal. This makes the semi-conductor lasers different from other lasers in that the optical feedback from the cleaved ends is very small, with a typical power reflection coefficient R of about 30%. The optical gain for semiconductor lasers, on the other hand, is extremely high, with a typical gain coefficient $g \simeq 50$ cm^{-1}.

The laser threshold is defined as a condition in which the loop gain is equal to the total cavity losses α_T. This condition can be established by considering a multiple reflection of a plane wave between two partially reflecting surfaces with reflectivity R_1 and R_2 of a Fabry–Perot cavity. If the gain coefficient is not saturated, the radiation intensity can grow exponentially with distance, as indicated by Equation (7.50). The power at a distance z between the mirrors can be expressed by

$$P(z) = P(0) \exp[g(\varepsilon) - \alpha(\varepsilon)]z$$

Making a round trip of $2L$ inside the cavity, we write

$$P(2L) = P(0) R_1 R_2 \exp[2L(g - \alpha)]$$

At threshold, $P(2L) = P(0)$. This leads to the condition for laser oscillation,

$$R_1 R_2 \exp(g_{th} - \alpha_T)2L = 1 \qquad (7.72)$$

The most significant losses are the absorption by free carriers and the scattering by imperfections. As long as the condition of Equation (7.72) is satisfied, laser oscillation will be self-sustained. The only complication of a low-feedback laser system is that the power distribution along the laser cavity length may not be uniform. Therefore, analysis of this type of laser must take the nonuniformity condition into account. In addition, the spatial effect involving confinement of the injection current is also a very important consideration and is discussed in Chapter 8.

Equation (7.72) can be written as

$$g_{th} = \alpha_T + \frac{1}{2L} \ln \frac{1}{R_1 R_2} \qquad (7.73)$$

Because the propagating mode spreads outside the active layer, g_{th} is reduced by the confinement factor Γ, as given by Equation (3.43). Substituting Equation (7.73) into Equation (7.69), we obtain an expression for the threshold current density:

$$J_{th}(A/cm^2) = \frac{8\pi ed \times 10^{-4} \, n_0^2 \varepsilon_m^2 \gamma \, \Delta\varepsilon}{\eta c^2 h^3 \Gamma} \left(\alpha_T + \frac{1}{2L} \ln \frac{1}{R_1 R_2} \right) \quad (7.74)$$

For p-type GaAs with $p_0 = 1 \times 10^{18}$ cm^{-3}, $R_1 = R_2 = 0.3$, $n_0 = 3.6$, $\alpha_T \simeq 10$ cm^{-1}, $\gamma = 9.2$, $\varepsilon_m = 1.4$ eV, and $\Delta\varepsilon = 0.06$ eV. Equation (7.74) gives a value for J_{th}/d at 300°K to be 3.5×10^3 A/cm^2-μm, assuming that $L = 400$ μm and $\eta = \Gamma = 1$. This calculated threshold value represents the theoretical limit for GaAs laser diodes, and was found to be in good agreement with measured values.

In many laser applications, it is desirable to choose an emitter with a low threshold current density so that the operating current for the device can be kept at a low level. Operating at low currents, energy dissipation and other deleterious effects that could lead to a gradual degradation of the emitter can be greatly reduced. According to the factors included in the Equation (7.74), a lower threshold current density can be obtained by eliminating the leakage of injection currents and optical waves from the active layer and by reducing various losses. Several techniques have been developed to confine the currents and optical waves and are discussed in Chapter 8. By improving the quality and the doping level of epitaxial layers in a heterostructure and by optimizing the end-face reflectivity and the cavity length, a threshold current of as low as 500 A/cm^2 at 300°K has been obtained by many workers.

7.9 OPTICAL MODES

The mode pattern of a semiconductor laser is determined primarily by the geometry of the laser. A typical heterostructure semiconductor laser is shown in Figure 7.11. It has a rectangular shape with multiple epitaxial layers grown on a crystalline substrate. The optical feedback is provided by the reflectivity of cleavage planes, which forms a Fabry–Perot cavity. Unlike microwave resonators, optical resonators are open cavities in which only the axial and small off-axis transverse modes can survive. Assuming that the cavity mirrors are flats and separated by a spacing L, the reflected and the transmitted field amplitudes, A_r and A_t, from an incident plane wave having an amplitude A_i at the two surfaces, as shown in Figure 7.12, can be obtained by summing all amplitudes resulting from multiple reflections. We shall write

$$\begin{aligned} A_r &= A_1 + A_2 + A_3 + \cdots \\ &= A_i(r + tt'r'e^{i\delta} + tt'r'^3 e^{2i\delta} + \cdots) \\ &= A_i \left(r + tt'r'e^{i\delta} \frac{1}{1 - r'^2 e^{i\delta}} \right) \end{aligned} \quad (7.75)$$

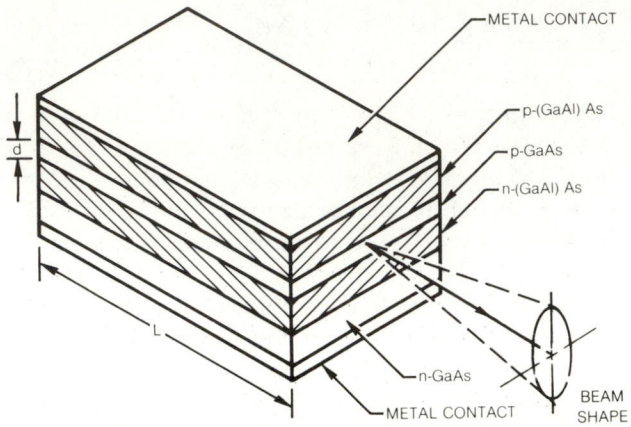

Figure 7.11   Structural diagram of a double-heterostructure AlGaAs laser diode showing the output radiation pattern emitted from cleaved end mirror along the junction plane.

where r, r', t, and t' are reflection and transmission coefficients for the field amplitudes at appropriate boundaries. δ is the phase difference of two waves separated by a distance of one round trip and can be written as

$$\delta = \frac{4\pi n_0 L}{\lambda} \cos \theta \qquad (7.76)$$

Since $\nu = c/\lambda$, the resonant frequency ν can be written as

$$\nu = \frac{c\delta}{4\pi n_0 L \cos \theta} \qquad (7.77)$$

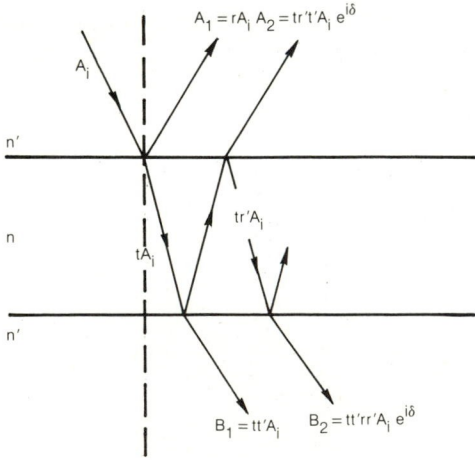

Figure 7.12   Multiple reflection and transmission from two planar surfaces.

We now define the power reflection and transmission coefficient R and T as $R = r^2$ and $T = t^2$. For simplicity, we shall assume that $r = r'$ and $t = t'$. We can rewrite Equation (7.75) in terms of R as

$$A_r = \frac{A_i(1 - e^{i\delta})\sqrt{R}}{1 - Re^{i\delta}} \tag{7.78}$$

In deriving Equation (7.78), we have assumed that $R + T = 1$. Similarly, we have

$$
\begin{aligned}
A_t &= B_1 + B_2 + B_3 + \cdots \\
&= \frac{A_i T}{1 - Re^{i\delta}}
\end{aligned}
\tag{7.79}
$$

The reflected power is proportional to the product of complex conjugate of the amplitude. Therefore, we write

$$\frac{P_r}{P_i} = \frac{A_r^* A_r}{A_i^* A_i} = \frac{4R \sin^2(\delta/2)}{(1 - R)^2 + 4R \sin^2(\delta/2)} \tag{7.80}$$

Similarly, for the power transmitted, we obtain

$$\frac{P_t}{P_i} = \frac{(1 - R)^2}{(1 - R)^2 + 4R \sin^2(\delta/2)} \tag{7.81}$$

From Equation (7.81), we see that if $\delta = 2m\pi$, the power transmitted is 100%. Substituting this value into Equation (7.77), we obtain resonance frequencies at which maximum transmission occurs to be

$$\nu_{max} = m \frac{c}{2n_0 L \cos\theta} \tag{7.82}$$

By letting $\theta = 0°$, the spacing between two axial modes can be obtained from Equation (7.82) as

$$\Delta\nu = \frac{c}{2n_0 L} \tag{7.83}$$

Figure 7.13 shows plots of P_t/P_i as a function of ν for various values of R. As values of R increase, the spectral width of the transmission decreases. If we denote $\Delta\nu_{1/2}$ to be the full width at the half-power points on the transmission curve, it can be shown that

$$\Delta\nu_{1/2} = \frac{\Delta\nu}{F} \tag{7.84}$$

where F is the finesse of the Fabry–Perot cavity and is given by

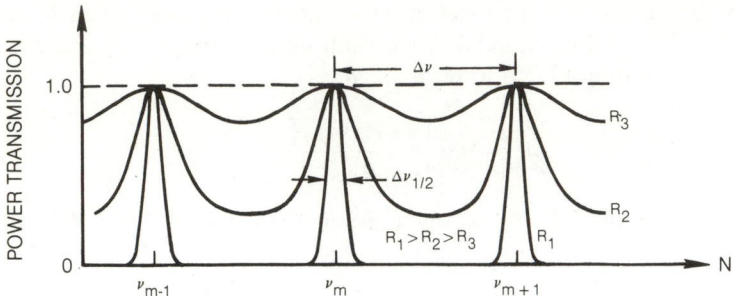

Figure 7.13   Transmission spectral characteristics of a Fabry–Perot interferometer.

$$F = \frac{\pi R^{1/2}}{1 - R} \tag{7.85}$$

Equation (7.85) indicates that as R increases, finesse increases and the cavity resonances are sharply peaked, as shown in Figure 7.13. For a finesse of 100, a mirror reflectivity of 97% is required. In practice, the F value depends not only on R but also on mirror flatness as well as the angular spreading of the beam. Because of these limitations, a finesse of 100 is not very easy to achieve even though $R > 97\%$.

For a typical GaAs laser whose cavity is formed by the two cleaved ends, the axial mode spacing is about 2 Å for $L = 400$ μm. Figure 7.14 shows a typical emission spectrum of a room-temperature continuous-wave (CW) AlGaAs laser diode with an output that consists of many longitudinal or axial modes. An increase in the injection current above the threshold corresponds to a sharp increase in the power output. Below the threshold, only the spontaneous radiation is emitted and the spectrum is very broad, with a spectral width FWHM $\simeq$ 200 Å. As threshold is approached, the spectrum begins to narrow by virtue of the stimulated emission through amplification. At the same time, the emitted beam width is narrowed in the junction plane. As the current increases further, the spontaneous emission approaches a saturation value and stimulated emission dominates. The beam is then fully narrowed, as shown in Figure 7.14 by the near-field distribution. The output is usually shared among many modes, which are distributed within the line width of the gain medium. In semiconductor lasers, the gain width is homogeneously broadened primarily by inelastic collision with phonons having a line shape usually represented by a Lorentzian function.

Lasers, either with narrow stripe width (<10 μm) or utilizing distributed feedback rather than the cleaved ends, can be made to operate in a single longitudinal and transverse mode over significant current ranges above threshold. The theoretical limit on the spectral broadening of a single mode can be estimated from the spontaneous emission rate R_{sp} or the photon lifetime τ_p.

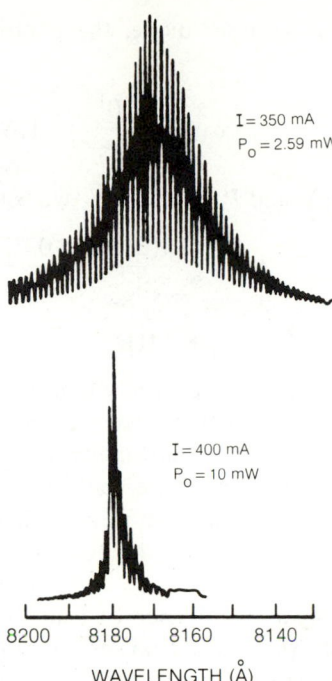

I = 350 mA
P_o = 2.59 mW

I = 400 mA
P_o = 10 mW

Figure 7.14 Output characteristics of a CW AlGaAs laser diode at different injection currents.

8200   8180   8160   8140

WAVELENGTH (Å)

The line width $\delta\nu_m$ of the mode at FWHM increases with increasing photons number N_p, and can be expressed as

$$\delta\nu_m = \frac{\Delta\nu\, N_p}{2} \tag{7.86}$$

where $\Delta\nu$ is the spectral width of the emission line. The value of $\delta\nu_m$ is dependent on the quality factor of the laser resonator Q, and is related to the photon lifetime τ_p as

$$Q = \frac{\nu_0}{\delta\nu_m} = 2\pi\nu_0\tau_p \tag{7.87}$$

Since the power in this laser mode P_m is

$$P_m = \frac{N_p h\nu_0}{\tau_p} \tag{7.88}$$

we obtain by combining Equations (7.86), (7.87), and (7.88) that

$$\Delta\nu = \frac{4\pi h\nu_0 (\delta\nu_m)^2}{P_m} \tag{7.89}$$

In a typical GaAs laser diode, the photon lifetime is of the order 10^{-12} s; therefore,

$$\delta v_m \simeq \frac{10^{12}}{2\pi} = 1.6 \times 10^{11} \text{ s}^{-1}$$

At $h v_0 = 1.5$ eV and $P_m = 10$ mW, we estimate that

$$\Delta v \simeq \frac{4\pi(2.4 \times 10^{-19})(1.6 \times 10^{11})^2}{10^{-2}}$$

$$\simeq 8 \text{ MHz}$$

which corresponds to a spectral width of 1.8×10^{-4} Å. Such a narrow width has never been achieved in practice, because the temperature fluctuation in this type of device introduces instability in cavity dimensions and noises in the spectrum. At room temperature, the lowest spectral width $\Delta\lambda \simeq 0.2$ Å has been achieved by operating a laser in a fundamental transverse mode. The value becomes much closer to the theoretical value by operating the laser at low temperatures.

The physical significance of Equation (7.89) is the predictable inverse dependence of Δv on laser power. A more refined expression for Equation (7.89) can be written by taking into account the effect of the spontaneous emission noise power, which is proportional to the population of the upper laser level N_2. In this case we rewrite Equation (7.89) in the form

$$\Delta v = \frac{4\pi h v (\delta v_m)^2 \mu}{P} \qquad (7.90)$$

where

$$\mu = \frac{N_2}{N_2 - N_1(g_1/g_2)} \qquad (7.91)$$

Since both N_1 and N_2 are power dependent, it can be shown (Ref. 7.4) that as $P \to \infty$, there remains a residual laser line width, which is power independent. This is due to the fact that as P increases, both N_1 and N_2 must increase. At sufficiently high values of P, the ratio N_2/P approaches a constant value, leading to a residual power-independent line width.

PROBLEMS

7.1. If the periodic boundary condition is imposed on the one-dimensional wave function of a system containing N atoms, show that the possible k_n values are $2\pi n/Na$, where $n = 0, 1, \ldots, N$.

7.2. Show that in the free-electron approximation, dE/dk vanishes at band edges, $k = n\pi/a$.

7.3. Show that the density of states in the valence band is given by the expression

$$\rho(E) = \frac{8\sqrt{2}}{h^3} \pi m_h^{3/2} \sqrt{E_v - E}$$

7.4. Derive Equation (7.42).

7.5. Calculate the Fermi level in an n-type material when n/N_c is approximately equal to 0.1.

7.6. Calculate N_c in a p-type AlGaAs at 300°K where $m_e/m = 0.092$.

7.7. Show that the ratio of the density of states

$$\frac{f_2(1 - f_1)}{f_1 - f_2} = \left\{ \exp \left[\frac{\varepsilon - (F_c - F_v)}{kT} \right] - 1 \right\}^{-1}$$

7.8. Show that the spacing of adjacent longitudinal modes in a semiconductor laser with a cavity length L is

$$\Delta\lambda = \frac{\lambda^2}{2n_0 L[1 - (\lambda/n_0) \, dn_0/d\lambda]}$$

7.9. Derive the threshold condition for laser oscillation as given by Equation (7.72).

7.10. Calculate the photon lifetime in a Fabry–Perot cavity in which a photon is lost by either absorption or transmission through the facets. Assume that a total loss coefficient for the cavity is 50 cm^{-1}.

7.11. Plot P_t/P_i as given by Equation (7.81) as a function of δ for $R = 90\%$.

REFERENCES

7.1. J. P. McKelvey, *Solid State and Semiconductor Physics*, Harper & Row, Publishers, Inc., New York, 1966.

7.2. G. H. B. Thompson, *Physics of Semiconductor Laser Devices*, John Wiley & Sons, Inc., New York, 1980.

7.3. H. C. Casey, Jr., and M. B. Panish, *Heterostructure Lasers*, Part A: *Fundamental Principles*, Academic Press, Inc., New York, 1978.

7.4. A. Yariv and K. Vahala, *IEEE J. Quantum Electron.*, QE-19, 889 (1983).

Properties and Growth of Semiconductor Heterojunctions

8.1 INTRODUCTION

In this chapter we make use of the radiative recombination property of semiconductors to generate a class of light-emitting devices commonly known as light-emitting diodes (LEDs) and laser diodes (LDs) and to examine various structural properties and operating parameters of these devices. Presently, device structures of most LEDs and almost all LDs are made of heterojunctions, which are multilayered dissimilar semiconductors. The major distinction between a heterojunction and a homojunction is that two different bandgap materials are used at the junction. Because two dissimilar materials are involved in such a structure, a mismatch in lattice constants can lead to a large number of nonradiative recombination centers at the junction and consequently reduce the light emission probability. Therefore, it is very important to select materials that have nearly identical lattice constants. GaAs and AlGaAs are the most widely used heterostructures and their properties have been studied extensively (Ref. 8.1) and are well understood. Another heterostructure consists of InGaAsP, and it has also received considerable attention recently because its emission spectra fall in the region of least material dispersion. In this chapter we first review the fundamentals of a *pn* junction in a simple binary structure and discuss the effects of forward bias on the potential barrier at the junction and on the distribution of electrons and holes in the bands. Similar considerations will be extended to a class of heterostructures which are made of ternary and quaternary compounds. The methods of epitaxial growth of these materials is then introduced. The relationships between the injection current and voltage

154

and the confinement factors for currents and optical waves in these structures will be derived. To meet the requirements of most fiber optical systems, LED is an adequate source based on its available output power and speed of response. For long-distance and extremely high-data-rate systems, the use of laser diodes should definitely be considered. The trade-off for laser power and bandwidth is not just in the cost but most important, in the reliability of the devices. In general, lasers degrade much more rapidly than LEDs and the deterioration originates primarily from material imperfections. Improving the reliability of laser sources constitutes one of the active areas of research at present.

8.2 THE *pn* JUNCTION

Intrinsic semiconductors can be made into *n*-type and *p*-type materials by introducing slight deviations from stoichiometry with dopants or impurities. For the *n*-type semiconductor, the material is doped with a donor impurity, and for the *p*-type the dopant is called an acceptor. In the case of GaAs, elements such as S, Se, Te, Si, Ge, and Sn can be used to form the donor and Be, Mg, Zn, Cd, C, and also Si and Ge can be the acceptor. In heavily doped *n*-type materials, all valence-band states and to some extent conduction band states are filled up to the Fermi level F as shown in Figure 8.1(a). On the other hand, a heavily doped *p*-type semiconductor exhibits a downward shift of the Fermi energy into the valence band so that all the conduction band and some states lying near the top of the valence band are empty, as shown in Figure 8.1(b).

At the *pn* junction in a semiconductor, the electron concentration on the *n* side is much larger than that on the *p* side. This difference in concentration produces a depletion region in which space-charged layers are formed to sustain high electric fields. These space-charged layers arrange themselves in an electric dipole configuration with uncompensated donor ions on the *n* side and uncompensated acceptor ions on the *p* side. This charge distribution causes a potential difference ϕ at the junction. Beyond the depletion region, the potential remains at a constant level. The energy of an electron at the bottom of the conduction band is lower by $e\phi$ on the *n* side of the junction than that on the *p* side. The same is true for a hole in the valence band. The energy diagram of a *pn* homojunction without a forward bias is depicted in Figure 8.2.

The depletion region in the vicinity of the junction has a much higher resistivity than any other part of the semiconductor. When an external bias voltage V is applied to the crystal, most of the voltage drop will occur across this region. A forward bias, in effect, will reduce the potential barrier height and will inject excess electrons into the *p* region. The state of the system will no longer be in thermal equilibrium, or in other words, a unique Fermi level for the system is no longer defined. Two quasi-Fermi levels F_c and F_v are assigned to *n* and *p* regions, respectively, as shown in Figure 8.3(a). The concentrations of injection currents in both *n* and *p* regions contiguous to the space-changed layers are

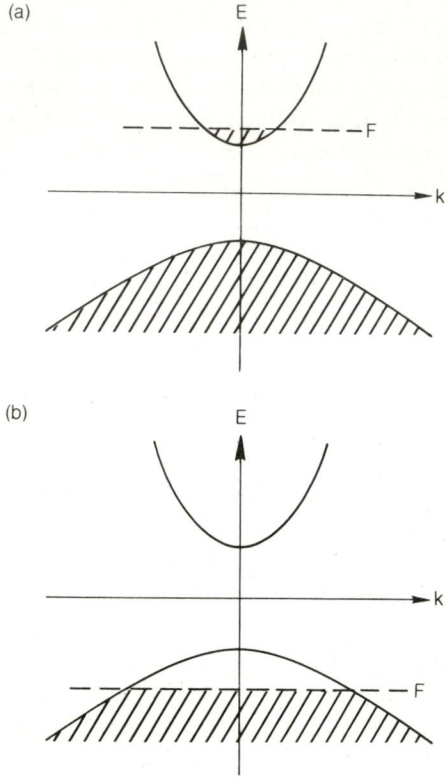

Figure 8.1 Electron energy diagram as a function of density of states for heavily doped (a) n-type and (b) p-type materials.

greatly perturbed from their equilibrium values, whereas the concentrations of majority carriers are not much affected. The electrons from the n region and holes from the p region can spill over the junction and become minority carriers on the opposite side. A condition known as population inversion can be established on the p side of the junction, as illustrated in Figure 8.3(a), by a penetration depth L of typically a few micrometers, which is characteristic of the diffusion length $L_e + L_h$ as determined by the doping gradient. Figure 8.3(b) shows a small downward step in the refractive index on the n side of the junction. This effect, commonly known as the free carrier depression, can provide a weak confinement for the optical beam. A much stronger effect can be obtained by varying the material composition in a ternary compound, and is discussed in Section 8.4.

The potential barrier at the interface as shown in Figures 8.2 and 8.3 can be deduced from Poisson's equation for charge density, which consists of free electrons, holes, and impurities. The differential equation for $\phi(x)$ over the width of the junction is given by

$$\frac{d^2\phi}{dx^2} = -\frac{e}{\varepsilon}(p - n + \Delta N) \tag{8.1}$$

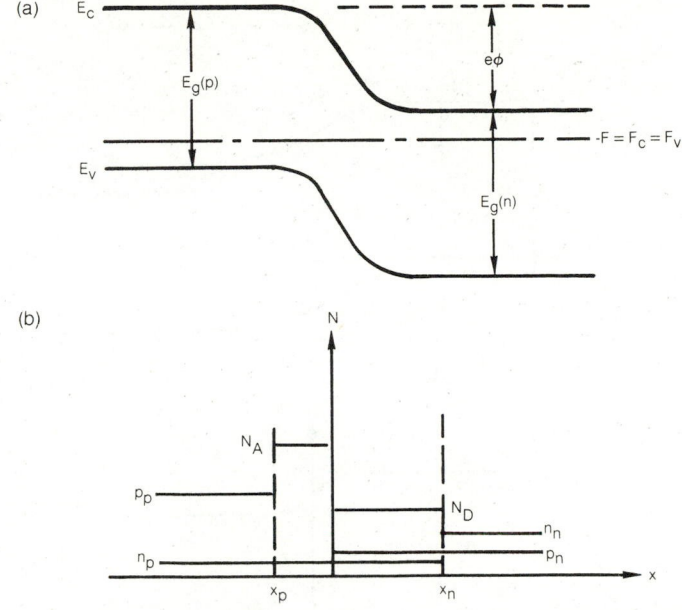

Figure 8.2 Abrupt *pn* homojunction without a bias: (a) energy-band diagram; (b) relative electron and hole concentration and also impurity levels in the *p* and the *n* regions.

where ΔN is the difference in acceptor N_A and donor N_D concentrations, and n and p are electrons and hole concentrations as given by Equations (7.22) and (7.23). In equilibrium, the Fermi level must be equal on both sides of the junction. When the *pn* junction is formed, a displacement in the energy-band diagram, as shown in Figure 8.2, occurs. The bands must be continuous by connecting with concave-upward and concave-downward parabolic functions of x in the form

$$\phi(x) = B + C(x - x_0)^2 \tag{8.2}$$

where the constants B, C, and x_0 can be determined by the boundary conditions. The potential barrier height is determined by initial positions of the Fermi levels relative to band edges and bandgap energy. Assuming nondegenerate materials (e.g., the Fermi levels occur outside conduction and valence bands), the barrier height is given by [see Figure 8.2(a)]

$$e\phi_D = E_g - \delta_c - \delta_v \tag{8.3}$$

where

$$\delta_c = E_c - F_c \quad \text{and} \quad \delta_v = F_v - E_v \tag{8.4}$$

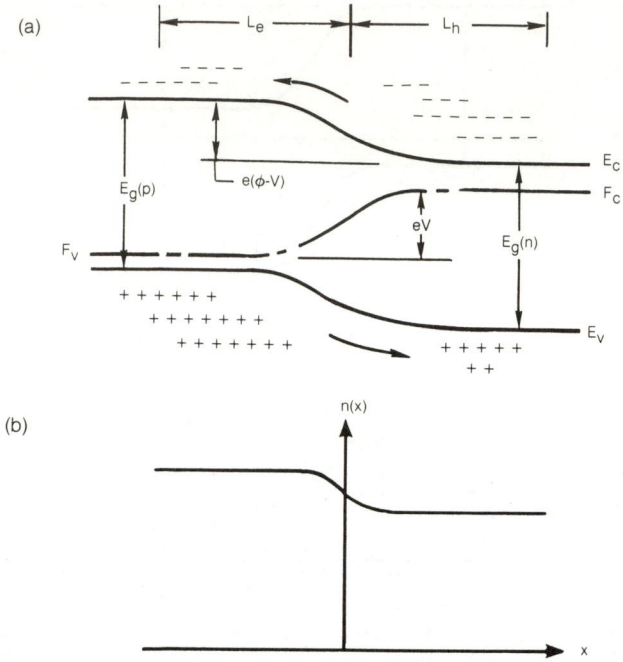

Figure 8.3 Abrupt *pn* junction under a forward-bias *V*: (a) energy-band diagram; (b) refractive index as a function of distance across the junction.

When a bias voltage V is applied to the p region, the junction is said to be forward biased and the effective barrier height is reduced. Electrons in the n region will flow into the p region. The excess hole density Δp at the edge of the depletion region in the n side of the junction can be shown to be

$$(\Delta p)_{x_n} = p_n \left[\exp \left(\frac{eV}{kT} \right) - 1 \right] \tag{8.5}$$

where p_n is the hole density in the n region with no bias [see Figure 8.2(b)]. Similarly, the excess electron density Δn in the p side is given by

$$(\Delta n)_{x_p} = n_p \left[\exp \left(\frac{eV}{kT} \right) - 1 \right] \tag{8.6}$$

where n_p is the electron density in the p region without bias. If we assume that the excess electron and hole distributions are exponential functions of the distance from the junction with a diffusion length L_e and L_h, respectively [see Figure 8.3(a)], we write

$$\Delta p = (\Delta p)_{x_n} \exp \left(-\frac{x}{L_h} \right)$$

$$\Delta n = (\Delta n)_{x_p} \exp\left(-\frac{x}{L_e}\right) \tag{8.7}$$

The current density can be calculated by taking the gradient of the distribution functions evaluated at the edges of the depletion region. Thus we obtain the current densities of holes J_h and electrons J_e at x_n and x_p, given by

$$(J_h)_{x_n} = -eD_h \left(\frac{\partial \Delta p}{\partial x}\right)_{x_n}$$

$$\tag{8.8}$$

$$(J_e)_{x_p} = -eD_e \left(\frac{\partial \Delta n}{\partial x}\right)_{x_p}$$

where D_h and D_e are the diffusion coefficients of the holes and the electrons. The total current density due to both carriers can be obtained by using Equations (8.5) to (8.8) and expressed as

$$J = e\left(\frac{D_h p_n}{L_h} + \frac{D_e n_p}{L_c}\right)\left[\exp\left(\frac{eV}{kT}\right) - 1\right] \tag{8.9}$$

As V increases, a breakdown value V_B will quickly be reached. At $n = 10^{18}$ cm^{-3}, the typical V_B is about 2.8 V. The junction capacitance per unit area due to voltage induced charges in the depletion region is given simply by

$$C = \frac{\varepsilon}{x_n + x_p} \tag{8.10}$$

where $(x_n + x_p)$ is the width of the depletion region and can be obtained by solving Equation (8.1). Upon two successive integrations, we obtain

$$x_n + x_p = \left(\frac{2\varepsilon(\phi - V)(N_A + N_D)}{eN_A N_D}\right)^{1/2} \tag{8.11}$$

The result of Equation (8.11) is obtained by using the boundary condition at which the neutrality of change has been imposed on either side of the junction.

The typical diffusion length in a GaAs *pn* junction device is on the order of several micrometers. Even at this short distance, it is already too wide to establish a good current confinement in order to achieve a low current threshold laser device operating at room temperature. Typical threshold currents of *pn* homojunction GaAs laser diodes are of the order 10^3 A/cm^2 at 77°K but increase exponentially with temperature and reach values above 10^5 A/cm^2 quickly as the temperature approaches room temperature. Therefore, no homojunction laser diodes have been successfully made to operate at room temperature. However, low-cost room-temperature and CW LEDs can be made from *pn* GaAs. Figure 8.4 shows such a structure in the form of a parallelopiped with a planar diffused *pn* junction lying a distance of about a few micrometers below the top surface. Metal electrodes were deposited on top and bottom surfaces [100] of the wafer, and the cleaved end faces [110] of the crystal were used as the cavity mirrors. With this simple structure, room-temperature continuous-wave operation was not possible even though a considerable amount of effort has been put forth into the development of this structure by providing adequate heat sink and current confinement with the help of the strip-geometry electrode. The only mode of operation of the homostruc-ture laser is in very short pulses with a pulse width of the order of 0.1 μs at low duty cycles of less than 0.1%.

Since the demonstration of the first working injection semiconductor laser, laser technology has undergone several stages of development before reaching its maturity as a practical source for optical fiber systems. All lasers today are made in much more complex configurations than the homostructure. However, many useful and economical LEDs are still made in this simple configuration by using zinc diffusion into *n*-type GaAs to form the *pn* junction. This simple device competes very strongly with the more sophisticated but considerably more efficient double-heterostructure AlGaAs laser diodes because of low cost and reliability. The most efficient geometry for the LED is the one that allows the radiation to be emitted from its surface, as shown in Figure 8.5. This geometry is commonly referred to as the Burrus type. Because *n*-type GaAs material is very conductive, it can be used as an electrode on one side of the junction without using a metallic layer, which would otherwise block the radiation emitting from the junction. It is necessary to remove as much *n*-type

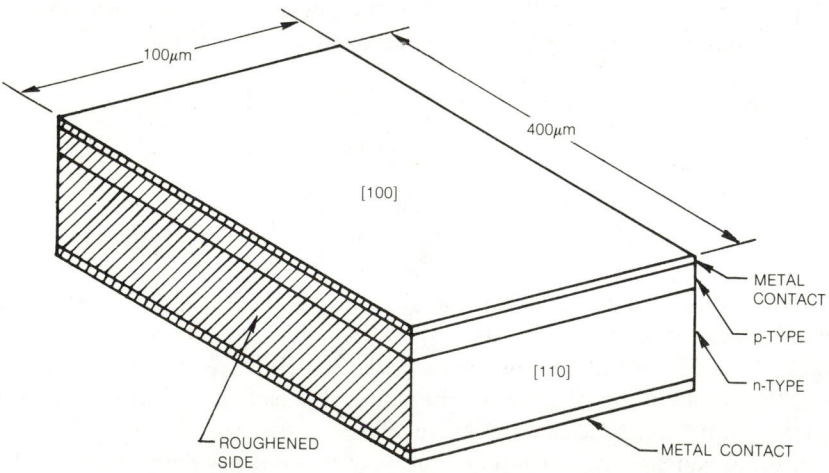

Figure 8.4   Schematic of a *pn* homostructure semiconductor laser.

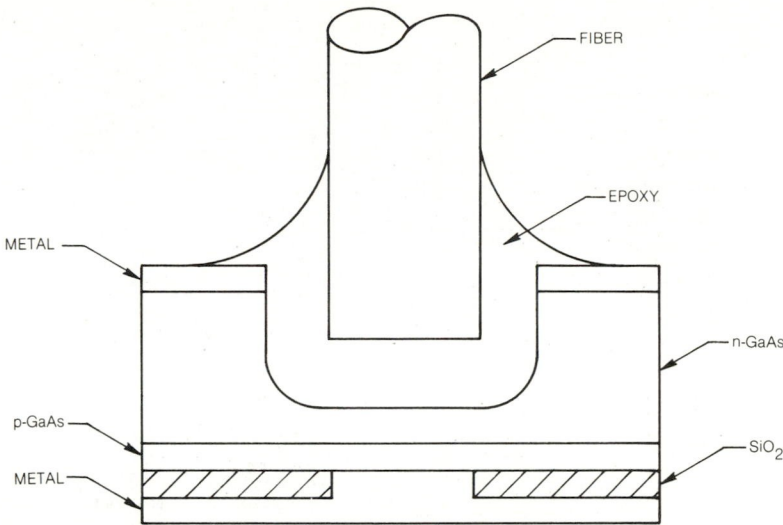

Figure 8.5   Schematic of a small-area Burrus-type LED-fiber coupling.

material as possible away from the region above the junction by using a stop-etching technique to reduce the optical loss due to the absorption by the free carriers. The emitting aperture of the Burrus-type LED is determined by the size of the opening hole in a SiO_2 layer, and can be made to match the aperture of the fiber core. Therefore, with this geometry, not only a very high radiance LED can be formed by restricting the emission to a small area but also very efficient coupling between the emitter and the fiber can be obtained. The major disadvantages of the LED are (1) the wide spectral width $\Delta\lambda \simeq 200$ Å, which is more than one order of magnitude wider than that of an LD, and (2) the low switching rate, which is governed by the spontaneous lifetime which is considerably longer than the stimulated lifetime. Both of these factors can affect the system bandwidth and impose a limit on the information-carrying capability.

8.3 SINGLE HETEROJUNCTIONS

Both injection currents and refractive indices can be controlled by varying either the doping or material composition. To enhance current and optical confine-ment several heterojunctions, such as *pN, nP, nN,* and *pP* in addition to the *pn* junction are often used in device topology. The capital letters are used to denote large bandgap materials in forming dissimilar semiconducting junctions. In the cases of *nN* and *pP*, where the carrier concentration differs only very slightly from the net impurity doping level, the spatial variation of the potential curves for the conduction bands are relatively constant on either side of the junction and differ by a finite step with a height given by ΔE_c. The energy diagrams for

(a)

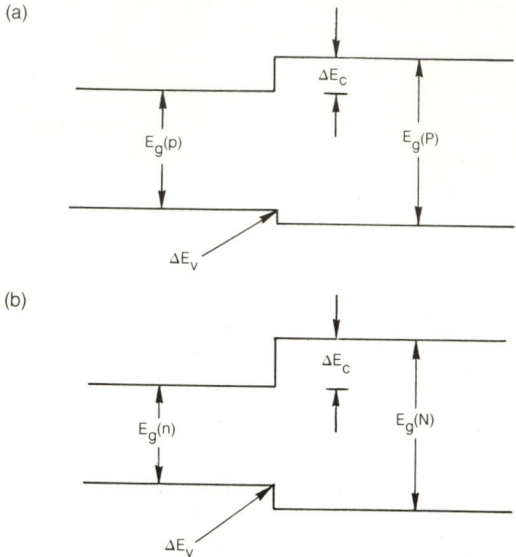

(b)

Figure 8.6 Energy-band diagram of two isotype heterojunctions: (a) pP and (b) nN heterostructures either with or without a bias.

these two isotype heterojunctions are shown in Figure 8.6. In these cases, the forward bias has very little effect on the band structure. Figure 8.7 shows the spatial variation of the band energies for a pN heterojunction (a) at thermal equilibrium and (b) under forward bias. For all heterojunctions, the effect of forward bias is to reduce the barrier height by an amount equal to eV. The band structure for an abrupt heterojunction can be constructed by solving Poisson's equation using appropriate charge distributions. As an example, we shall examine some important features of a pN heterojunction using Anderson's model (Ref. 8.2), in which the conduction band discontinuity at the interface was assumed to be the difference between the electron affinities. The electron affinity is the energy difference between the conduction band and the vacuum level at the free surface. Recent development from a more fundamental approach could lead to a modification of the simple Anderson model; nevertheless, the concept presented here provides a useful background for the understanding of heterojunctions.

As shown in Figure 8.8, energy band relations for p-GaAs and N-AlGaAs materials before the formation of a heterojunction are given by

$$E_g(N) = E_g(p) + \Delta E_c + \Delta E_v \tag{8.12}$$

where

$$\Delta E_c = \chi_p - \chi_N \tag{8.13}$$

and

$$\Delta E_v = \chi_N + E_g(N) - \chi_p - E_g(p)$$
$$= \Delta E_g - \Delta E_c \tag{8.14}$$

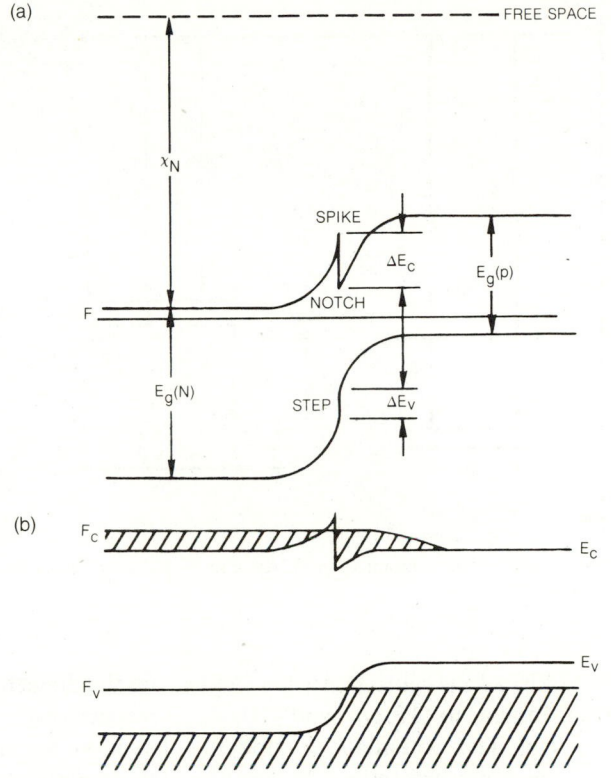

Figure 8.7 Energy-band diagram of an abrupt pN heterojunction: (a) at thermal equilibrium; (b) under a forward bias.

In Equations (8.13) and (8.14), χ is the electron affinity, which is the energy required to liberate an electron from the band edge to free space.

If the electron concentration in the depletion region is negligible, Equation (8.1) can be separated into two independent regions. If we let ϕ_p be the potential on the p side and ϕ_N be the potential on the N side, we can write two independent equations as follows:

$$\frac{d^2\phi_p}{dx^2} = \frac{e}{\varepsilon_p}\,\Delta N_p \qquad -X_p \le x < 0 \tag{8.15}$$

and

$$\frac{d^2\phi_N}{dx^2} = \frac{-e}{\varepsilon_N}\,\Delta N_N \qquad 0 < x \le x_N \tag{8.16}$$

where ΔN_p is the difference in the number of acceptors and donors $(N_A - N_D)_P$ in p-GaAs, and ΔN_N is the difference in the number of donors and acceptors

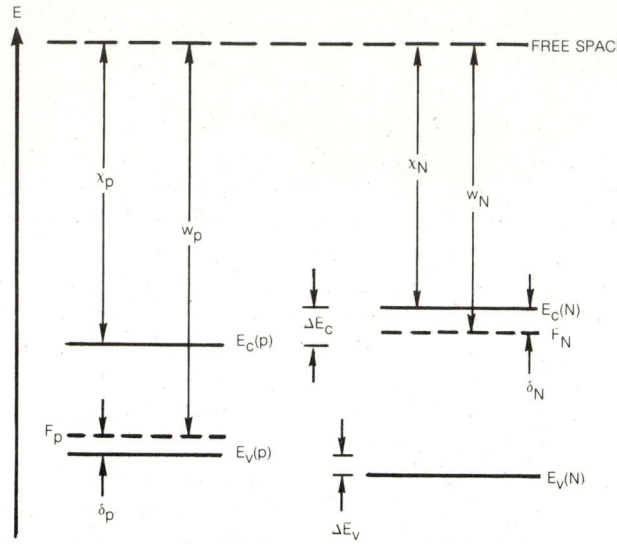

Figure 8.8 Energy-level diagram of p-GaAs and N-AlGaAs before forming an abrupt junction.

$(N_D - N_A)_N$ in N-AlGaAs [Figure 8.9(a)]. ε_p and ε_N are the dielectric constants and have values equal to $13.1\varepsilon_0$ and $10.06\varepsilon_0$, respectively, where $\varepsilon_0 = 8.85 \times 10^{-14}$ F/cm. The potentials outside the depletion region remain at constant values, which are determined by the boundary conditions. Integrating Equations (8.15) and (8.16), we obtain the field expressions

$$\mathscr{E}_p = -\frac{d\phi_p}{dx} = -\frac{e}{\varepsilon_p}\Delta N_p x + A_p \tag{8.17}$$

and

$$\mathscr{E}_N = -\frac{d\phi_N}{dx} = +\frac{e}{\varepsilon_N}\Delta N_N x + A_N \tag{8.18}$$

At two edges of the depletion region, the field vanishes. Therefore, we obtain

$$A_p = -\frac{e}{\varepsilon_p}\Delta N_p x_p \tag{8.19}$$

and

$$A_N = -\frac{e}{\varepsilon_N}\Delta N_N x_N \tag{8.20}$$

Substituting Equations (8.19) and (8.20) into (8.14) and (8.18) and integrating again, we obtain

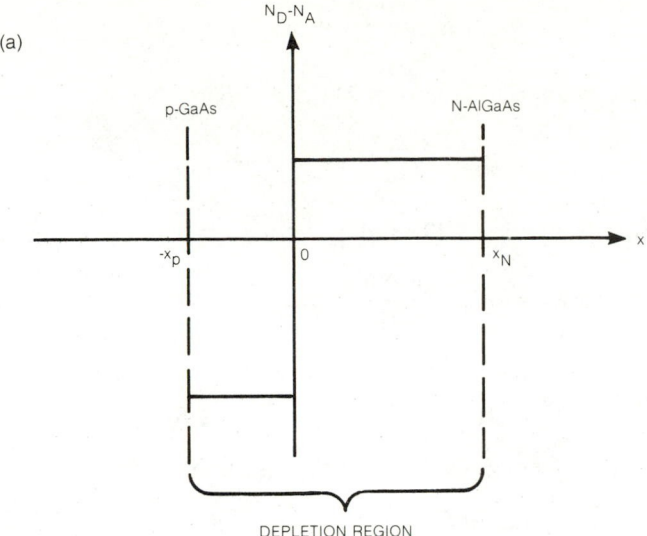

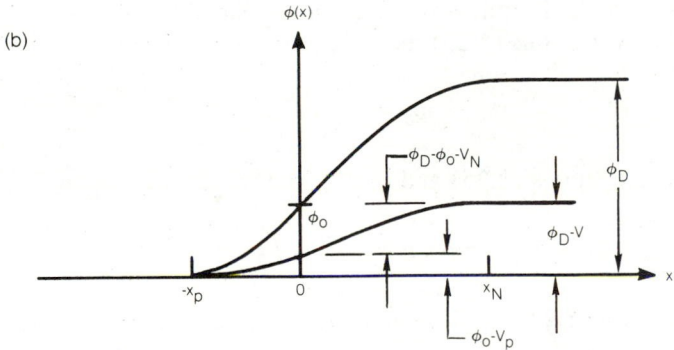

Figure 8.9   Abrupt pN heterojunction: (a) impurity distribution on the p side and on the n side; (b) potential variation under zero bias and a forward bias.

$$\phi_p = \frac{e}{\varepsilon_p}\,\Delta N_p\left(\frac{x^2}{2} + x_p x\right) + B_p \qquad (8.21)$$

and

$$\phi_N = \frac{e}{\varepsilon_N}\,\Delta N_N\left(x_N x - \frac{x^2}{2}\right) + B_N \qquad (8.22)$$

By choosing the ordinate as shown in Figure 8.9(b), such that $\phi_p = 0$ at $x = -x_p$, we obtain from Equation (8.21) that

$$B_p = \frac{e\,\Delta N_p}{2\varepsilon_p}\,x_p^2 \tag{8.23}$$

substituting Equation (8.23) into (8.21), we obtain

$$\phi_p = \frac{e\,\Delta N_p}{2\varepsilon_p}\,(x + x_p)^2 \tag{8.24}$$

Similarly, from Equation (8.22) and by letting $x = x_N$, we obtain

$$B_N = \phi_D - \frac{e\,\Delta N_N}{2\varepsilon_N}\,x_N^2 \tag{8.25}$$

where

$$\begin{aligned}
e\phi_D &= W_p - W_N \\
&= E_g(p) + \chi_p - \delta_p - \chi_N - \delta_N \\
&= E_g(p) - \Delta E_c - (\delta_p + \delta_N)
\end{aligned} \tag{8.26}$$

In Equation (8.26), W is the work function, which is the energy required to free an electron from the Fermi level to free space. Since $\phi_N = \phi_0$ at $x = 0$, we get by combining Equations (8.22) and (8.25) that

$$\phi_D - \phi_0 = \frac{e\,\Delta N_N}{2\varepsilon_N}\,x_N^2 \tag{8.27}$$

Substituting Equations (8.25) and (8.27) into (8.22), we obtain

$$\phi_N = \phi_0 + \frac{e}{2\varepsilon_N}\,\Delta N_N[x_N^2 - (x_N - x)^2] \tag{8.28}$$

From Equations (8.21), (8.23), and (8.27) we can determine the depletion depth, as

$$x_d = x_p + x_N \tag{8.29}$$

where

$$x_p = \left[\frac{2\varepsilon_p \phi_0}{e\,\Delta N_p}\right]^{1/2} \tag{8.30}$$

and

$$x_N = \left[\frac{2\varepsilon_N(\phi_D - \phi_0)}{e\,\Delta N_N}\right]^{1/2} \tag{8.31}$$

Another relationship between x_p and x_N can be derived from the continuity condition of the field at $x = 0$ (e.g., $\varepsilon_p E_p = \varepsilon_N E_N$). From Equations (8.17) to (8.20), we obtain

$$\frac{x_p}{x_N} = \frac{\Delta N_N}{\Delta N_p} \tag{8.32}$$

The striking feature of the band structure of the pN heterojunction is the formation of interfacial energy spikes, notches, and steps, as shown in Figure 8.7. The appearance of these abrupt interfaces is a consequence of joining two dissimilar compounds. Under equilibrium condition, the Fermi level must be the same in both compounds. This requirement and the energy relationship of Equation (8.26) lead to the formation of a spike and a notch in the conduction band. Because of a large difference in the valence-band energies, a step is needed to connect the two bands. The magnitude of the interfacial spikes can be reduced by composite grading at the heterojunction interface. In practice, composite grading may occur during epitaxial growth. The wider the graded region at the junction, the greater the reduction of spiking for most structures.

Under forward bias, the voltage V is assumed to be distributed between the p and the N side, as

$$V = V_p + V_N \tag{8.33}$$

The built-in potentials must be modified by

$$\phi_D \rightarrow \phi_D - V$$
$$\phi_N \rightarrow \phi_N - V_N \tag{8.34}$$
$$\phi_p \rightarrow \phi_p - V_p$$

Consequently, the relations between energies in the energy-band diagram must also be modified. On the p side, we have

$$E_v = e(\phi_D - V_p) = +\Delta E_v - \frac{e^2}{2\varepsilon_p} \Delta N_p (x_p + x)^2$$
$$\text{for } -x_p \leq x < 0 \tag{8.35}$$

and

$$E_c = E_v + E_g(p) \tag{8.36}$$

On the N side, we have

$$E_v = e(\phi_D - \phi_0 - V_N) - \frac{e^2}{2\varepsilon_N} \Delta N_N [x_N^2 - (x_N - x)^2]$$
$$\text{for } 0 < x \leq x_N \tag{8.37}$$

and

$$E_c = E_v + E_g(N) \tag{8.38}$$

These results are shown in Figure 8.7(b).

Even though the accuracy of the foregoing model for an abrupt heterojunction is questionable at present, the current confinement, as shown in Figure 8.7(b), has been confirmed experimentally. The expressions for the depletion width and the built-in potential are nevertheless useful in describing the important features of the heterojunction. The treatment for abrupt junctions assumes a more or less constant net concentration of acceptor and donor impurities, and thus becomes considerably simpler to carry out. For graded junctions, these concentrations are functions of distance across the junction. For example, N_A increases and N_D decreases gradually as the junction is approached from the N side and the two curves will cross over in the p region. Mathematical treatment for the graded junction is much more complex and will not be considered here.

Under a forward bias, the injection current flows from the wider bandgap into the narrow bandgap material through the junction. The current and voltage relationship can be derived by using the diffusion equation. Considering excess electrons that diffuse into the p region, we write the steady-state situation as

$$\frac{d^2n}{dx^2} - \frac{n - n_p}{L_e^2} = 0 \tag{8.39}$$

where L_e is the diffusion length for the electrons. It is defined by

$$L_e = \sqrt{D_e \tau_e} \tag{8.40}$$

where D_e and τ_e are the diffusion coefficient and electron recombination lifetime, respectively. The term n_p in Equation (8.39) is the excess electron concentration in the p region at thermal equilibrium. The solution of Equation (8.39) can be obtained by two successive integrations. We write

$$n = C_1 \exp\left(\frac{x}{L_e}\right) + C_2 \exp\left(-\frac{x}{L_e}\right) + n_p \tag{8.41}$$

At $x = -\infty$, $n = n_p$. This implies that $C_2 = 0$. At $x = x_p$, $n = n_p \exp(e\psi/kT)$, where $\psi = \phi_D - V$, so that

$$C_1 = n_p\left[\exp\left(\frac{e\psi}{kT}\right) - 1\right]\exp\left(\frac{x_p}{L_e}\right) \tag{8.42}$$

Substituting Equation (8.42) into (8.41), we have

$$n = n_p\left[\exp\left(\frac{e\psi}{kT}\right) - 1\right]\exp\left(\frac{x_p + x}{L_e}\right) + n_p \tag{8.43}$$

Equation (8.43) indicates that under a forward bias, the injected minority carriers concentration decreases exponentially with decreasing x from the junction. At $x = -x_p$, the current can be evaluated by the equation

$$i_e = eD_e \frac{dn}{dx}\bigg|_{x=-x_p} = -\frac{eD_e n_p}{L_e}\left[\exp\left(\frac{e\psi}{kT}\right)-1\right] \qquad (8.44)$$

Similarly, diffusion of holes into the N side is given by

$$i_h = \frac{eD_h n_N}{L_h}\left[\exp\left(\frac{e\psi}{kT}\right)-1\right] \qquad (8.45)$$

At room temperature, $n_p \simeq 1 \times 10^{18}$ cm^{-3} and $n_N \simeq 1.5 \times 10^{17}$ cm^{-3}. The ratio $i_e/i_h \gg 1$. Therefore, the injection current flows from the N side into the p side and can simply be expressed by

$$I = I_0 \exp\left(\frac{e\psi}{kT}\right) \qquad (8.46)$$

Another single heterostructure is the nP junction, as shown in Figure 8.10. Figure 8.10(a) shows the energy-band structure of the nP abrupt junction with zero bias. The marked difference between this nP structure and the pN structure (Figure 8.7) is caused by the difference in Fermi levels for these materials. Figure 8.10(b) shows the band energies of an nP junction under forward bias, where the separation between F_c and F_v is equal to eV. In the nP case, $i_p \gg i_N$. The following numerical examples are useful for constructing energy-band diagrams. We shall consider only the heterojunction between GaAs and

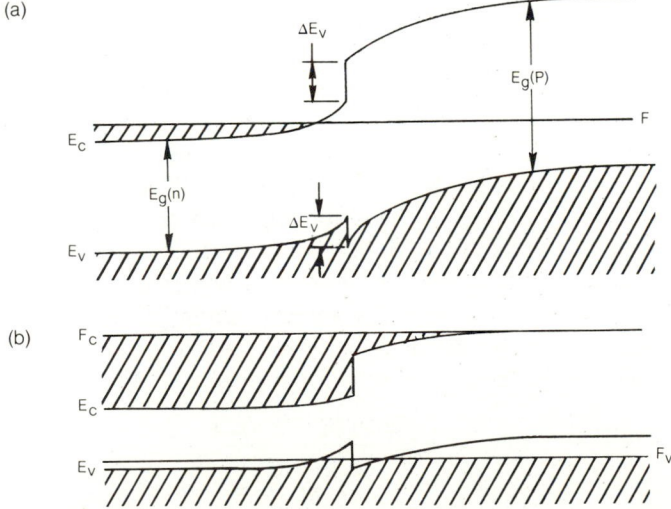

Figure 8.10 Abrupt nP heterojunction: (a) without a forward bias; (b) with a forward bias.

$Al_x Ga_{1-x}As$, where x is a fraction of the composition. The energy gap E_g is a function of x. For $0 < x < 0.45$,

$$E_g(eV) = 1.424 + 1.247x \tag{8.47}$$

for $0.45 < x < 1$,

$$E_g(eV) = 1.424 + 1.147(x - 0.45)^2 \tag{8.48}$$

For a GaAs-$Al_{0.3}Ga_{0.7}As$ nP heterojunction at room temperature (297°K), $(N_D - N_A)_n = 1 \times 10^{18}$ cm^{-3} and $E_c - F_c = 0.042$ eV on the n side; $(N_A - N_D)_p = 2.3 \times 10^{17}$ cm^{-3} and $F_v - E_v = 0.109$ eV on the P side. For a pN heterojunction of the same composition ($x = 0.3$) at 297°K, $(N_A - N_D)_p = 1.1 \times 10^{18}$ cm^{-3} and $F_v - E_v = 0.053$ eV on the p side; $(N_D - N_A)_N = 3 \times 10^{17}$ cm^{-3} and $E_c - F_c = 0.046$ eV on the N side. The expressions for built-in potentials, depletion width, and injection currents under forward bias in a nP heterostructure can be obtained by following a treatment similar to the one already discussed in detail for the pN heterostructure.

The simplest single heterostructure SH laser has the npP topology, which consists of three layers: n-GaAs, p-GaAs, and P-AlGaAs. It can be fabricated by liquid-phase expitaxial LPE growth of a P layer of AlGaAs on a heavily doped n-type GaAs substrate. Either during the growth or the annealing cycle, a sufficient amount of Zn can be diffused into the n-GaAs substrate to form the pn junction. The optimum thickness of the active p layer in this case is about 1 to 2 μm. The measured threshold current density of this laser as a function of the active layer thickness d is shown in Figure 8.11 for two different laser cavity lengths: $L = 250$ μm and $L = 400$ to 500 μm. The lowest J_{th} (300°K) value is about 8.5×10^3 A/cm^2, which represents a significant improvement from the pn homostructure laser. Even then, CW operation of this laser at room temperature is very difficult. As mentioned before, a reduction of J_{th} values for the SH lasers can be accomplished by a closer optical and current confinement. By introducing optical confinement in SH lasers, the total cavity loss can be reduced from a typical value of 100 cm^{-1} down to a value between 20 and 40 cm^{-1}. This is simply a result of the difference in refractive index Δn at the pP heterojunction, where the Δn value is about a factor of 5 times that at the pn junction. Furthermore, the injection current is confined in the active p layer because of the potential barrier at the heterojunction. Within this active layer, the gain profile is a slowly varying function of x. At the pP junction, $i_e = 0$. This boundary condition leads to a solution for Equation (8.44) in the form

$$n(x) = n_p \frac{\cosh[(d - x)/L_e]}{\cosh(d/L_e)} \tag{8.49}$$

for $d \ll L_e$, $n \simeq n_p$. More discussion on leakage currents and charge distribution involving diffusion is given in Chapter 9. The value for n_p can be calculated by using the Fermi–Dirac integral. As d decreases below 2 μm, the optical confinement factor Γ decreases rapidly and consequently, J_{th} increases. Another

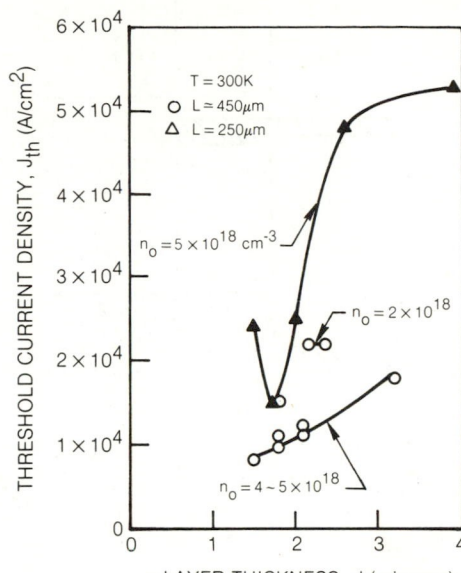

Figure 8.11 Variation of J_{th} at $300°K$ as a function of an active p-layer thickness of npP SH lasers. (From Ref. 8.2. Reprinted with permission of Academic Press, Inc., New York.)

point to be noted (Figure 8.11) is that higher electron concentrations in the n layer yield lower threshold currents, and consequently higher emission power.

8.4 DOUBLE HETEROJUNCTIONS

Early work on semiconductor lasers dealt mostly with homostructures at liquid-nitrogen temperature ($77°K$). The threshold current density for these lasers operating at room temperature was very high ($\geq 5 \times 10^4$ A/cm²). With single heterostructures, the current threshold can be reduced by about a factor of 5 or more. Further reduction of the current threshold is possible by constructing the laser with double or multiple heterojunctions, because a further reduction of the thickness of the active layer can be activated with these structures without decreasing the optical confinement. Because of strong current and optical confinement, significantly higher efficiency can be obtained from double-heterojunction (DH) lasers. The most commonly used double heterostructure consists of a very thin p-type GaAs layer which is sandwiched between an N-type and a P-type AlGaAs, as shown in Figure 8.12(a). The conduction-band discontinuity ΔE_c provides a barrier to the electrons at the pP heterojunction and thus confines the injected electrons to the p-GaAs active layer, where the radiative recombination takes place. The thickness d of this p layer is usually much shorter than the diffusion length and is typically on the order of 0.1 to 0.3 μm. The spatial variation of bands for a NpP structure as

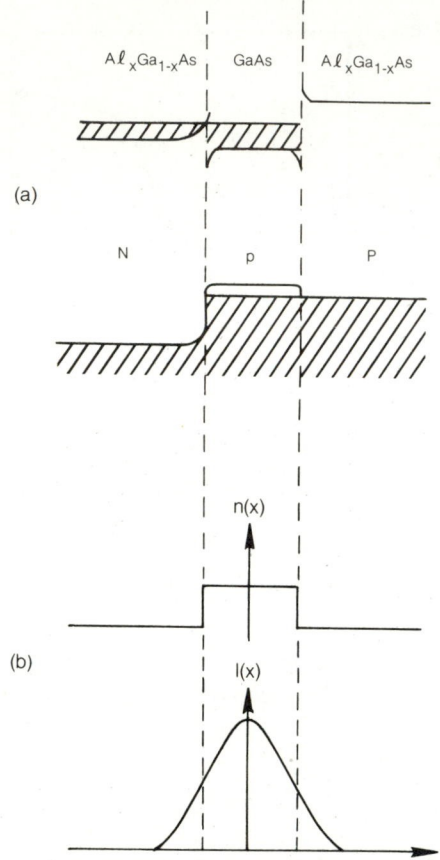

Figure 8.12 (a) Abrupt *NpP* double-heterostructure AlGaAs laser under high injection currents; (b) refractive index, and light intensity profiles as a function of heterostructural thickness.

shown in Figure 8.12(a) is under a forward bias at high currents. The valence-band discontinuity ΔE_v plus the built-in potential $e(\phi_D - V)$ provides a potential barrier for holes at the Np heterojunction, and prevents hole injection into the N region. In this case, the confinement of holes is as effective as that created for the confinement of electrons by the discontinuities in the bands. Within the thin active p layer, a constant level of the injection carrier concentration can be assumed.

The refractive index profile and the light intensity distribution of this NpP DH laser are shown in Figure 8.12(b). A higher refractive index for the active p layer than that for the P and N cladding layers is achieved in this case by choosing a different material composition factor x. For low-doped N- or P-type $Al_xGa_{1-x}As$, the refractive index decreases linearly with increasing x from $n = 3.59$ at $x = 0$ to $n = 3.32$ at $x = 0.4$. The relation can be approximated by the equation

$$n(x) = 3.59 - 0.71x \tag{8.50}$$

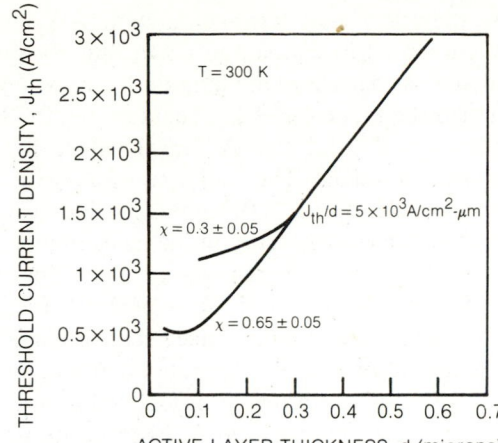

Figure 8.13   Variation of J_{th} at 300°K as a function of an active layer thickness for GaAs-AlGaAs DH lasers having a cavity length of 500 μm. (From Ref. 8.2. Reprinted with permission of Academic Press, Inc., New York.)

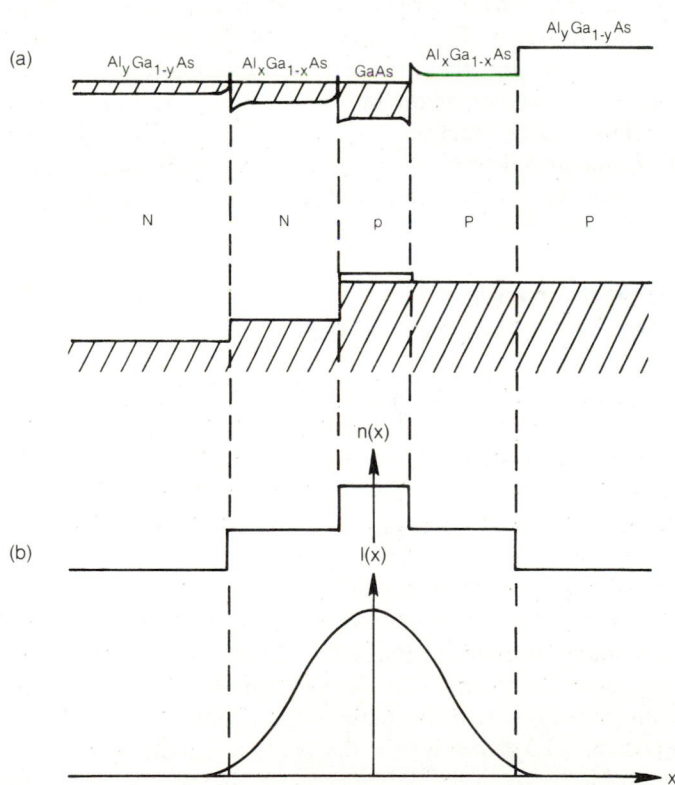

Figure 8.14   (a) Energy-band diagram (b) refractive index and light intensity as a function of heterostructural thickness for a LOC laser.

A dielectric slab waveguide with a symmetric index profile as shown in Figure 8.12(b) can support two linearly polarized guided TE and TM wave modes, as discussed in Chapter 3. For a symmetric DH laser with $x = 0.3$, the first-order mode ($m = 1$) occurs at $d = 0.38$ μm and the second mode ($m = 2$) occurs at $d > 1.0$ μm. At larger values of x, the fundamental mode occurs at smaller d values. The variation of J_{th} with d for several x values is shown in Figure 8.13. For $d > 0.3$ μm, J_{th} is independent of x. For $d < 0.3$ μm, there is considerable uncertainty in the measurements; nevertheless, J_{th} values decrease consistently with increasing x value. For $x = 0.3$, $J_{th} \sim 1000$ A/cm^2 at $d \geq 0.15$ μm. For $x = 0.65$, $J_{th} \simeq 500$ A/cm^2 at $d \leq 0.1$ μm. The GaAs-AlGaAs DH laser is the most advanced semiconductor laser system at present and it was the first laser that was operated continuously (CW) at room temperature.

One major problem associated with DH lasers having a very thin active layer is the deterioration of laser performance as a result of optical damage gradually occurring within the active region. This is basically a material-related problem; however, one way to reduce the severity of the problem is to spread the light into a wider region without affecting the carrier confinement. Separate confinements for light and carriers are possible but require additional epi-layers to the double heterostructure. Figure 8.14 shows the energy diagram of the bands, the index profile, and the light-intensity distribution for a five-layer system. This structure is known as the large optical confinement (LOC) and is basically an extension of the DH structure by adding to its outer layers two wider-bandgap N-type and P-type AlGaAs layers. With this type of structure a threshold current density of less than 500 A/cm^2 has been achieved.

8.5 MATERIAL PROPERTIES
AND GROWTH OF SEMICONDUCTORS

Laser actions have been obtained from a variety of semiconductor compounds with threshold currents varying from less than 0.5 kA/cm^2 to greater than 100 kA/cm^2 at room temperature or below. This wide variation in performance is due to many factors, most of which can be related to the material properties. The most efficient LD must be made of a direct bandgap material for its active region and have a multilayer heterostructure as discussed in Section 8.4. The major problem associated with heterostructures arises from the interfacial lattice mismatch, which creates internal strain and dislocations commonly referred to as the material imperfections. To increase the performance and in particular, the life expectancy of a LD, material imperfections must be kept to a minimum. For this reason, lattice constants of these materials must be matched as perfectly as possible. To determine the matching compounds, it is convenient to plot the relationship between the lattice constant and the bandgap energy for a number of binary and ternary compounds. In Figure 8.15, the curves joining the binary compounds give the values for the energy gap and lattice constant of

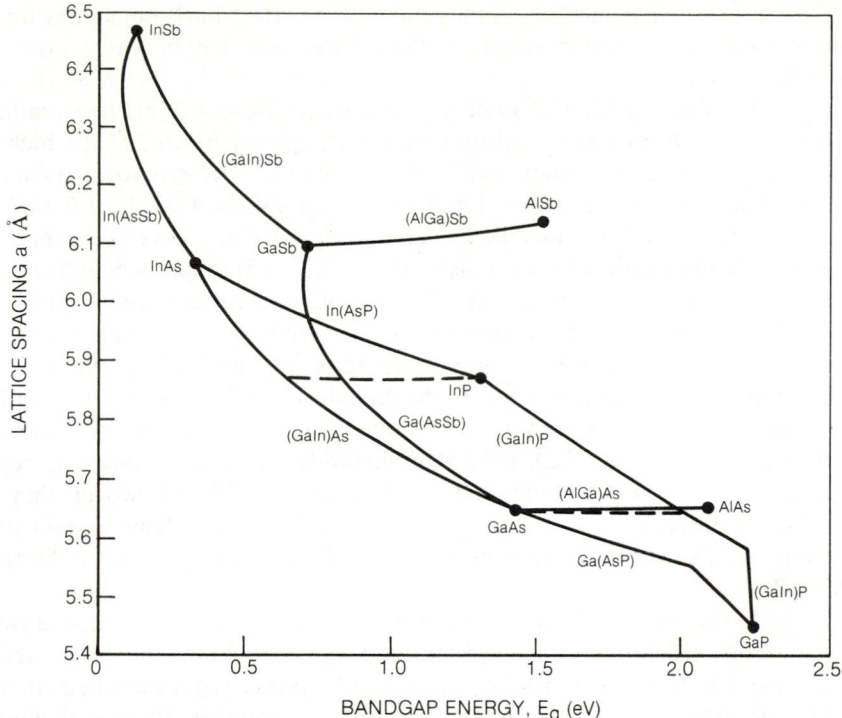

Figure 8.15 Lattice constant versus bandgap energy for semiconductor compounds.

the ternary compounds. The behavior of the quaternaries can be interpolated from these ternaries. AlGaAs is the simplest system, for which a nearly perfect lattice match exists between GaAs and AlAs independent of the value of x. In contrast to this system, the lattice constant for all other ternaries varies appreciably with the composition. Therefore, to make a perfect lattice match for systems other than AlGaAs, it is necessary to form a quaternary in order to gain the extra degree of freedom.

One interesting quaternary which has been studied extensively is the InGaAsP system emitting at wavelengths in the range 1.1 to 1.6 μm. An epitaxial growth of this system must start with a substrate from which the lattice constant of the epitaxial quaternary layer must be matched to that of the substrate. From Figure 8.15 we see that two possible choices of the substrate are GaAs and InP. If GaAs is used as a substrate, the bandgap energy of the epitaxial layer can only follow the dashed line toward a higher value and terminate at the boundary curve represented by the GaInP solid solution. On the other hand, the InP substrate allows the growth of lower-bandgap quaternary material by following the dashed curve that terminates at the InGaAs solid solution. Clearly, the latter system is of more practical interest and, in principle, a quaternary compound with a perfect lattice matching for the active layer can be grown to emit at the desired wavelength with a minimum material dispersion.

The ternary $Al_x Ga_{1-x} As$ compound changes from a direct to an indirect bandgap material as the composition factor, x, approaches 0.37, at which the bandgap energy of this compound is about 1.96 eV. The crossover points for other compounds are: 2.33 eV for AlInP, 2.25 eV for GaInP, 2.04 eV for AlInAs, and 1.97 eV for GaAsP. The composition factor x governs not only the radiative recombination rate but also the index of refraction, which decreases with increasing bandgap energy in the range of practical interest.

There are three methods commonly used for epitaxial growth of semiconductors: namely, (1) liquid-phase epitaxy (LPE), (2) chemical vapor deposition (CVD), and (3) molecular beam epitaxy (MBE). At present, LPE is the simplest and most widely used technique for producing the most efficient light-emitting devices. CVD and MBE methods are more accurate in reproducing layer thickness and material composition than LPE; however, they are more complicated, and in some cases many technical problems remain to be solved before superior-quality materials can be made in large quantity by these methods.

LPE may be defined in general terms as the growth of an oriented single crystal from a saturated or supersaturated liquid solution onto a crystalline substrate. This process is usually carried out by producing a saturated solution with appropriate composition at high temperatures and then allowing this liquid to be in contact with the substrate by some mechanical means at a lower temperature, during which epitaxial growth is instigated. To grow the pure binary compound, it is necessary to provide one element from group III and one element from group V. The melting points for group III elements such as Ga and In are near room temperature; the melting points for group V elements are at much higher temperatures. Therefore, it is only necessary to provide a sufficient amount of the group III element to be dissolved in the group V material at an elevated temperature that is adjusted to a growth rate. A variety of dopants can be added directly to the solution to yield the appropriate concentration. Some dopants can also be diffused into the III–V compounds in the solid phase at the growth temperature.

Ternary and quaternary semiconductors can be grown in a similar way as the binaries, from solutions containing the appropriate constituents. To grow multilayer LPE, a common technique involves sliding a graphite boat, as shown in Figure 8.16, with multiple compartments containing different melts on top of a graphite plate with a recessed space for the substrate wafer. The boat is enclosed in a silica chamber placed within a furnace in an extremely pure hydrogen atmosphere. During the growth the melt, which is composed of accurately weighted constituent elements, must remain in the saturated condition at the growth temperature. Alternatively, a certain amount of supersaturation can be induced in each of the melts by passing them

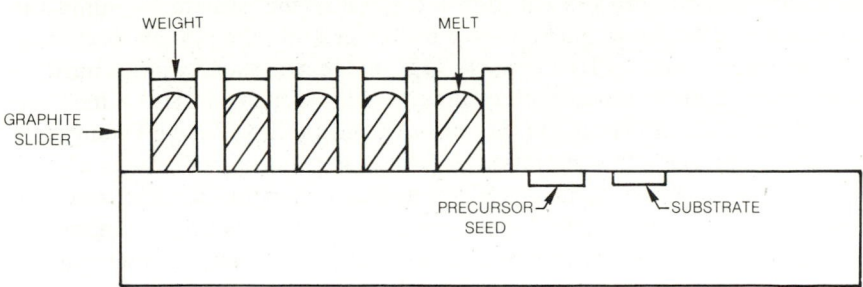

Figure 8.16   Sliding boat apparatus for liquid-phase epitaxy.

successively over the substrate at temperatures slightly (about 10 to 20°C) below the normal growth temperature to ensure nucleation of epitaxial growth of a thin layer. Often a weight is applied against the upper surface of the melt to overcome the balling effect of surface tension. With some care, a very thin layer ($\lesssim 0.1$ μm) of fairly uniform thickness can be grown.

Even though excellent laser materials have been obtained by using this simple method, it is, nevertheless, difficult to reproduce reliably the surface quality and thickness uniformity, because very small thermal gradient and fluctuation can induce imperfections as well as strains and dislocations in epitaxial layers. Another problem associated with the LPE process is the occurrence of excess oxidation of melting compound when the solution is kept at equilibrium for a very long time. This event is particularly deleterious and can affect the life of the lasers. For these reasons, considerable efforts have been made to develop alternative techniques for a more reliable epitaxial growth of large area substrates.

Vapor-phase epitaxy (VPE) or chemical vapor deposition (CVD) is also a widely used technique for growing single crystals. This method is based on the principle of chemical reactions among various constituent elements being transported by gases over the heated substrate. There are three main variants often used for growing GaAs, AlGaAs, and InGaAsP compounds: (1) the trichloride, (2) the hydride, and (3) the metallorganic systems. With the trichloride system single crystals of GaAs can be grown by passing arsenic trichloride gas over the heated metallic gallium. The volatile gallium chloride and arsenic so formed are being transported in pure hydrogen gas over the substrate at a lower temperature to instigate the epitaxial growth. For this system the relative fractions of Ga and As in the vapor phase are determined mainly by the temperature of the Ga solution, which must be accurately controlled. To provide better control of the concentration of the constituent elements, the hydride process can be used because this system bypasses the liquid phase and the major reactants are introduced into the system directly in the gas phase. The group III elements Ga and In are usually introduced as monochlorides, and the group V elements are hybrids such as PH_3, AsH_3, and SbH_3. Using either LPE or CVD, both GaAs and InGaAsP have been grown successfully; however, the limitation of these systems is that they are not suitable for the growth of compounds containing aluminum. To accommodate this element, the metallorganic system has been introduced, whereby the metallic compounds can be transported in a hydrogen carrier gas as metal alkyls together with the arsine gas. The metals used in this system consist of trimethyl aluminum, trimethyl gallium, diethyl zinc, and so on. The room temperature AlGaAs/GaAs DH lasers having low threshold current have been fabricated by this technique with excellent control in thickness uniformity and concentrations.

Molecular beam epitaxy(MBE) is one of the newest methods and has gained considerable attention recently. Conceptually, it is the most direct method of material epitaxy utilizing ion beams directly from heated sources in an ultrahigh vacuum. Very accurate control over the alloy composition and deposition rate can be obtained with this technique. Low-threshold and single-transverse-mode DH lasers have been produced by this technique, and the performance has been brought up to the standards made by other methods. However, there are problems associated with MBE which are of a fundamental nature and involve kinetic effects such as surface migration and desorption under nonequilibrium growth conditions. By heating the substrate to a high temperature above 600°C, good single-crystal material with the desired composition can be grown with correct stoichiometry by controlling the relative arrival rates of the constituents from different ion beams. MBE is considered to be the most precisely controlled deposition process, which offers not only precise dimensional controls in the growth direction but can also achieve three-dimensional structures. To obtain low-threshold current density for DH lasers prepared by MBE, the growth conditions and substrate preparation must be handled with care. Of particular importance is the elimination of residual gases such as water vapor, CO, and O_2 in the growth chamber. To achieve this, an interlock chamber for sample exchange is used so that the growth chamber is always kept under ultrahigh vacuum. The threshold current density was found to be lowered significantly with increasing substrate temperature, T_s. The optimum T_s is around 650°C. Although MBE appears very attractive for present-day material research, the practicality of this technique for device production remains to be seen.

PROBLEMS

8.1. Calculate the depletion width of a *pn* junction under no bias voltage. *Hint:* Use the charge neutrality condition $N_D x_n = N_A x_p$.

8.2. Explain physically why the depletion width of an abrupt pn junction is proportional to the square root of the potential drop $\phi_D - V$.

8.3. Construct an energy-band diagram for an nP heterojunction at equilibrium by showing the spike and notch and the step in the band edges.

8.4. Calculate the optimum waveguide thickness for the fundamental mode in a symmetric DH laser with $\Delta n = 0.01$ and $\lambda = 0.9$ μm.

REFERENCES

8.1. H. C. Casey, Jr., and M. B. Panish, *Heterostructure Lasers,* Part A: *Fundamental Principles,* and Part B: *Materials and Operating Characteristics,* Academic Press, Inc., New York, 1978.

8.2. R. L. Anderson, *Solid-State Electron.,* 5, 341 (1962).

Semiconductor Lasers

9.1 INTRODUCTION

The sources for transmitters of optical fiber communication systems consist primarily of light-emitting diodes (LEDs) and semiconductor injection lasers. These devices emit in the wavelength range 0.75 to 1.6 μm. In the lower portion of the spectra (0.75 to 0.95 μm) AlGaAs heterojunction devices are the most widely used sources. In the longer-wavelength region (1 to 1.6 μm), InGaAsP devices are now available for system applications. The radiative recombination properties and the structural and material characteristics of these semiconductor devices have been discussed in previous chapters. In this chapter we make use of these properties to generate a class of light-emitting devices having a variety of features governing their operation and performance. For most fiber optical systems, LED is an adequate source based on its available output power and speed of response. For long-distance, extremely high-data-rate systems, the use of laser diodes (LDs) should definitely be considered. The trade-off for laser power and bandwidth is not only in the cost but most important, in the reliability of the devices. In general, lasers degrade much more rapidly than LEDs and the deterioration originates primarily from material imperfections. Improving the quality and reliability of laser sources constitutes one of the active areas of research at present.

9.2 STRIPE-GEOMETRY LASERS

There are a variety of semiconductor lasers with either symmetric or asymmetric multiple-heterojunction topology to provide separate optical and carrier confinements in the transverse direction. For lateral confinement in the junction plane, either a stripe or the buried geometry shown in Figure 9.1 is commonly used. The simplest configuration is that shown in Figure 9.1(a), where the optical gain region is confined by a stripe electrode with width typically 5 to 30 μm. In this case, the P-AlGaAs layer of the heterostructure with a relatively low doping concentration of 5×10^{17} cm^{-3} has been used to reduce the current spreading in a thickness direction. The top GaAs layer is necessary to provide a good electrical contact with the electrode, which is formed by first opening a narrow stripe in the SiO$_2$ layer and then diffusing a shallowed depth of Zn through this window before metallic deposition. Alternative methods, which have also been used to provide the stripe geometry, consist of either etching the top GaAs layer to form a shallow mesa, or ion implantation to reduce the carrier mobility in the region outside the electrode.

A more sophisticated structure is made by forming a buried p-GaAs active layer in a stripe configuration completely surrounded by AlGaAs, as shown in Figure 9.1(b). Although the fabrication of this device is complicated, it basically involves preferential etching and LPE growth around the mesa. Such a structure can provide very stable output power in a single transverse mode when the width of the active layer in the junction plane is made as narrow as 2 to 3 μm. Table 9.1 summarizes performance characteristics of a representative class of stripe geometry AlGaAs double-heterostructure lasers. In general, the performance of these lasers depends mainly on the stripe width, the length, and the amount of current spreading. For a stripe width greater than 10 μm, the laser output usually consists of many transverse modes, in some cases a nonuniform distribution, and kinks can be developed in the output which are usually unstable. By reducing the width down below 5 μm, these nonlinearities can be

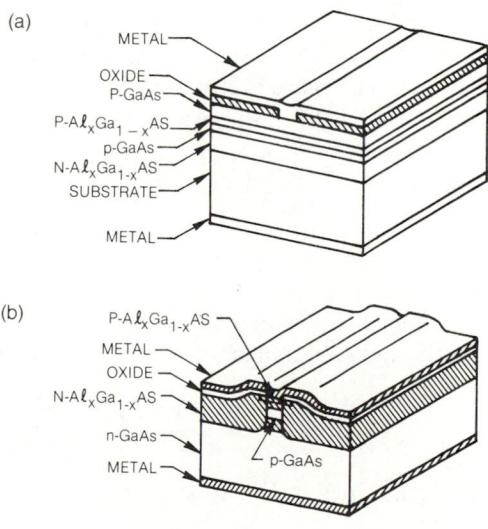

(a)

METAL
OXIDE
P-GaAs
P-Al$_x$Ga$_{1-x}$AS
p-GaAs
N-Al$_x$Ga$_{1-x}$AS
SUBSTRATE
METAL

(b)

P-Al$_x$Ga$_{1-x}$AS
METAL
OXIDE
N-Al$_x$Ga$_{1-x}$AS
n-GaAs
METAL
p-GaAs

Figure 9.1 Topology of semiconductor heterojunction lasers: (a) stripe-geometry DH laser; (b) buried heterostructure BH laser.

TABLE 9.1   Typical Output Characteristics of Some Semiconductor Lasers

Laser Type	Stripe Width (μm)	I_{th} (mA)	Peak Power (mW)	Output
Planar	10	100–125	10–50	Multimode
Ion implanted	12	100–150	5–10	Multimode
Zn diffused	5	30–70	5–10	Single mode
Buried	2	5–50	1–5	Single mode

avoided; however, threshold currents become poorly defined. Stripe lasers with a narrow width differ in another respect from those having wider stripes. When a fast-rise current pulse is applied to wider-stripe heterostructure lasers, the laser output usually exhibits a transient oscillation, commonly known as relaxation oscillation. This kind of pulsation can be suppressed by using very narrow stripe geometry.

Although the stripe-geometry laser has many attractive features, its threshold current density increases rapidly when the stripe width is reduced below 10 μm. Also, the astigmatic effect, which will be discussed later, increases rapidly with decreasing stripe width. These effects can be explained by the spatial variations of carrier concentration and gain along the junction plane. This situation can be analyzed by using a model in which electrons diffuse away from the active region along the junction plane defined by the stripe. From this diffusion model, one can obtain an expression for the carrier concentration profile. One can compare the calculated profile with the measurements of the spatial variation of spontaneous emission along the junction plane. Excellent agreement has been found to exist between the computed carrier concentration profile and the spontaneous emission measurements (Ref. 9.1). Figure 9.2 shows the spatial variation of current density, carrier concentration, refraction index, and gain coefficient of a stripe-geometry laser. The current density is assumed to be constant over the stripe width, as depicted in Figure 9.2(b). Furthermore, no variation is assumed to take place along the length of the electrode in the z direction.

The leakage current density outside the electrode can be expressed by a simple exponential decay function as

$$J(y) = J_0 \exp \left[-\frac{|y| - S/2}{l_0} \right] \qquad \text{for } |y| > \frac{S}{2} \qquad (9.1)$$

where J_0 is the current density inside the electrode, S is the stripe electrode width, and l_0 is a characteristic length as defined by

$$l_0 = \frac{2L}{\beta r I_0} \qquad (9.2)$$

The quantities in Equation (9.2) are the junction parameter β, the resistivity r,

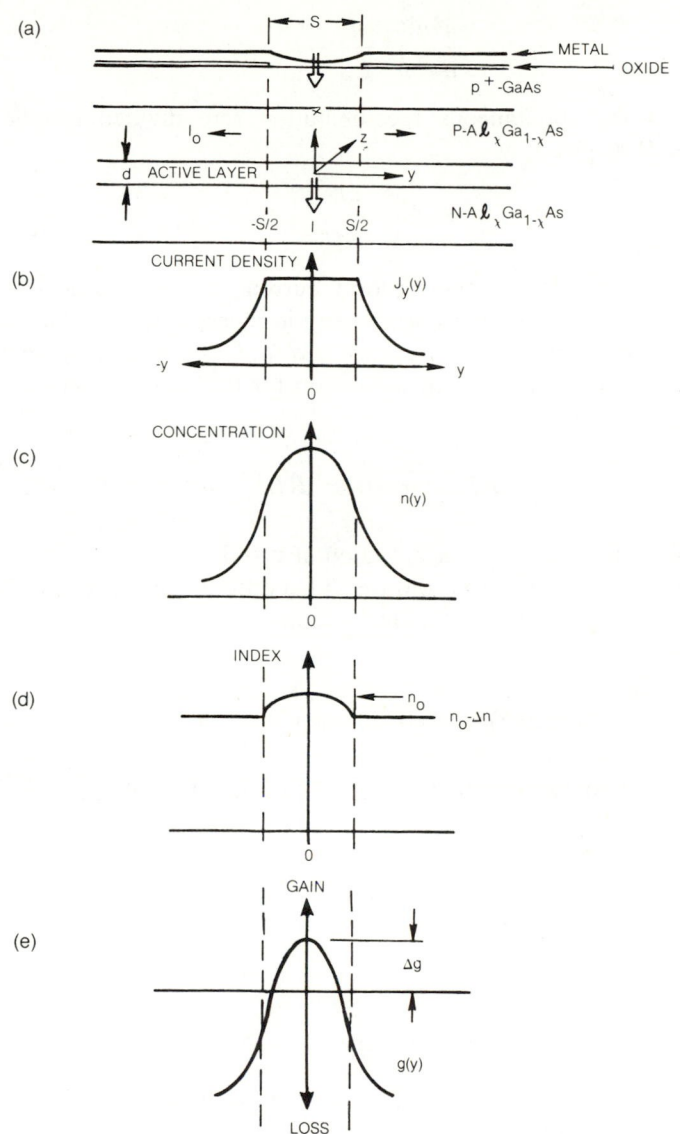

Figure 9.2 Characteristics of a stripe-geometry DH laser: (a) laser configuration; (b) lateral current density distribution; (c) carrier concentration profile; (d) index profile; (e) gain and loss profile.

the electrode length L, and the leakage current I_0 at the edge of the electrode, respectively. Since the thickness of the active layer d is usually very small in comparison with the stripe width S and the diffusion length L_e, a one-dimensional diffusion equation along the y axis is sufficient to describe the lateral carrier concentration profile. We write

$$\frac{d^2n}{dy^2} = \frac{n}{L_e^2} - R \tag{9.3}$$

where R is the spontaneous recombination rate divided by the diffusion coefficient D as given by

$$R = \frac{J_0}{edD} \tag{9.4}$$

In Equation (9.4), J_0 is the uniform current density inside the electrode $(-S/2 < y < S/2)$, and e and d are electronic charge and active-layer thickness, respectively. Assuming that the leakage current I_0 is very small compared to the injection current, I, one obtains a solution for the electron density profile for $|y| < S/2$ as given by

$$n(y) = RL_e^2 + [n(0) - RL_e^2] \cosh \frac{y}{L_e} \tag{9.5}$$

where $n(0)$ is the electron concentration at $y = 0$.

For $|y| > S/2$, electrons diffuse with a different diffusion constant D' and diffusion length L_e' into a profile that is characterized by an exponential decay function as

$$n(y) = n\,(S/2)\exp\left(-\frac{y}{L_e'}\right) \tag{9.6}$$

Using the boundary condition that electron densities at $|y| = S/2$ must be equal, we obtain

$$n(0) = RL_e^2\left[1 - \left(\cosh\frac{S}{2L_e} + \zeta \sinh\frac{S}{2L_e}\right)^{-1}\right] \tag{9.7}$$

where

$$\zeta = \frac{DL_e'}{D'L_e} \tag{9.8}$$

Substituting Equation (9.7) into (9.5), we write for the carrier concentration inside the electrode ($|y| < S/2$),

$$n(y) = RL_e^2\left[1 - \frac{\cosh(y/L_e)}{\cosh(S/2L_e + \zeta \sinh(S/2L_e)}\right] \tag{9.9}$$

Evaluating Equation (9.9) at $|y| = S/2$, we get

$$n(S/2) = RL_e^2\,\zeta\,\sinh\frac{S}{2L_e}\exp\left(\frac{S}{2L_e'}\right)\left(\cosh\frac{S}{2L_e} + \zeta\sinh\frac{S}{2L_e}\right)^{-1} \tag{9.10}$$

From the measurement of spontaneous emission along the lateral junction plane, one can compute the diffusion lengths L_e and L'_e assuming that the local spontaneous emission intensity is a linear function of local carrier concentration. This assumption is found to be valid for cases both below and above threshold. A quantity that is useful for determining the diffusion length from local measurements of spontaneous emission is the ratio $n(S/2)/n(0)$. From Equations (9.7) and (9.10), we get

$$\frac{n(S/2)}{n(0)} = \zeta \sinh\frac{S}{2L_e}\left[\cosh\frac{S}{2L_e} + \zeta \sinh\frac{S}{2L_e} - 1\right]^{-1} \tag{9.11}$$

The local gain must obey the same functional relationship for the electron concentration. With this assumption we can therefore relate the gain profile to the electron concentration profile by a simple transformation:

$$g(y) = an(y) - b \qquad \text{for } |y| < \frac{S}{2} \tag{9.12}$$

where $n(y)$ is given by Equation (9.9) and a and b are parameters dependent on injection currents. From Equation (9.12) we can express

$$\Delta g = g(0) - g\,(S/2) = a\,\exp\left(-\frac{S}{2L_e}\right)\left(\cosh\frac{S}{2L_e} - 1\right) \tag{9.13}$$

At a given current, the calculated gain width is in good agreement with the measured beam width obtained by an extrapolation from the far-field beam divergence.

The far-field patterns of a stripe-geometry DH laser are shown in Figure 9.3. As the width of the stripe S is increased beyond 10 μm, higher-order transverse modes supersede the lower-order modes even if the injection current is low. At high currents, the laser output usually consists of multimodes. The field distribution of these transverse modes can be generated by using the following expression:

$$H_n(y) = (-1)^n \exp(y^2)\frac{d^n}{dy^n}\exp(-y^2) \tag{9.14}$$

where H_n is the Hermite polynomial of order n. The first three Hermite polynomials are $H_0(y) = 1$, $H_1(y) = 2y$, and $H_2(y) = 2y^2 - 2$. To calculate the far-field intensity distribution, one must take the square of the fields and multiply that by the gain profile. These mode patterns are governed to a large extent by the properties of the laser resonator. The resonator can be designed to limit the number of possible modes. In the thickness direction, it is normally done by reducing the active-layer thickness to a size equivalent to only one-half period of the wave. In the lateral direction along the junction plane, the width of the electrode is important, because the oblique angles permit excitation of higher-order modes. Therefore, the width must be kept below 10 μm. There are

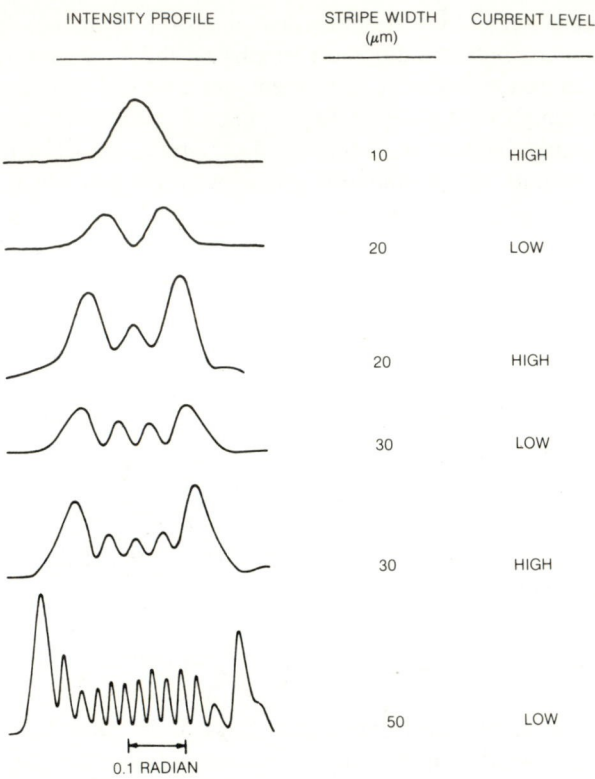

INTENSITY PROFILE	STRIPE WIDTH (μm)	CURRENT LEVEL
	10	HIGH
	20	LOW
	20	HIGH
	30	LOW
	30	HIGH
	50	LOW

0.1 RADIAN

Figure 9.3   Far-field radiation patterns of a stripe-geometry DH laser.

other ways to achieve single-transverse-mode operation by using other structures, such as the buried or constricted double-heterojunction CDH configuration, which will be discussed later.

9.3 CURRENT THRESHOLD AND GAIN GUIDING IN STRIPE-GEOMETRY LASERS

To treat a laser system in a planar geometry, we make use of the time-independent wave equation for a two-dimensional waveguide as given by

$$\nabla^2 E_n(x, y) + \omega^2 \mu \varepsilon(x, y) E_n(x, y) = 0 \qquad (9.15)$$

In Equation (9.15) the dielectric medium is characterized by the function $\varepsilon(x, y)$, which is a function of not only the dielectric constant but also the gain g of the medium. We write

$$\varepsilon(x, y) = \begin{cases} \varepsilon_1 - s^2 g^2(y) & \text{for } |y| < \dfrac{S}{2} \\[4mm] \varepsilon_2 & \text{for } |y| > \dfrac{S}{2} \end{cases} \tag{9.16}$$

$E_n(x, y)$ in Equation (9.15) is the E field for the nth mode, which varies not only along the waveguide thickness (x direction), but also along the junction plane (y direction). In this case, the modal gain is defined by

$$G_n = \frac{\displaystyle\int_{-\infty}^{\infty} g(y) E_n^2(x, y)\, dy}{\displaystyle\int_{-\infty}^{\infty} E_n^2(x, y)\, dy} \tag{9.17}$$

To a first approximation we assume that

$$E_n(x, y) = E_n(x) E_n(y) \exp(-i\beta z) \tag{9.18}$$

Because $E_n(y)$ is a slowly varying function, it is reasonable to assume that $E_n(x)$ is not significantly affected by the confinement along the y direction; therefore, $E_n(x)$ satisfies the one-dimensional wave equation

$$\frac{\partial^2 E_n(x)}{\partial x^2} + \beta_x^2 E_n(x) = 0 \tag{9.19}$$

It should be noted that because both gain and loss are involved within the active layer, the constant s must be represented by a complex quantity associated with a complex index of refraction, which will be elaborated in Section 9.5. Substituting Equations (9.16) and (9.18) into (9.15) yields

$$E_n(x)\frac{d^2 E_n(y)}{dy^2} + (\omega^2 \mu\varepsilon - \beta_x^2 - \beta^2) E_n(x) E_n(y) = 0 \tag{9.20}$$

Multiplying Equation (9.20) by $E_n^*(x)$ and integrating over x yields

$$\frac{d^2 E_n(y)}{dy^2} + [\omega^2\mu(\varepsilon_1 - s^2 g^2)\Gamma + \mu^2\mu\varepsilon_2(1 - \Gamma) - \beta_x^2 - \beta^2] E_n(y) = 0 \tag{9.21}$$

where Γ is the optical confinement factor along the thickness direction as defined by Equation (3.43). Solutions of Equation (9.21) are well-known Hermite–Gaussian polynomials, as given by Equation (9.14). As an example we shall compute G_n for the fundamental mode ($n = 0$). In this case, $E_0(x) = \cos \beta_x x$ and $E_0(y) = 1$, so

$$E_0(x, y) = \cos \beta_x x \tag{9.22}$$

Substituting Equations (9.9), (9.12), and (9.22) into (9.17) yields

$$G_0 = aRL_e^2 \left[1 - \frac{2L_e/S}{(S/2\pi L_e)^2 + 1} \frac{\tanh(S/2L_e)}{1 + \zeta\tanh(S/2L_e)} \right] - b \qquad (9.23)$$

For high-order modes ($n \neq 0$), the modal gain is given by

$$G_n = aRL_e^2 \left[1 - \frac{2(n+1)^2 L_e/C}{(S/2L_e)^2 + (n+1)^2} \frac{\tanh(S/2L_e)}{1 + \zeta \tanh S/2L_e} \right] - b \qquad (9.24)$$

where ζ is defined by Equation (9.8).

For a lightly doped p layer at 10^{17} cm^{-3} with $S/L_c \simeq 2$, the parameters a and b are found (Ref. 9.1) to be $(1.08 \pm 0.06) \times 10^{-16}$ cm^2 and 146 cm^{-1}, respectively. At threshold, the current density in this case is 4.45×10^3 A/cm^2. The gain profile at threshold for $|y| < S/2$ is

$$g_{th}(y) = 146 \left[1 - 0.72 \cosh\frac{y}{L_e} \right] \qquad \text{cm}^{-1}$$

and for $|y| > S/2$

$$g_{th}(y) = 322 \exp\left(-\frac{y}{L_e'} \right) - 146 \qquad \text{cm}^{-1}$$

A plot of g_{th} is shown in Figure 9.4.

It is interesting to see how the spatial distribution of modal gain can affect the current threshold. From Equation (9.23) we see that the threshold modal gain G_{th} corresponds to a threshold recombination rate R_{th}. If we take the ratios

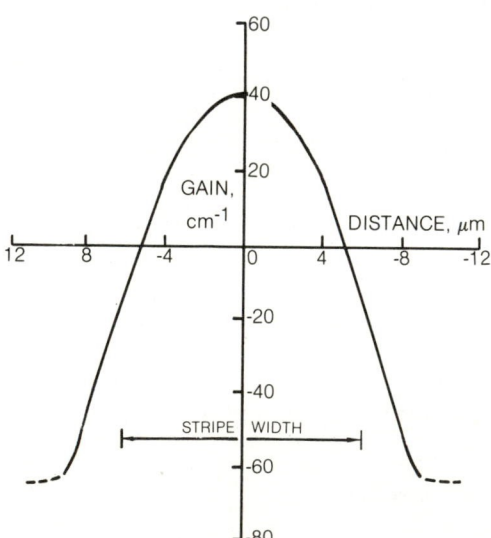

Figure 9.4 Threshold gain as a function of distance for a lightly doped stripe-geometry AlGaAs DH laser. (From Ref. 9.1.)

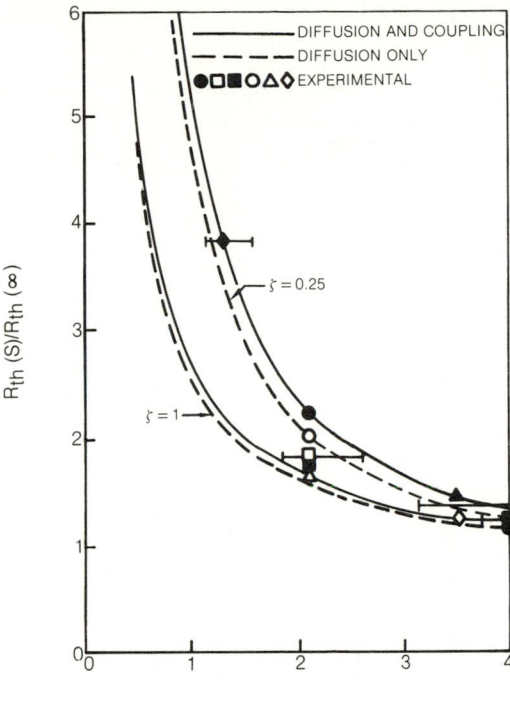

Figure 9.5 Threshold recombination rate as a function of normalized stripe-electrode width for a stripe-geometry laser normalized to that of an infinitely wide electrode. (From Ref. 9.1.)

of the threshold current density for a stripe-geometry laser having a stripe width S to a laser with an infinite electrode width, we get

$$\frac{R_{\mathrm{th}}(S)}{R_{\mathrm{th}}(\infty)} = \left[1 - \frac{2L_e/S}{(S/2\pi L_e)^2 + 1} \frac{\tanh(S/2L_e)}{1 + \zeta \tanh(S/2L_e)} \right]^{-1} \tag{9.25}$$

Figure 9.5 shows a plot of the injection rate required for lasing in the fundamental mode as a function of the stripe width normalized to a diffusion length for two cases: $\zeta = 1$ and $\zeta = 0.25$. It is clear that in both cases the threshold increases rapidly with decreasing stripe width. The increase in threshold is due to two major factors: (1) there is a loss of electrons due to out diffusion, and (2) there is a reduction in mode coupling to a nonuniform gain profile.

Another interesting observation from the output of stripe-geometry laser is the nonplanar wavefront, which is cylindrical concave in the direction of laser propagation. This is completely different from the planar wavefront for the case of a planar waveguide of an infinite extent. This behavior can be attributed to the spatial variation of gain medium, as shown in Figure 9.2(e), and is commonly referred to as gain guiding, a phenomenon in which the field distribution is dependent more on the gain of the medium than on the usual

(a) SIDE VIEW

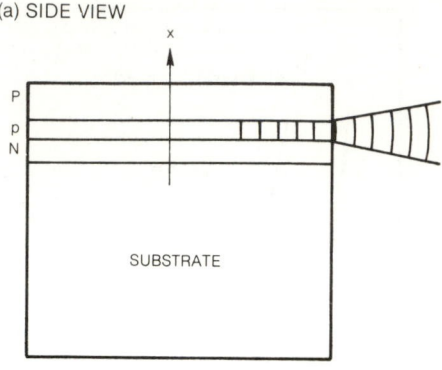

(b) TOP VIEW

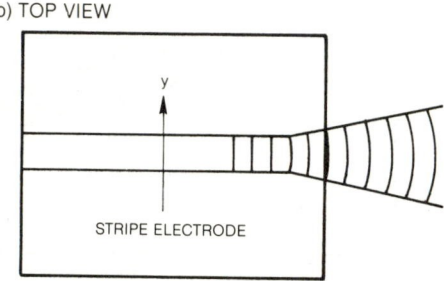

Figure 9.6 Schematic diagrams showing astigmatic wavefronts from (a) a side view and (b) a top view of a stripe-geometry laser.

refractive index. The wavefronts for stripe-geometry lasers are illustrated in Figure 9.6. The phase fronts for $E_n(x)$ as shown in Figure 9.6(a) are determined by a constant Δn across the junctions. However, the field $E_n(y)$ is influenced by gain guiding and has a cylindrical phase front as shown in Figure 9.6(b), resulting from a virtual beam waist occurring behind the facet. This output is therefore astigmatic because for the field confined perpendicular to the junction plane, the beam waist is at the facet, and it is located behind the facet for the field confined along the junction plane. Measurements (Ref. 9.2) of the far-field beam divergence and beam width confirmed the fact that the optical confinement along the junction plane is caused by the gain profile alone. Below threshold, the measured intensity profile of spontaneous emission corresponds well with the carrier distribution $n(y)$, as given by Equations (9.9) and (9.6).

The strength of astigmatism can be expressed in terms of a K factor as defined by

$$K = \frac{[\int |F(y)|^2\, dy]^2}{|\int F^2(y)\, dy|^2} \tag{9.26}$$

where $F(y)$ denotes the complex modal field distribution in the junction plane. If the wave is index guided, $F(y)$ is real; therefore, $F^2(y) = |F(y)|^2$. In this case it is clear that $K = 1$. K exceeds unity when $F(y)$ is complex, as for the case of gain-guided lasers. The astigmatism is shown (Ref. 9.3) to have a great

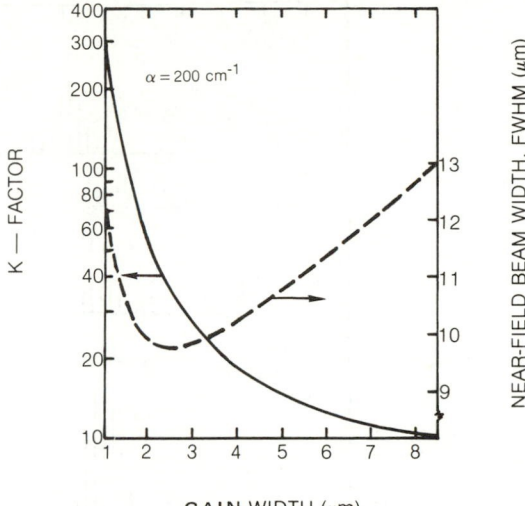

Figure 9.7 Variation of astigmatic K factor and beam width as a function of gain width. (From Ref. 9.4.)

influence on the spontaneous emission factor, which is defined as a ratio of the rate of spontaneous emission into one oscillating mode to the total emission rate R_{sp}. It has been shown (Ref. 9.3) that by narrowing the stripe width, the spontaneous emission factor increases much faster than that expected from the corresponding decrease of the active-layer volume. As a result, narrow stripe-geometry lasers exhibit a much broader spectrum than index-guiding lasers with a comparable active volume.

To extend this concept further, Streifer, Scifres, and Burnham (Ref. 9.4) utilized a more accurate representation of the lateral mode distribution to calculate the K factor. Without going through the details, the numerical results for K are plotted in Figure 9.7 as a function of the lateral gain width W_g, where W_g is determined by a combined effect of stripe width, current spreading, and lateral charge diffusion. Roughly speaking, $W_g \simeq 2S$, where S is the stripe width. These results were obtained by assuming a band-to-band absorption coefficient $\alpha = 200$ cm^{-1}. Also shown in Figure 9.7 is the near-field beam width as a function of the gain width.

9.4 POWER SPECTRUM OF DH LASERS

The spectral characteristics of a 4-μm stripe-geometry DH laser are shown in Figure 9.8. Because of a narrow electrode width, laser oscillation is limited to only a single transverse TE$_0$ mode. However, there are still many longitudinal modes with a spectral separation $\Delta\nu \simeq 2$ to 5 Å, as indicated by Equation (7.83). At low power ($P = 0.82$ mW), the spectral envelope λ_s is approximately equal to the width of the homogeneously broadened spontaneous emission line

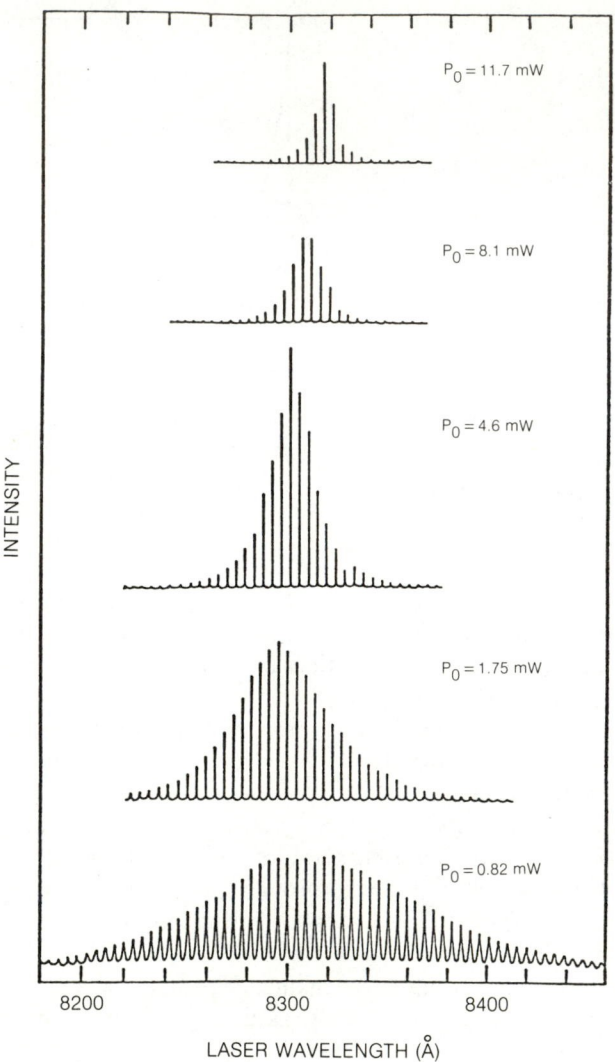

Figure 9.8 Spectral characteristics of a 4-μm stripe-geometry DH laser at several injection current levels. (From Ref. 9.4.)

λ_h, which can be characterized by a Lorentzian profile. Within this width, many longitudinal modes oscillate simultaneously. As power increases, the spectral width decreases and its shape is no longer symmetric, with a downshift toward the shorter wavelengths.

To explain some of these behaviors, Streifer, Scifres, and Burnham (Ref. 9.4) have presented a model based on spontaneous emission coupling into longitudinal modes. The results of their analysis are in excellent agreement with experimental data for both gain-guided and index-guided DH lasers. The theory assumes that the homogeneously broadened spontaneous emission is coupled to all the longitudinal modes by a different amount depending on the emission line shape. Below threshold, the emission envelope is very broad and is equivalent to that of an LED. Above threshold, the stimulated emission causes the injected carriers to recombine, hence the gain begins to saturate and the spectral envelope narrows significantly. Using a self-consistent approach, requiring that the field must reproduce itself in a round trip within the cavity, a simple result has been obtained from a rather complicated analysis for the dependence of the full width λ_s of the longitudinal spectral envelope at half-maximum power (FWHM) on the output power P_T. It is given by (Ref. 9.4)

$$\lambda_s = \lambda_h \left[\left(\frac{P_T^2}{4P_i^2} + 1 \right)^{1/2} - \frac{P_T}{2P_i} \right] \tag{9.27}$$

where λ_h is the homogeneous spontaneous line width, P_T is the total power of the fundamental mode transmitted through facet 1, and P_i is the internal circulating power of the mode just below threshold. They are related by the expression

$$P_T = 2P_i(1 - R_1)\sqrt{R_2} \, \frac{(1 + R_1)\sqrt{R_2} + (1 + R_2)\sqrt{R_1} - 4\sqrt{R_1 R_2}}{(1 + R_2^2)R_2 + (1 + R_2^2)R_1 - 4R_1 R_2} \tag{9.28}$$

where

$$P_i = \frac{1 - R_1}{R_1} \, \frac{\sqrt{R_1} + \sqrt{R_2}}{2\sqrt{R_2}} \, \frac{1 - \sqrt{R_1 R_2}}{\ln(1/R_1 R_2)} \, \frac{hc\lambda_0 K}{4\pi n^2 A} \tag{9.29}$$

R_1 and R_2 are the facet power reflectivities, λ_0 is the free-space wavelength, and

$$A = \frac{ed}{\Gamma} \, \frac{dg}{dJ} \tag{9.30}$$

In Equation (9.30), dg/dJ is the slope of the gain versus current density below threshold and is approximately a constant for most semiconductor lasers. As an example: If $\Gamma = 0.3$, $d = 0.1$ μm, and $dg/dJ \simeq 100$ cm^{-1}/kA/cm^2, then $A \simeq 0.6 \times 10^{-24}$ cm^2-s.

At low output levels (i.e., $P_T < P_i$), we obtain from Equation (9.27) that

$$\lambda_s \simeq \lambda_h \left(1 - \frac{1}{2} \frac{P_T}{P_i} \right) \tag{9.31}$$

Above threshold, we can write λ_s in another form as

$$\lambda_s \simeq \frac{\lambda_h P_i}{P_T} \tag{9.32}$$

Equation (9.32) indicates that in the lasing regime, the spectral envelope varies inversely with the output power. Since the expression above is independent of lasing parameters, it is applicable to all semiconductor lasers regardless of the guiding mechanism.

Figure 9.9 shows the variations of λ_s as a function of P_T for a number of K values. The measured λ_s values for the 4- and 8-μm stripe lasers agree quite well with the calculated values, $K = 30$ and 20, respectively. At higher power levels, the spectra are no longer symmetric, and longitudinal modes tend to lase on the short-wavelength side of the dominant mode. Such behavior is probably caused by nonlinear effects, which will not be treated here. Also shown in Figure 9.9 are measurements of an index-guided channeled substrate planar CSP laser. Since $K = 1$ for this device, these measurements are in good agreement with the theory. A slight phase-front distortion in the index-guided lasers could result in larger K values and broaden the spectrum.

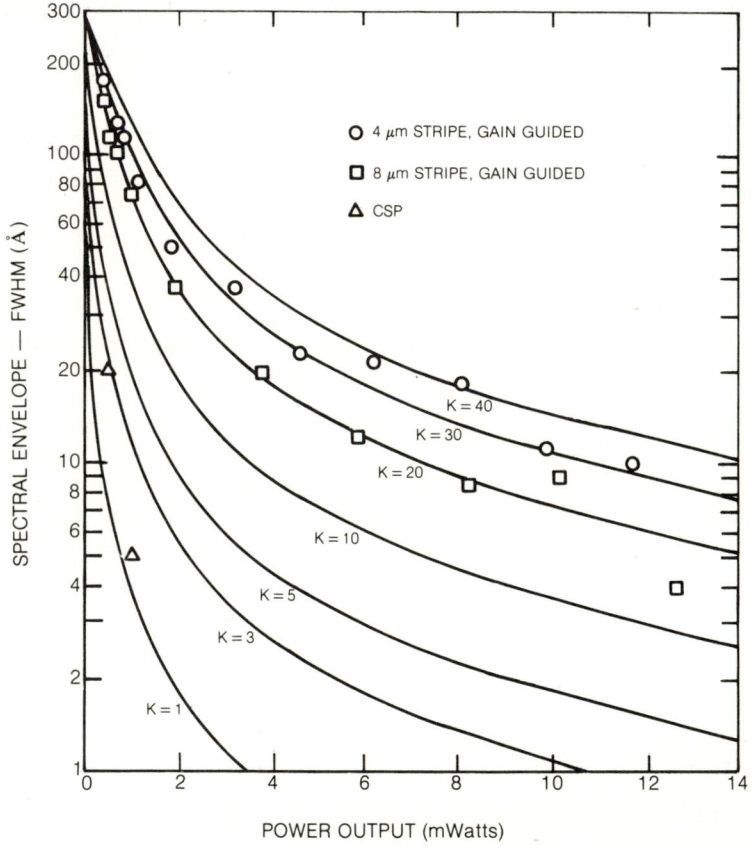

Figure 9.9 Threoretical and experimental results of spectral envelope width (FWHM) versus output power. (From Ref. 9.4.)

9.5 DISTRIBUTED FEEDBACK LASERS

The most advanced semiconductor laser structure is the one utilizing a periodic corrugation spatially distributed along the length of a gain medium to produce a feedback mechanism for laser oscillation. Figure 9.10 illustrates the structural configuration of a typical distributed feedback (DFB) laser. The fabrication of a desired corrugating structure commonly known as phase grating on a semiconducting layer between heterojunctions usually involves a very high-resolution material-processing technology. Even though the physical model of this laser has been established for a long time, it is still not available commercially. We shall follow the treatment by Kogelnik and Shank (Ref. 9.5), who analyzed this laser system using a coupled-wave theory. A simple physical model was established by assuming a coupling between two counter-running waves caused by an interaction known as backward Bragg scattering. Therefore, this laser is referred to sometimes as a distributed Bragg reflection (DBR) laser. In this model, the field amplitude of each of the two oppositely traveling waves is superimposed with a backward Bragg scattered component of the other wave of a particular order that must satisfy the Bragg condition. Since the medium provides optical gain along its path length, such an interaction is capable of generating sufficient feedback to sustain laser oscillation. The major advantage of this laser structure is the high degree of spectral selectivity by the phase grating.

For simplicity, the analysis deals only with a linear system in which any nonlinear effect such as gain saturation is ignored. Therefore, this model is valid only near threshold. For a medium having gain or loss, the wave equation as given by Equation (2.9) must be modified to include the effects of source or sink. This is accomplished by adding a term involving the electrical conductivity, $\mu\sigma(\partial E/\partial t)$ into the wave equation for a passive medium. In this case the wave equation takes the form

$$\nabla^2 \mathbf{E} = \mu\sigma \frac{\partial \mathbf{E}}{\partial t} + \mu\varepsilon \frac{\partial^2 \mathbf{E}}{\partial t^2} \tag{9.33}$$

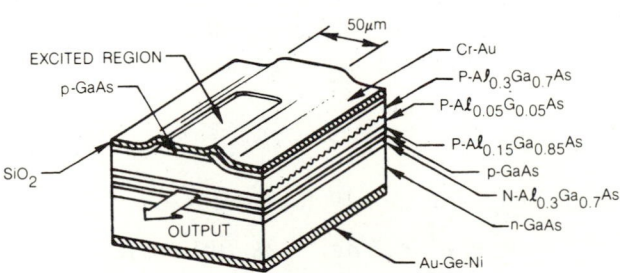

Figure 9.10   Topology of a AlGaAs DH DFB laser.

For a plane wave polarized in the x direction and its spatial variations occurring only in the z direction, $\partial/\partial x = \partial/\partial y = 0$. $\mathbf{E}$ can be replaced by E_x. Equation (9.33) reduces to

$$\frac{\partial^2 E_x}{\partial z^2} = \mu\sigma \frac{\partial E_x}{\partial t} + \mu\varepsilon \frac{\partial^2 E_x}{\partial t^2} \tag{9.34}$$

Substituting the time-dependent term, $e^{-i\omega t}$, into Equation (9.34), we obtain a scalar wave equation of the form

$$\frac{\partial^2 E_x}{\partial z^2} + k^2 E_x = 0 \tag{9.35}$$

where

$$k^2 = i\omega\mu(\sigma + i\omega\varepsilon) \tag{9.36}$$

It is customary to express the propagation constant k in terms of a complex refractive index $\mathcal{N}$. We let

$$k^2 = \mathcal{N}^2 k_0^2 \tag{9.37}$$

where $\mathcal{N}$ has a real part n_r and an imaginary part n_i. We now assume that the real part has a sinusoidal variation with z as

$$n_r(z) = n + n_m \cos Kz \tag{9.38}$$

where $K = 2\pi/\Lambda$, with Λ the period of the phase grating, n_m the maximum modulation index, and n the real index of the material. The imaginary part is related to the field gain coefficient, g_F, where g_F must be distinguished from the usual power gain coefficient g (i.e., $g = 2g_F$) by the expression

$$n_i = -g_F k_0 \tag{9.39}$$

Substituting Equations (9.38) and (9.39) into (9.37), we obtain

$$k^2 = k_0^2 [(n + n_m \cos Kz)^2 - (k_0 g_F)^2 + 2(g_F k_0)i(n + n_m \cos Kz)] \tag{9.40}$$

For all practical purposes, we shall assume that the gain is small over a distance of the order of a wavelength, λ_0, and also that the perturbation of refractive index n_m is very small compared with n. In other words,

$$g_F \ll \frac{2\pi n}{\lambda_0} = \beta \qquad n_m \ll n \tag{9.41}$$

With these assumptions, Equation (9.40) reduces to

$$k^2 = \beta^2 + 2i\beta g_F + 2k_0 \beta n_m \cos Kz \tag{9.42}$$

Substituting Equation (9.42) into (9.35), we write

$$\frac{\partial^2 E_x}{\partial z^2} + (\beta^2 + 2i\beta g_F)E_x = -2k_0\beta \cos(2\beta_B z)E_x \qquad (9.43)$$

where

$$\beta_B = \frac{2\pi n}{\lambda_B} \qquad (9.44)$$

is a number that corresponds to half Bragg wavelengths in the medium. If we select the periodicity of the phase grating Λ to be $\lambda_B/2n$, so that the Bragg condition is satisfied, only two opposite waves are synchronized in phase, while all other diffraction orders can be neglected in the coupled-wave model. We shall denote these two counter-running waves by $\mathscr{E}_+(z)\, \exp(i\beta_B z)$ and $\mathscr{E}_-(z)\, \exp(-i\beta_B z)$. The total electric field is the sum of these two waves as given by

$$E_x(z) = \mathscr{E}_+(z)\, \exp(i\beta_B z) + \mathscr{E}_-(z)\, \exp(-i\beta_B z) \qquad (9.45)$$

where $\mathscr{E}_+$ and $\mathscr{E}_-$ are complex amplitudes. To simplify the analysis further, we shall assume that the transfer of power between these two waves occurs slowly and therefore the second derivatives $\partial^2 \mathscr{E}_-/\partial z^2$, $\partial^2 \mathscr{E}_+/\partial z^2$ can be neglected. In this approximation we obtain upon substituting Equation (9.45) into (9.43), a pair of coupled-wave equations of the form

$$-\frac{\partial \mathscr{E}_-}{\partial z} + (g_F - i\delta) = iK_c\mathscr{E}_+$$

$$\frac{\partial \mathscr{E}_+}{\partial z} + (g_F - i\delta) = iK_c\mathscr{E}_- \qquad (9.46)$$

where the parameter δ is a normalized frequency defined by

$$\delta \equiv \frac{\beta^2 - \beta_B^2}{2\beta_B} \simeq \beta - \beta_B = \frac{n}{c}(\omega - \omega_B) \qquad (9.47)$$

and the parameter K_c is the coupling coefficient, defined by

$$K_c = \frac{\pi n_m}{\lambda_0} \qquad (9.48)$$

The coupled-wave equations (9.46) describe wave propagation in the DFB structure in the presence of gain and a periodic perturbation in the refractive index. As illustrated by Figure 9.11, laser oscillation builds up from zero amplitudes with waves reflected at the device boundaries. The boundary conditions for the wave amplitudes are

$$\mathscr{E}_-\left(-\frac{L}{2}\right) = \mathscr{E}_+\left(\frac{L}{2}\right) = 0 \qquad (9.49)$$

(a)

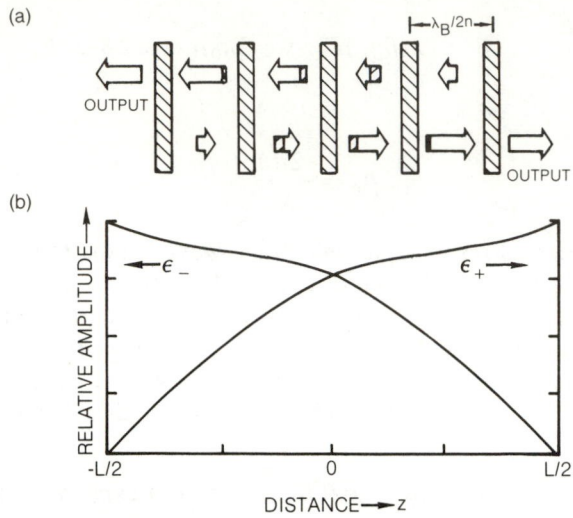

(b)

Figure 9.11   (a) Illustration of laser oscillation in a periodic structure; (b) field amplitudes of a left-traveling wave $\mathcal{E}_-$ and a right-traveling wave $\mathcal{E}_+$ versus distance. (From Ref. 9.5.)

To solve the coupled-wave equations, we shall assume a general solution of the form

$$
\begin{aligned}
\mathcal{E}_- &= C_1 e^{\gamma z} + C_2 e^{-\gamma z} \\
\mathcal{E}_+ &= d_1 e^{\gamma z} + d_2 e^{-\gamma z}
\end{aligned}
\tag{9.50}
$$

The symmetric and antisymmetric requirements [i.e., $\mathcal{E}_-(z) = \pm \mathcal{E}_+(-z)$] lead to the following relations:

$$
C_1 = \pm d_2, \qquad C_2 = \pm d_1
\tag{9.51}
$$

Furthermore, the boundary conditions require that

$$
\frac{C_1}{C_2} = \frac{d_2}{d_1} = -e^{\gamma L}
\tag{9.52}
$$

Using these relations, we can rewrite Equation (9.50) in the form

$$
\mathcal{E}_- = \sinh \gamma \left(z + \frac{L}{2} \right)
$$

$$
\mathcal{E}_+ = \pm \sinh \left(z - \frac{L}{2} \right)
\tag{9.53}
$$

To determine the allowed eigenvalues corresponding to the longitudinal modes in the DFB structure, we substitute the expressions of Equation (9.53) into the coupled-wave equations (9.46) and obtain a pair of transcendental equations

$$-\gamma \sinh\frac{\gamma L}{2} + (g_F - i\delta)\cosh\frac{\gamma L}{2} = \pm\, iK_c \cosh\frac{\gamma L}{2}$$

$$-\gamma \cosh\frac{\gamma L}{2} + (g_F - i\delta)\sinh\frac{\gamma L}{2} = \mp iK_c \sinh\frac{\gamma L}{2} \tag{9.54}$$

Performing the sum and difference of the equations above, we obtain

$$\gamma + (g_F - i\delta) = \pm\, iK_c e^{\gamma L}$$

$$\gamma - (g_F - i\delta) = \mp iK_c e^{-\gamma L} \tag{9.55}$$

Multiplying these two equations, we obtain a dispersion relation for the complex propagation constant γ:

$$\gamma^2 = K_c^2 + (g_F - i\delta)^2 \tag{9.56}$$

Adding these two equations, we obtain an eigenvalue equation for γ:

$$K_c = \pm\,\frac{i\gamma}{\sinh \gamma L} \tag{9.57}$$

Equation (9.57) indicates that in general values of γ are complex and each corresponds to a different branch of the complex hyperbolic sine functions. Furthermore, each eigenvalue γ associates with a threshold gain constant and a resonant frequency δ. The values for g_F and δ can be evaluated from the following relation, which is obtained by taking the difference of Equation (9.55):

$$g_F - i\delta = \gamma \coth \gamma L \tag{9.58}$$

Because of the dispersive nature of the DFB structure, there exist "stop bands" of frequencies in which mode propagation is forbidden. From Equations (9.45) and (9.50) and the dispersion relation as given by Equation (9.56), we see that there exist four waves in the structure. For each traveling wave $\exp(i\beta_B \pm \gamma)z$ there are two γ values, which are complex quantities, determined by the dispersion relation in terms of K_c, g_F, and δ. The eigenvalues γ can be obtained only by solving the complex transcendental equation (9.57) numerically, with the help of a computer. It is possible, however, to obtain approximations in the limits of high and low threshold gain. In the high-gain limit, where $g_F \gg K_c$, the complex propagation constant γ as given by Equation (9.56) becomes

$$\gamma \simeq g_F - i\delta \tag{9.59}$$

and the eigenvalue equation (9.55) can be written as

$$2(g_F - i\delta) \simeq \pm iK_c \exp(g_F - i\delta)L \tag{9.60}$$

Multiplying the complex conjugate, we obtain from Equation (9.60) that

$$\frac{(\exp 2g_F L)K_c^2}{4(g_F^2 + \delta^2)} = 1 \tag{9.61}$$

Equation (9.61) is equivalent to the oscillation condition for semiconductor lasers with end mirrors as given by Equation (7.72), where $g - \alpha$ corresponds to $2g_F$ and the product of mirror reflectivities corresponds to $K_c^2/4(g_F^2 + \delta^2)$. By equating the phase of Equation (9.60), we get

$$(m + \tfrac{1}{2})\pi + \tan^{-1}\frac{\delta}{g_F} = \delta L \tag{9.62}$$

where $m = 0, \pm 1, \ldots$. Near the Bragg frequency, we obtain from Equation (9.48) that $K_c = \pi n_m / \lambda_0$. Substituting these K_c values into Equations (9.61) and (9.62) and letting $\delta = 0$ yields

$$\left(\frac{\pi n_m}{\lambda_0}\right)^2 \exp(2g_F L) = (2g_F)^2 \tag{9.63}$$

and

$$\delta = \frac{(m + \tfrac{1}{2})\pi}{L} \tag{9.64}$$

From Equation (9.64) we obtain the expression for the resonance frequencies

$$\nu = \nu_B \pm \frac{c}{2nL}(m + \tfrac{1}{2}) \tag{9.65}$$

Equation (9.65) shows that the resonances are spaced approximately $c/2nL$ apart, which is the same as in a usual two-mirror laser cavity of length L. It should be noted that there is no resonance at ν_B, the Bragg frequency, and the width of the stop band is $2K_c$. Within the stop band, the waves are evanescent. The threshold gain for a given resonance frequency can be calculated from Equation (9.64). To verify the spectral selectivity of a DFB laser, we observe from Equation (9.61) that for a wavelength that is deviated from the Bragg wavelength by $\delta = g_F$, K_c^2 must be doubled in order to keep the threshold gain the same. This strong spectral selectivity is a direct consequence of the dispersive property of the DFB laser structure.

    To obtain approximate formulas for the low-gain limiting case, where $g_F \ll K_c$, we shall take only the real part of Equations (9.57) and (9.58) and expand in a power series near $g_F = 0$. We get

$$\delta \simeq K_c \tag{9.66}$$

for the first resonance that occurs again just outside the stop band. The threshold condition in this case is

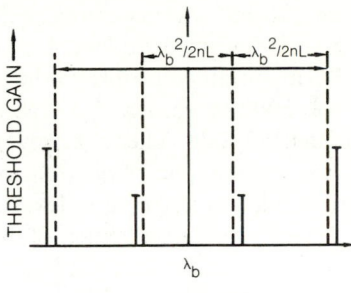

Figure 9.12 Calculated mode spectrum and threshold gain of a DFB laser. (From Ref. 9.5.)

$$g_F L \simeq \left(\frac{\lambda}{n_m L} \right)^2 \tag{9.67}$$

Figure 9.12 shows the numerical results for the resonance spectrum and the threshold gain of the periodic structure obtained by Kogelnik and Shank (Ref. 9.5). The dashed lines indicate the basic frequency spacing $\Delta \nu = c/2nL$. In terms of the Bragg wavelength, we write

$$\Delta \lambda = \frac{\lambda_B^2}{2nL} \tag{9.68}$$

This is the exact result that one gets in the high-gain limit. Figure 9.12 indicates that the mode spectrum is symmetric with respect to the Bragg frequency ν_B and there is no resonance at ν_B. Within the stop band, all oscillations cease. The stop band increases with increasing K_c, and eventually becomes comparable to $2nL$. At this point it starts to push the resonances away from ν_B. The threshold gain as shown in Figure 9.12 also increases with the frequency spacing from ν_B. As a result, the DFB laser structure provides the highest spectral selectivity.

9.6 SINGLE-MODE
AND HIGH-POWER SEMICONDUCTOR LASERS

Although the simplest type of practical laser diode is the stripe-geometry DH laser with standard stripe widths varied in the range 10 to 20 μm, the output often contains a series of instabilities in the form of nonlinear "kinks" in the light–current characteristics. These pulsations cause excess noise and mode shifts in the power spectrum. To eliminate these problems, the electrode width must be reduced to below 10 μm. As a result, the threshold currents and astigmatic effect increase significantly. To avoid these problems, one can use the distributed feedback laser; however, such a laser is difficult to fabricate and is not available commercially. Other approaches have been pursued by making three-dimensional structures that can provide a light–current confinement based on a built-in real-index waveguiding. Light–current confinement to a narrow channel (3 to 7 μm) eliminates kinks in the output at high injection levels. Built-in real-index waveguiding results in a stable fundamental spatial mode as well as a single-frequency mode. There are several types of waveguiding structures, depending on how the real part of the bulk index is altered in the lateral plane. It can be a positive-index, a negative-index, or a semileaky guide. The amount of index change Δn and the spatial extent of the index variation must be chosen so that the structure can support only the fundamental mode, and is predominantly index guiding rather than gain guiding.

Figure 9.13 shows a number of index-guided single-mode CW lasers. In addition to the buried-heterostructure BH laser, which was already introduced previously, Figure 9.13(a) shows a transverse junction stripe TJS laser, and Figure 9.13(b) shows a plane-convex waveguide PCW laser. Figure 9.13(c) shows a channeled substrate planar CSP laser, and Figure 9.13(d) shows a constricted double-heterostructure CDH laser. The output of these lasers is typically in the range 1 to 7 mW. With the exception of the TJS laser, all the lasers noted above are fabricated by LPE growth of heterojunctions over a channeled or ridged substrate with a width varied from 1 to 5 μm. The lateral variation in thickness is equivalent to a lateral variation in refractive index that provides the optical confinement along the junction plane. For the TJS laser, the lateral waveguiding is obtained by two consecutive Zn diffusions that produce a variation in the index between p and p^+ material; therefore, both the cathode and the anode can be placed on the top surface. The CDH laser is grown on a "double-dovetail" channel configuration. By placing a 10-μm-wide electrode on the top layer, a constricted active region is formed that resembles a leaky waveguide. Figure 9.14 shows the power spectrum of a CDH laser. At a low injection level, the output consists of a number of longitudinal modes. As the injection current increases above 100 mA, a single longitudinal mode at a power of $\sim$10 mW is obtained with its wavelength shifting toward the longer region. This is attributed to Joule heating in the device junction. The CW power–current (P–I) characteristics of a CDH laser operated at various temperatures between 20 and 70°C are shown in Figure 9.15. At all temperatures the curves are linear and "kinkless." As the temperature is increased from 20°C to 70°C, the threshold current increases by only 22%. Higher outputs can be obtained from these lasers by changing the laser structures to the LOC configurations shown in Figure 9.16. This involves the growth of additional cladding layers of intermediate index. LOC CDH lasers have been operated successfully to yield single-longitudinal-mode CW output at 50 mW per facet.

9.7 LONG-WAVELENGTH SOURCES

The interest in obtaining low material dispersion has led to rather significant development in recent years of the InGaAsP system, which emits at long wavelengths ranging from 1.1 to 1.6 μm. In this section several low-threshold

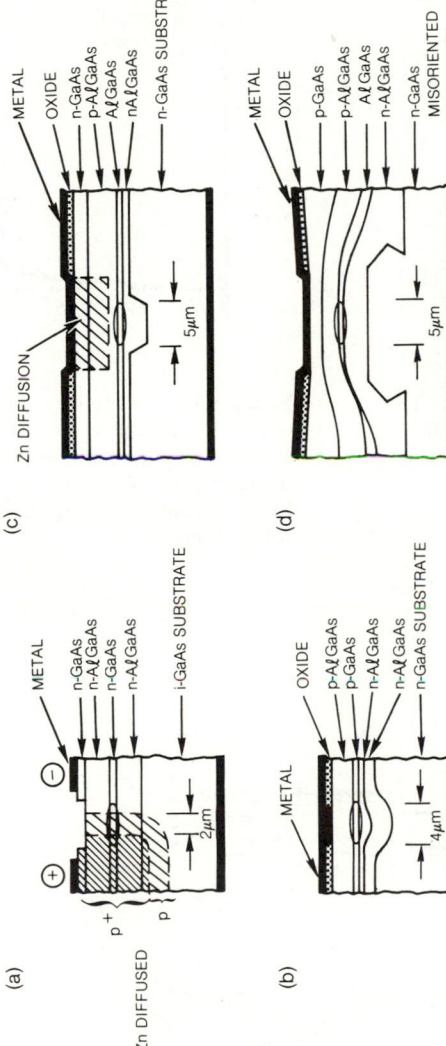

Figure 9.13 Single-mode DH lasers: (a) transverse junction stripe TJS lasers; (b) planar convex waveguide PCW lasers; (c) channeled substrate planar CSP lasers; (d) constricted double-heterostructure CDH lasers. (From Ref. 9.6. Reprinted with permission of North-Holland Publishing Company, Amsterdam.)

203

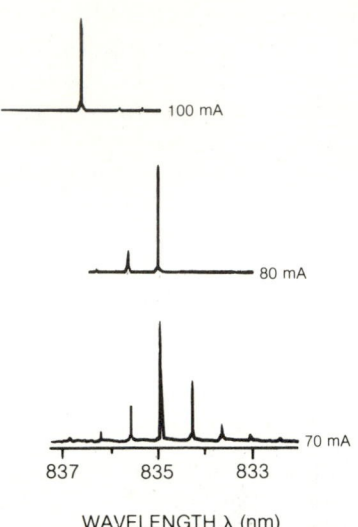

WAVELENGTH λ (nm)

Figure 9.14 Power spectrum of a CDH laser. (From Ref. 9.6. Reprinted with permission of North-Holland Publishing Company, Amsterdam.)

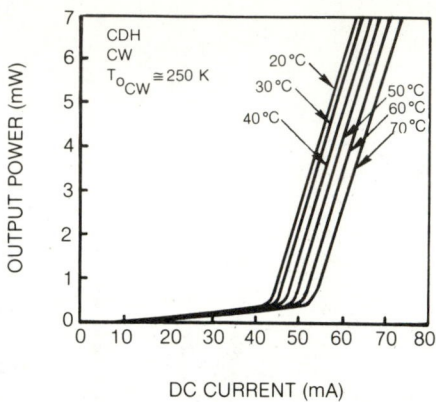

DC CURRENT (mA)

Figure 9.15 CW power output as a function of injection current for a CDH laser operated at temperatures varied from 20 to 70°C. (From Ref. 9.6. Reprinted with permission of North-Holland Publishing Company, Amsterdam.)

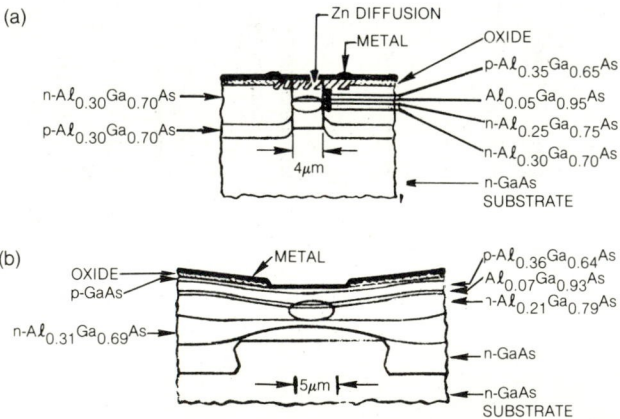

Figure 9.16 Single-mode and high-power large-optical-cavity DH lasers: (a) LOC BH laser topology; (b) LOC CDH laser topology. (From Ref. 9.6. Reprinted with permission of North-Holland Publishing Company, Amsterdam.)

InGaAsP/InP lasers and their performance characteristics are described. With the conventional liquid-phase epitaxy methods, the growth of an InGaAsP active layer encounters a serious problem, due to the fact that the active layer can easily be melted back into the In solution during the subsequent growth of the cladding layers. To prevent meltback, subsequent growth must be carried out at relatively low temperatures. Several anti-meltback methods have been introduced: for example, the use of an anti-meltback layer between the active and InP cladding layers. Such a structure is shown in Figure 9.17(a). The InGaAsP layer is grown on a Sn-doped InP substrate oriented in the (100) plane with a carrier concentration of 2×10^{18} cm^{-3}. The growth temperature and rate must be controlled very carefully to prevent the loss of the active layer. The lowest threshold current density of approximately 1.2 kA/cm^2 has been obtained (Ref. 9.7) from an active-layer thickness of 0.2 μm. A similar technique (Ref. 9.7) has been used to fabricate InGaAsP/InP buried hetero-structure lasers, as shown in Figure 9.17(b). These lasers can be operated continuously at room temperature with single-transverse-mode output up to 10 mW, at a differential quantum efficiency as high as 43%. Operating in the 1.3-μm region, a crescent-shaped active layer, as shown in Figure 9.18, has been

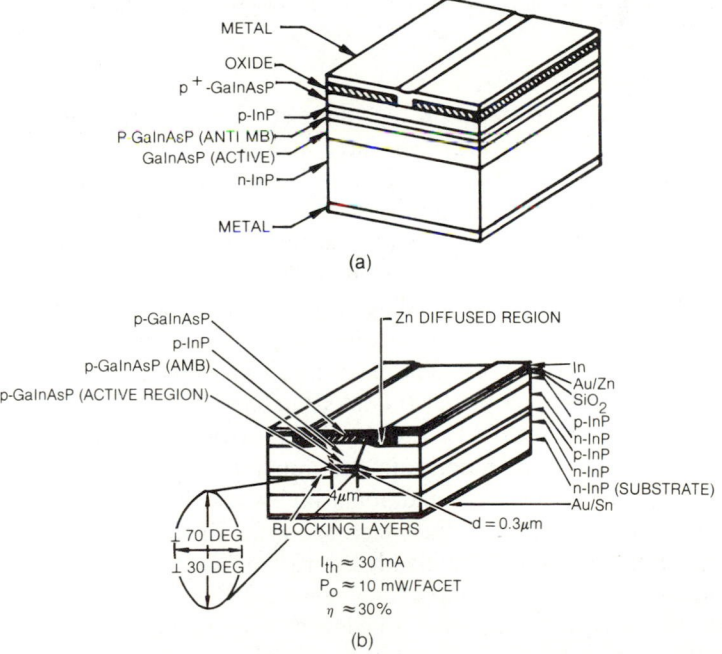

(a)

(b)

Figure 9.17   Topology of (a) semiconductor stripe-geometry DH GaAsP/InP laser, and (b) buried DH InGaAsP/InP laser. (From Ref. 9.7. Reprinted with permission of IEEE, © 1981.)

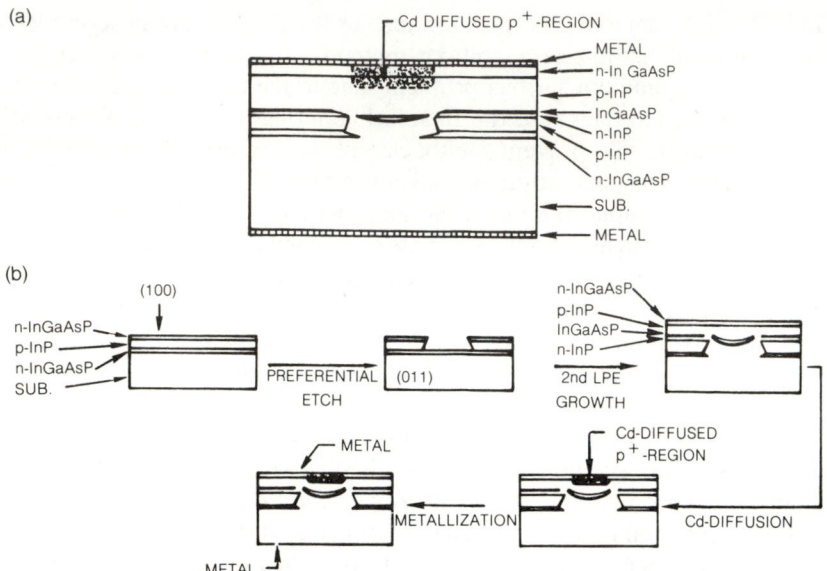

Figure 9.18   (a) Topology of a buried crescent InGaAsP/InP laser; (b) fabrication procedure of a BC laser. (From Ref. 9.8. Reprinted with permission of IEEE, © 1981.)

fabricated by using a two-step LPE technique (Ref. 9.7). Low threshold current and single-transverse-mode operation have also been obtained from this laser. As shown in Figure 9.18(a), a crescent-shaped InGaAsP active region is embedded in InP by LPE growth on a dovetail channeled substrate. A double-current confinement scheme is incorporated with two reverse-biased *pn* junctions at both sides of the active layer. The reverse-biased InP junction serves as a barrier against the spreading of current, even when it is very close to the active region. The active region has a parabolic cross section and is completely surrounded by InP. The fabrication procedure of this structure is shown in Figure 9.18(b). A planar InGaAsP heterostructure is first grown on a (100) *n*-InP substrate by LPE. Subsequently, a dovetail-shaped channel is etched along the (011) direction into the upper two layers, with a depth of approximately 2 μm. Then four additional layers, as shown in Figure 9.18(b), are grown successively on the channeled substrate. The crescent region has a refractive index equivalent to a cladded parabolic index waveguide. With an active width of 2 μm and an active thickness of 0.1 μm, only the fundamental transverse mode is allowed, while all higher-order modes are beyond cutoff. Figure 9.19(a) depicts the *L–I* curve and Figure 9.19(b) shows the output spectral characteristics of a crescent-shaped InGaAsP laser. The I_{th} of this laser is about 20 mA and the output increases with increasing currents and exhibits no "kinks." At high currents, the *L–I* curve saturates due to the temperature

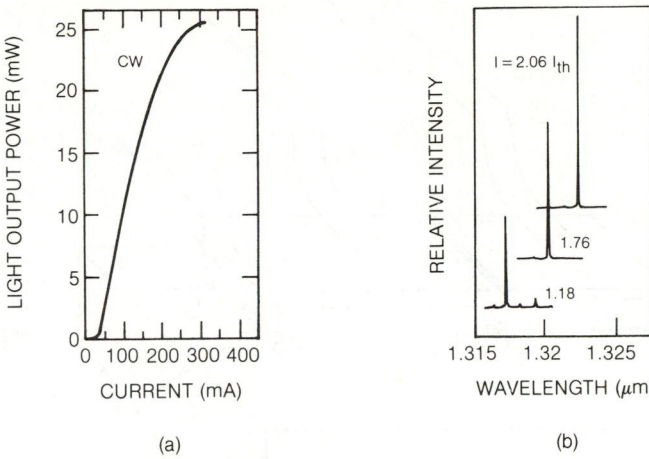

(a)   (b)

Figure 9.19 (a) L–I curve for a BC InGaAsP/InP laser; (b) output spectral characteristics of a BC InGaAsP/InP laser. (From Ref. 9.8. Reprinted with permission of IEEE, © 1981.)

effect. The maximum CW power obtainable from this laser without causing any catastrophic damage is about 25 mW per facet. At low injection levels, there are several longitudinal modes; however, as the current level increases, the output power begins to concentrate in one mode as the current reaches the $1.1I_{th}$ level. With a further increase in the injection current, the output spectrum shifts toward the longer wavelength, again due to the temperature effect; however, the output remains in a single longitudinal mode for currents up to twice the threshold value.

9.8 CLEAVED COUPLED-CAVITY SEMICONDUCTOR LASERS

The most recent development in semiconductor laser technology is a new electronically tunable and single frequency laser source, commonly called the cleaved coupled-cavity (C^3) laser. This relatively new laser consists of two optically interacting cavities and can be batch manufactured by cutting a planar heterojunction device along its cleavage plane and realigning the two parts. Each part is electronically controlled and independent of the other. Because of the spectral purity and extremely broad frequency tunability of its output, this laser has the potential to increase the data rate and the transmission length, or both, significantly. Using this laser, it has been shown (Ref. 9.9) that single frequency operation can be maintained under 2 Gb/s direct modulation with error rates less than 10^{-10}. It has also been demonstrated that a C^3 laser can transmit information at 420 Mb/s through a 119 km unrepeated length of fiber with a BER $<10^{-9}$. In this case the C^3 laser was made to operate at 1.55 μm

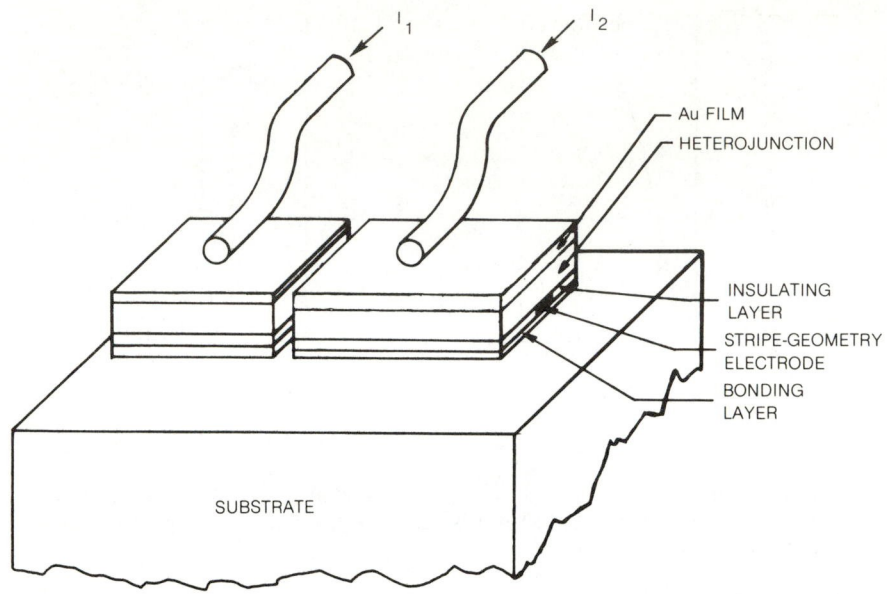

Figure 9.20   A schematic diagram of a cleaved coupled-cavity laser.

with a chromatic dispersion of 2.08×10^{-3} ps/km. Since the frequency of this laser can be tuned, it is then possible to design a system involving multichannel optical frequency shift-keying and a new approach to optical data switching and routing.

Figure 9.20 is a schematic of a C^3 laser. The basic laser material can be either a gain-guided or index-guided heterojunction. Two standard Fabry-Perot (FP) cavities with two completely separate stripe-geometry electrodes are involved. These two laser diodes have slightly different lengths, typically about 125 μm, and are strongly coupled optically with each other through a separation of less than 5 μm. All the reflecting facets are formed by cleaving along crystallographic planes so that they are perfectly parallel. The electrodes that are buried underneath the laser material must be perfectly aligned with respect to each other, and must be electrically isolated from each other. To achieve the above, a heterojunction epilayer is cleaved at two ends to form a standard FP laser cavity of approximately 250 μm length. It is then covered with a thick ($\sim$5 μm) electroplated Au layer. The device is recleaved near the middle to form two separate FP diodes. Because of the thick Au layer, these two diodes remain hinged together and are bonded with indium upside down on a Cu heat sink. Two stripe-geometry electrodes have to be deposited on the epilayer before bonding. The separation between the two diodes is only a few microns, nevertheless it can be varied by stretching the thick Au film. This fabrication technique can be extended to manufacture an array of coupled-cavity laser diodes.

The C^3 laser can be operated in either a tuning or an untuning mode. In either case, the first diode is always operated above the threshold. The second diode can either be above or below threshold. In a tuning mode, the laser wavelength can be controlled by varying the injection current level, which is always kept below the threshold. A change in injection currents causes a change in carrier density, which affects the refractive index of the medium and causes a shift in the cavity modes in the second diode. Typically, a frequency excursion of 150 Å and a tuning rate of 10 Å/mA can be achieved. Let n_1 and n_2 be the effective refractive indices in the first and the second diodes, respectively. Then the mode spacings for the two diodes in terms of wavelength can be expressed as

$$\Delta\lambda_1 = \frac{\lambda_0^2}{2n_1L_1}$$

and

$$\Delta\lambda_2 = \frac{\lambda_0^2}{2n_2L_2}$$

Figure 9.21(a) shows the allowed Fabry-Perot modes for the two independent diodes. Since these two diodes are strongly coupled the modes from each cavity that coincide spectrally will interfere constructively to form the resultant modes of the coupled cavity, while all others interfere destructively and are suppressed. The spectral spacing $\Delta\lambda_C$ of the coupled cavity modes, therefore, is significantly larger than those of the two individual diodes and can be approximated by

$$\Delta\lambda_C = \frac{\Delta\lambda_1\Delta\lambda_2}{|\Delta\lambda_1 - \Delta\lambda_2|} = \frac{\lambda_0^2}{2|n_1L_1 - n_2L_2|}$$

As a result, the spectrum of the coupled cavity laser becomes extremely pure and only the one that occurs near the peak of the gain, as shown in Figure 9.21(b), can survive. As the injection current in diode 2 varies by a small amount equivalent to a shift in wavelength by $\delta\lambda$, the wavelength of the C^3 laser will be shifted by exactly one FP mode spacing ($\Delta\lambda_1 \simeq \Delta\lambda_2$), which is about 15Å, depending on the cavity length. This type of stepwise tuning will continue over a very large spectral width of the gain profile ($\sim$150Å). The ability to tune over such a wide range of wavelengths naturally leads to system applications involving frequency shift-keying of wavelength division multiplexing.

In addition, the C^3 laser has the ability to perform a set of basic logic operations, including AND, OR, INVERT, and EXCLUSIVE OR. In these operations, both diodes are operated above threshold with a pulsed current superimposed with a dc bias. Therefore, the laser output spectra can be in either λ_1, λ_2, or $\lambda_1 + \lambda_2$ states. By a superposition of injection current pulses I_1 and I_2 with a varying pulse width and time delay, it is possible to achieve all of the above operations. For example, the overlap of the two current pulses produces

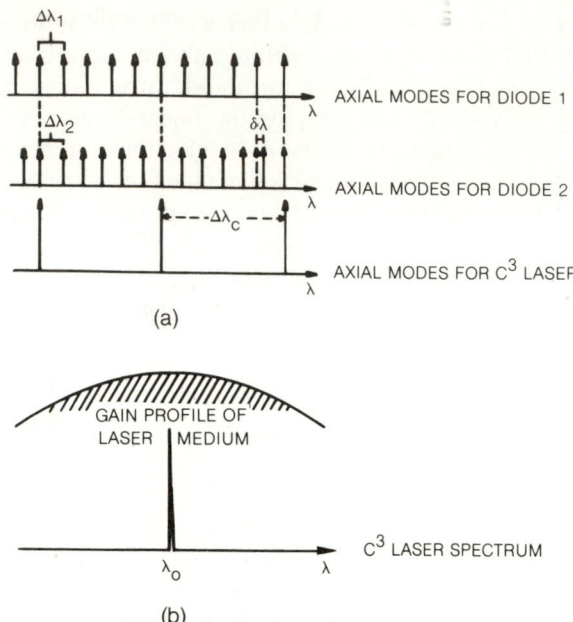

Figure 9.21 Spectral characteristics of cleaved coupled-cavity lasers. (a) Allowed Fabry-Perot modes for the first and the second diodes, and for the coupled-cavity. (b) The output spectrum of the C^3 laser.

an increase of the output power in the overlapping period. One simple method for obtaining AND or OR logic functions is to limit the levels in the detection channel that is sensitive to all wavelengths. Another method is to analyze the output spectra by using a diffraction grating with an array of detectors. This method provides the flexibility of obtaining a variety of logic functions by virtue of the fact that both the power and the wavelength are monitored independently.

In summary, this laser can perform not only the basic logic functions at an extremely fast switching rate (greater than 1 Gb/s), but can also offer a very broad frequency tuning range that is useful for very high data rate transmission systems. Because of this versatility, it is considered to be one of the most significant achievements in semiconductor laser technology since the discovery of single heterojunctions in 1970.

PROBLEMS

9.1. Calculate the center frequency of a light source emitting at $\lambda = 0.9$ μm.

9.2. Derive the carrier distribution function as given by Equation (9.5) under the stripe electrode.

9.3. If $n(S/2)/n(0) = 0.5$, compute L_c and L'_c.

9.4. For a stripe-geometry laser emitting at $\lambda_0 = 0.83$ μm having a gain width of about 4 μm, an active layer thickness of 0.1 μm, and assuming that $A = 10^{-24}$ cm^2-s, calculate the internal circulating power and the output power if $R_1 = R_2 = 0.3$.

9.5. Derive an oscillation condition for a DFB laser.

9.6. Show that the increase in threshold gain Δg_F for a DFB laser when the resonance wavelength λ is exceeded over one spectral bandwidth $\Delta\lambda/\lambda$ is $\Delta g_F = 2\pi n/\lambda_B$.

REFERENCES

9.1. B. W. Hakki, *J. Appl. Phys.*, *44*, 5021 (1973).

9.2. D. D. Cook and F. R. Nash, *J. Appl. Phys.*, *46*, 1660 (1975); T. L. Paoli, *IEEE J. Quantum Electron.*, *QE-13*, 662 (1977).

9.3. K. Petermann, *IEEE J. Quantum Electron.*, *QE-15*, 566 (1979).

9.4. W. Streifer, D. R. Scifres, and R. D. Burnham, *Appl. Phys. Lett.*, *40*, 305 (1982); *IEEE J. Quantum Electron.*, *QE-17*, 736 (1981); *Electron. Lett.*, *17*, 933 (1981).

9.5. H. Kogelnik and C. V. Shank, *J. Appl. Phys.*, *43*, 2327 (1972).

9.6. D. Botez, *J. Opt. Commun.*, *1*, 2 (1980).

9.7. S. Arai, M. Asada, T. Tanbunek, Y. Suematsu, Y. Itaya, and K. Kishino, *IEEE J. Quantum Electron.*, *QE-17*, 640 (1981).

9.8. E. Oomura, T. Murotani, H. Higuchi, H. Namizaki, and W. Susaki, *IEEE J. Quantum Electron.*, *QE-17*, 646 (1981).

9.9. W. T. Tsang, N. A. Olsson, and R. A. Logan, *IEEE J. Quantum Electron.*, *QE-19*, 1621 (1983).

10

Optical Transmitters

10.1 INTRODUCTION

The transmitter is one of the major components in an optical fiber system. It usually consists of a semiconductor light source, a driving circuit or an optical modulator, and an electronic interface with the input terminal. Its main function is to convert an input electrical waveform into an identical optical waveform at a high rate. Exact conversion may not be achievable because of the inadequacy of frequency response of the device. Besides the limitation of the transmitter, fiber dispersion and receiver noise can also cause a degradation in transmitted waveform and consequently, reduce the integrity of the input. In this section we discuss the limiting factors responsible for the frequency response of the transmitter and introduce electronic circuits commonly used to modulate the LED and the LD. There are a variety of signal coding schemes used in communication systems today. The performance of various optical fiber components dictates the choice of input waveform and the modulation scheme. Only a few simple digital and analog signal codes are considered. The modulation can be obtained either by direct switching of the light source or by using an external modulator whose refractive index can be changed in the presence of an applied electric field or acoustic wave. Since the primary concern here is to transmit information, some attention is given to the information theory of optical communication. Here we shall introduce only the theoretical aspects of the transmitter, leaving the receiver portion to Chapter 11.

10.2 FREQUENCY RESPONSE

For semiconductor light-emitting devices, the required voltage drop across the junction capacitance is only about 2 to 5 V, depending on the series resistance. If the space-charged capacitance of the junction is small compared with the diffusion capacitance C_d of the device, the frequency response $\mathscr{R}(\omega)$, which is limited mainly by the carrier lifetime τ, can be expressed in terms of the ratio of the time-dependent photon flux $\phi(\omega)$ to the carrier flux $J(\omega)/e$ as

$$\mathscr{R}(\omega) = \frac{e\phi(\omega)}{J(\omega)} \tag{10.1}$$

When the diffusion capacitance C_d dominates the frequency response, the rate of change of the number of injected electrons N for a LED is dictated by a diffusion equation

$$\frac{\partial N}{\partial t} = -R_{\mathrm{sp}} + \frac{D}{e}\frac{\partial^2 J}{\partial x^2} \tag{10.2}$$

where the first term on the right-hand side is the spontaneous recombination rate R_{sp}, and is equal to N/τ. The second term is caused by the diffusion current. By separating N into two parts, one steady-state and the other frequency-dependent, we can write

$$N = N(0) + N(\omega)e^{i\omega t} \tag{10.3}$$

Substituting Equation (10.3) into the rate equation (10.2) yields an equation for the steady-state case:

$$D\frac{\partial^2 N(0)}{\partial x^2} - \frac{N(0)}{\tau} = 0 \tag{10.4}$$

and for the frequency-dependent case we have

$$D\frac{\partial^2 N(\omega)}{\partial x^2} - \frac{N(\omega)(1 + i\omega\tau)}{\tau} = 0 \tag{10.5}$$

We now define two diffusion lengths:

$$L(0) = (D\tau)^{1/2} \qquad \text{(steady-state)} \tag{10.6}$$

and

$$L(\omega) = \left(\frac{D\tau}{1 + i\omega\tau}\right)^{1/2} \qquad \text{(frequency-dependent)} \tag{10.7}$$

Substituting L into Equations (10.4) and (10.5), these two equations converge to one as given by

$$\frac{\partial^2 N}{\partial x^2} - \frac{N}{L^2} = 0 \tag{10.8}$$

Subjecting N to the boundary conditions that at $x = 0$, $N = N(0)$ and at $x = d$, $N = -L \ \partial N/\partial x \,|_d$, the frequency-dependent solution of Equation (10.8) is

$$N(\omega) = N(0) \ \frac{\cosh\left[(d-x)/L(\omega)\right] + \left[L(\omega)/L(0)\right] \sinh\left[(d-x)/L(\omega)\right]}{\cosh\left[d/L(\omega)\right] + \left[L(\omega)/L(0)\right] \sinh\left[d/L(\omega)\right]} \tag{10.9}$$

To determine the photon flux ($\phi(\omega)$), we shall integrate the recombination rate $N(\omega)/\tau$ over the width of the recombination region, and write

$$\phi(\omega) = \frac{1}{\tau} \int_0^d N(\omega) dx \tag{10.10}$$

Substituting Equation (10.9) into (10.10), we obtain

$$\phi(\omega) = \frac{N(0)}{\tau} \ \frac{L(\omega) \sinh\left[d/L(\omega)\right] + \left[L^2(\omega)/L(0)\right]\left[\cosh d/L(\omega) - 1\right]}{\cosh\left[d/L(\omega)\right] + \left[L(\omega)/L(0)\right] \sinh\left[d/L(\omega)\right]} \tag{10.11}$$

Since the frequency-dependent current flow $J(\omega)$ across the junction is simply the product of eD and the concentration gradient at the junction, we write

$$J(\omega) = eD \left.\frac{\partial N}{\partial x}\right|_{x=0} \tag{10.12}$$

Substituting Equation (10.9) into (10.12), we obtain

$$J(\omega) = -eDN(0) \frac{\left[1/L(\omega)\right] \sinh\left[d/L(\omega)\right] + \left[L^2(\omega)/L(0)\right] \cosh\left[d/L(\omega)\right]}{\cosh\left[d/L(\omega)\right] + \left[L^2(\omega)/L(0)\right] \sinh\left[d/L(\omega)\right]} \tag{10.13}$$

Substituting Equations (10.11) and (10.13) into (10.1) yields the frequency-response function as

$$\mathscr{R}(\omega) = \frac{1}{D} \ \frac{L(\omega) \sinh\left[d/L(\omega)\right] + \left[L^2(\omega)/L(0)\right] \cosh\left[d/L(\omega) - 1\right]}{\left[1/L(\omega)\right] \sinh\left[d/L(\omega)\right] + \left[1/L(0)\right] \cosh\left[d/L(\omega)\right]} \tag{10.14}$$

If $d \gg L$, Equation (10.14) reduces to a simple form

$$\mathcal{R}(\omega) \simeq \frac{L^2(\omega)}{D\tau} = \frac{L^2(\omega)}{L^2(0)} = (1 + \omega^2\tau^2)^{-1/2} \qquad (10.15)$$

The results of Equation (10.15) is an overestimate if the space-change capacitance C_s is large compared with the diffusion capacitance of the device. In a forward-biased junction with a current density J across an area A, the diffusion capacitance C_d resulting from the storage of electrons within a diffusion length of the junction interface can be written as

$$C_d = e\frac{A}{kT}J \qquad (10.16)$$

In a junction where the injection current density is of the order of 40 kA/cm², the diffusion capacitance begins to dominate the space-charged capacitance. Under normal operating conditions, current densities are typically below 40 kA/cm², so that the space charge plays an important role. This can lead to an increased lifetime, and consequently the frequency response is reduced.

Two approaches commonly used to increase the frequency response are (1) operating the device at high current densities, or (2) using heavily doped material for the active layer. Highly doped active LED devices have produced the widest bandwidth characteristics (>1 GHz). However, these devices usually have low quantum efficiency, owing to the poorer quality of heavily doped material, in which there can be a large increase in nonradiative recombination centers. The treatment described above is applicable to lasers as well. In the case of lasers, we shall replace the spontaneous rate R_{sp} with the stimulated rate R_{st}. Since the stimulated lifetime is much shorter than the spontaneous lifetime, the laser is intrinsically a faster device than the LED.

10.3 BIAS AND CONTROL CIRCUITS

The light output and the bias voltage versus injection current for a typical LED and LD is shown in Figure 10.1. These light sources can be modulated either by applying current pulses directly through a simple circuit as shown in Figure 10.2(a), or using an external electro-optic modulator, which is discussed in Section 10.5. In the case of direct modulation, the spontaneous recombination time τ_{sp} is primarily the limiting factor on the speed of response or the bandwidth of the device. The distorted waveform is equivalent to a filter version of the input current $I(t)$ with a bandwidth B of $1/2\pi\tau_{sp}$. The output $P(t)$ can be expressed as

$$P(t) = \frac{\eta h \nu}{e}\int I(t')e^{-(t-t')/\tau_{sp}}dt' \qquad (10.17)$$

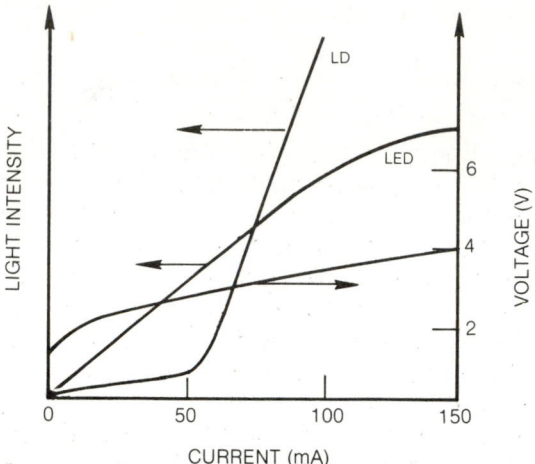

Figure 10.1   Typical L and V versus I curves for LEDs and LDs.

where η is the quantum efficiency, $h\nu$ is the photon energy, and e is the electronic charge. The use of a simple pulse network to switch a LED usually requires large currents. The currents can cause problems in the circuit design associated with the feedback noise that can interfere with the performance of other system components. Alternatively, one can use a biasing circuit as shown in Figure 10.2(b), where the LED is biased at a constant voltage, but can be switched with a transistor gate logic to turn the injection current on and off. To achieve high modulation rates, one must use a very low-impedance drive circuit to compensate for the LED capacitance, which is typically of the order of 100 pF at zero bias. Variations in the device capacitance and resistance with forward-biased currents must also be taken into account; otherwise, nonlinear effects can lead to problems associated with pattern-dependent pulse shapes and intersymbol interference. To reduce these problems, a pre-bias resistor circuit, as shown in Figure 10.2(b), has been introduced to provide a dc bias for the LED at a sacrifice of the extinction ratio.

To maintain a high degree of linearity in the output, various feedback and equilization techniques must be employed, especially for analog systems. Figure 10.2(c) shows a simple feedback control. In this circuit, a small portion of the output is captured by a local photodetector. The sampled signal is then amplified and compared with the input drive signal. With this technique it is important to assure a single-mode operation, because the $L–I$ characteristic is not necessarily the same for different modes. For this reason it is easier to adopt this technique for lasers than for LEDs.

One major difference between a laser and a LED is the threshold behavior as indicated by the light versus current curves. For this reason it is desirable to bias the laser near the threshold current to avoid the time delay necessary for

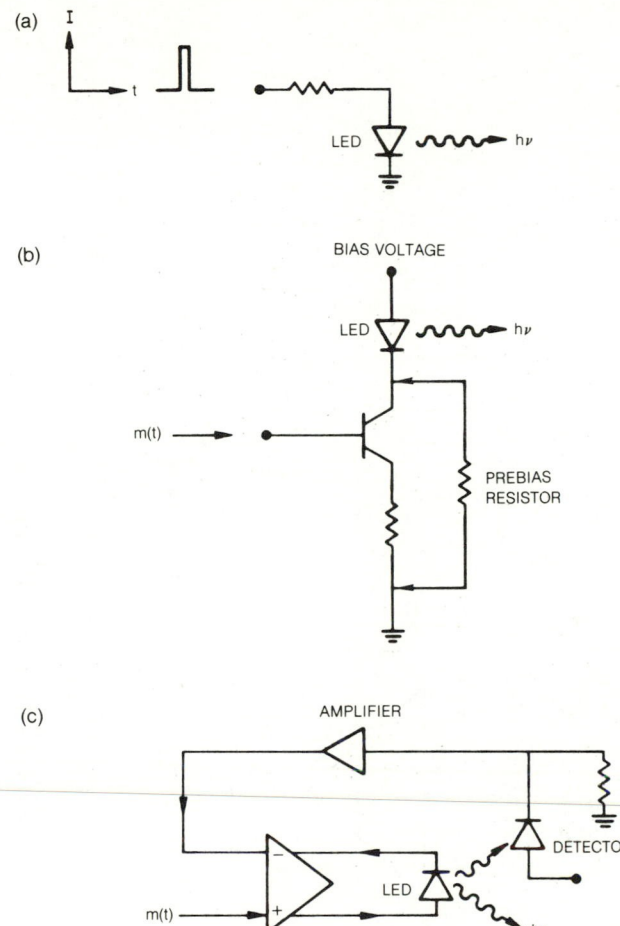

Figure 10.2 (a) Simple driving circuit for LEDs; (b) bias circuit using a transistor gate logic; (c) simple feedback stabilizing circuit.

building the current density from zero. An additional advantage of using a bias circuit is that the switching current can be made much smaller than the bias current, so that high-speed modulation drive circuits can readily be made available.

One problem associated with the pre-bias circuit for laser feedback control is the sudden surge in currents when the modulation signal is momentarily removed. As the modulation signal resumes, a temporary high level of the bias voltage could cause a catastrophic failure or burnout. To avoid this problem, a more complex circuitry involves the use of two comparators, as

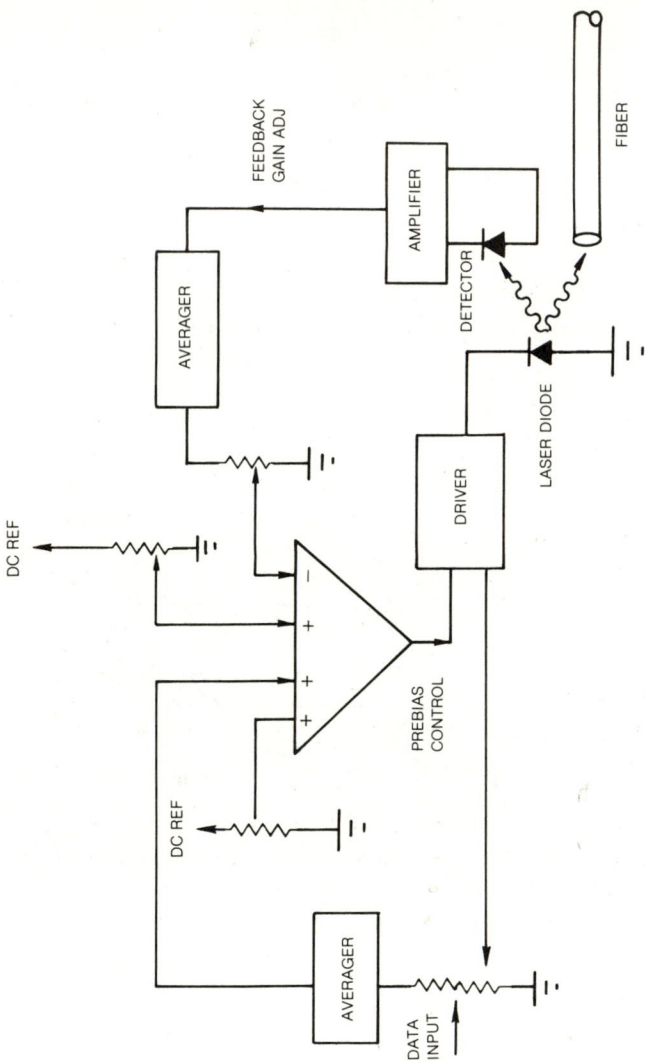

Figure 10.3 Pre-biased and compensated stabilization circuit.

shown in Figure 10.3, one monitoring the modulation signal level and the other regulating the feedback control loop. If the drive voltage is off, the average value of the signal voltage will go to zero and simultaneously an adjustment is made in the bias circuit to reduce the laser output. Such a circuit requires not only a large number of components but also very delicate balancing and calibration among various offsets, which are necessary for compensating the difference in modulation signal levels and in the local photodetector output waveforms. These offsets are interactive, resulting in a very complex alignment procedure.

10.4 DIGITAL AND ANALOG CODES

Either digital or analog signals can be used to code an optical carrier for optical fiber systems. The digital coding involves a variety of pulse-code modulation (PCM) formats and can be decoded simply by means of direct detection. For example, in a simple binary pulse code (0 or 1), the only requirement imposed on the receiver is to determine whether a signal is above or below threshold. This is not true in the case of an analog system, for which the receiver must reproduce as closely as possible the waveform, the frequency, or the phase of the input signal. Although direct detection can also be used for simple analog systems [e.g., the intensity modulation (IM)], heterodyne detection is often employed in systems where the modulation involves either a frequency or a phase-shift key. Generally speaking, analog systems require a higher degree of spectral purity and system linearity than do digital systems. Furthermore, system components required for a heterodyne receiver are considerably more complex than those used in a direct detection system. In this section we consider only the simplest digital and analog systems—the binary code and the IM—both of which can be regarded as some sort of intensity modulation. The former is usually clocked and the latter is unclocked. Because of the nature of their error occurrence, the fidelity of the transmitted signal for the former is dictated by the bit error rate (BER) and for the latter is determined by the signal-to-noise (S/N) ratio. These topics are discussed in detail in Chapter 11.

A binary signal waveform is shown in Figure 10.4(a), where a well-defined time slot Δt is assigned to a pulse that can be either present or absent. This represents a clocked binary signal which is synchronized with a constant periodic reference signal. If the pulse width is less than Δt, the signal is said to be "return to zero (RZ)." Otherwise, the signal is called "not return to zero (NRZ)." In practice, signals are always degraded and contain errors. For example, within a time slot, there can be a false alarm for the presence of an output that is not supposed to be there. These errors are introduced as a result of transmitter and receiver noise, fiber dispersion, and imperfections in electronic circuitry. The latter can result in time jitter [Figure 10.4(b)], waveform distortion [Figure 10.4(c)], baseline wander [Figure 10.4(d)], and so on. The rate at which these errors occur is called the bit error rate (BER). One way to

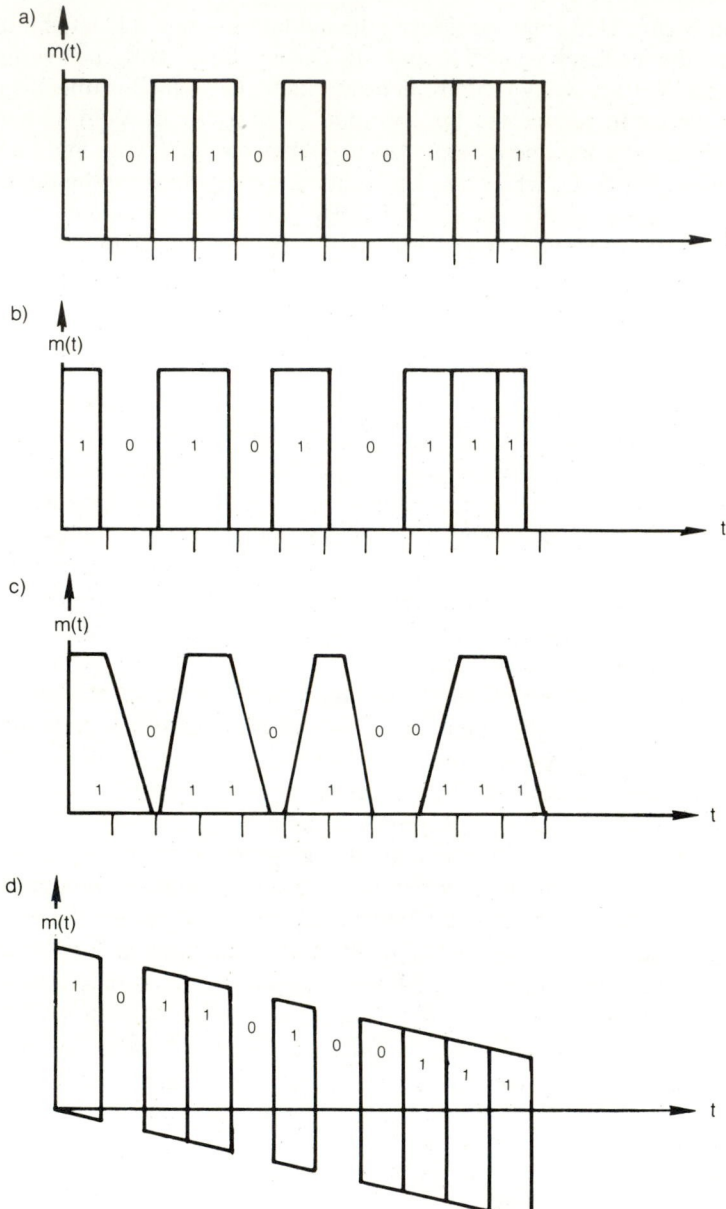

Figure 10.4 (a) Clocked binary signal; (b) binary signal with time jitter; (c) distorted waveforms; (d) waveform with baseline wander.

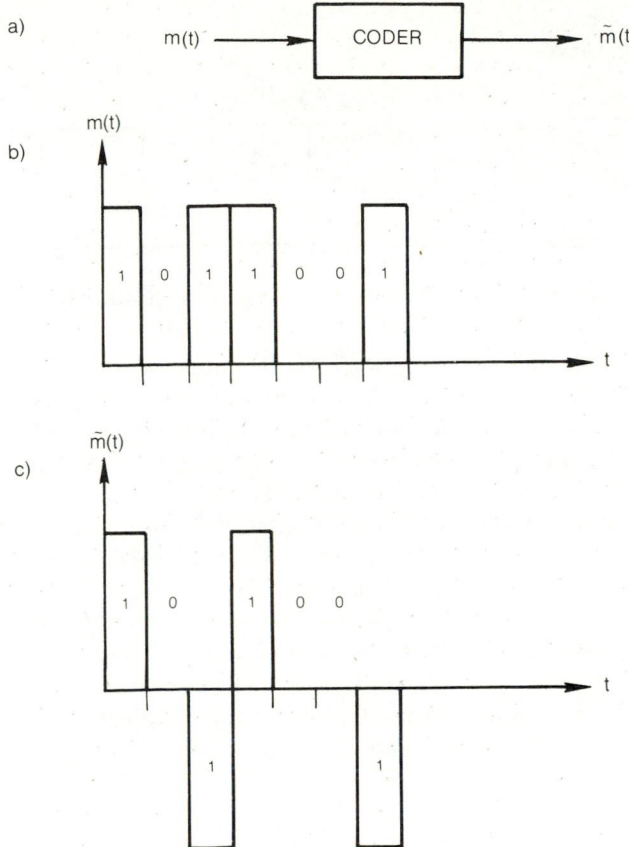

Figure 10.5   (a) Function of a bipolar coder that transforms $m(t)$ to $\tilde{m}(t)$; (b) input waveform $m(t)$; (c) output waveform $\tilde{m}(t)$.

reduce BER is to use a coder, whose function is to transform the input waveform into a form more suitable for transmission. In this case a decoder is required in the receiver system to convert the signal back to its original waveform. One example is the bipolar coder, illustrated in Figure 10.5. This coding scheme can alleviate the problem of baseline wander commonly existing in an ac coupled filter, through which a rectangular pulse is distorted with a long tail of opposite polarity. The bipolar coder converts the input pulses alternately into positive and negative pulses. As a result, this coding scheme creates a cancellation in the tail of ac coupled pulses with opposite polarity. It is rather easy to decode this format at the receiving end. There are many other coding schemes besides PCM [e.g., pulse position modulation (PPM), Manchester code, etc.]. The details and trade-offs of these schemes can be found in the literature. For many low-data-rate applications, unclocked PCM signals are found to be adequate.

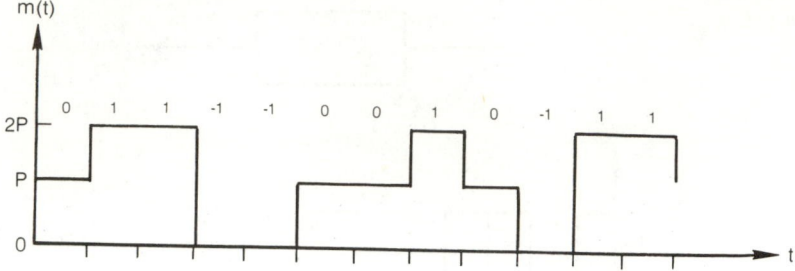

Figure 10.6 Typical alternative mark inversion AMI signal waveforms.

The basic requirement for a PCM system is the time synchronization of transmitted signals. This is usually done by providing timing information in the transmitted signals. However, such timing information can easily be lost, if, for example, the signal contains a long series of zeros. For this reason ternary line codes rather than binary are often used. The simplest code that can ease the clock extraction is the so-called "alternative mark inversion" AMI coding. In this code, a power level P represents zero, while $2P$ and 0 represent $+1$ and -1, respectively. The typical AMI signal waveform is shown in Figure 10.6. The main disadvantage of this coding is the relatively stringent stability requirement on the transmitter power. The next level of complexity in coding is the 1B2B (one bit represented by two bits) code. Examples of this code are the "complemented mark inversion" (CMI) and the "bi-polar phase" (BP), shown in Figure 10.7. All these codes are inherently well balanced in such a way that

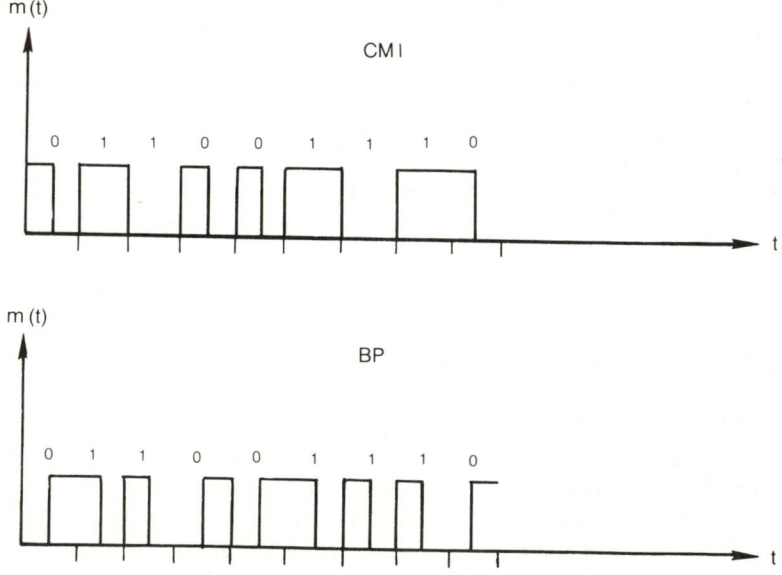

Figure 10.7 Typical complemented mark inversion CMI and bipolar BP waveforms.

there are no long runs of zeros or ones, as illustrated by these waveforms. The penalty for these codes is the doubling of the modulation rate, which raises the system bandwidth requirement.

The simplest analog modulation format is the direct intensity modulation (IM). The analog input $m(t)$ is used to modulate the source with an output $P_m(t)$ as given by

$$P_m(t) = P_0[1 + \gamma m(t)] \tag{10.18}$$

where P_0 is the average source output power and γ is the modulation index. A typical analog transmission link is shown in Figure 10.8. The modulation signal $m(t)$ is used as the input to drive the transmitter. The simplest way is to switch the light source directly. A more sophisticated but expensive way is to obtain $P_m(t)$ by using an external modulator, which can be either electro-optic, acousto-optic, or magneto-optic. The output of this transmission link typically contains the modulated signal plus an error signal which is caused by the noise and nonlinearities of the system. Therefore, the system fidelity is measured by the signal-to-noise ratio.

The rate for direct modulation depends on the response time of the source and also on the time constant of the circuit that provides the drive current. In the

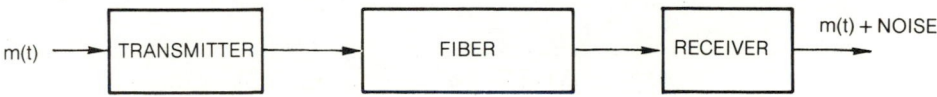

Figure 10.8   Typical analog transmission link.

case of LEDs, a modulation rate up to 20 megabits per second (Mb/s) is considered to be routine. With careful selection of LED devices and circuits, a modulation rate as high as 100 Mb/s can be achieved; it is ultimately limited by the spontaneous lifetime or the recombination time, as discussed in Section 10.2. In the case of LDs, modulation speeds beyond 1 GHz have been obtained. The capability of very high-data-rate communication systems with laser diodes over LEDs is derived from a faster time constant, which is a fundamental characteristic of stimulated emission rather than spontaneous emission.

10.5 MODULATION

The use of an external modulator in a transmitter has a distinct advantage over direct modulation because it relieves the burden imposed on the light source. By passing a CW source through a modulator, a subcarrier that contains information can be generated by applying a voltage on the modulator crystal which can be either electro-optic, acousto-optic, or magneto-optic. In this section we only discuss the electro-optic effect, which is the most widely used technique to produce intensity, frequency, and phase modulation. There are many electro-optic crystals useful in the wavelength region of interest: for example, potassium dihydrogen phosphate (KDP), ferroelectric peroskites such as $LiNbO_3$ and $LiTaO_3$, and cubic crystals such as GaAs and CdTe. For more information on electro-optic crystals, the review paper by Kaminow and Turner (Ref. 10.1) is very informative.

Electro-optic modulation of light is based on a linear electro-optic effect in crystals whose refractive index is changed upon application of an electric field. In general, the optical properties of a crystal are described in terms of an index ellipsoid or indicatrix. The equation of indicatrix along the principal axis x_i is

$$\sum_{i=1}^{3} \frac{x_i^2}{n_i^2} = 1 \tag{10.19}$$

where n_i is the principal refractive indices. For example, if an E field is applied along x_1 and the optical carrier is propagating along x_3, a change in refractive index occurs for both n_1 and n_2, depending on the direction of the polarization vector of the carrier. The new indices $(n_1 + \Delta n_1)$ and $(n_2 + \Delta n_2)$ become

$$\frac{1}{(n_1 + \Delta n_1)^2} = \frac{1}{n_1^2} + \gamma_{11} E_1 \tag{10.20}$$

$$\frac{1}{(n_2 + \Delta n_2)^2} = \frac{1}{n_2^2} + \gamma_{21} E_1 \tag{10.21}$$

where γ_{11} and γ_{21} are electro-optic coefficients associated with the orientation of the interacting fields. If $\Delta n \ll n$, Equations (10.20) and (10.21) reduce to

$$\Delta n_1 = -\tfrac{1}{2} \, n_1^3 \gamma_{11} E_1 \tag{10.22}$$

$$\Delta n_2 = -\tfrac{1}{2} \, n_2^3 \gamma_{21} E_1 \tag{10.23}$$

The situation becomes more complex if the directions of the optical polarization and the E field are not oriented along the principal axes of the crystal. In this case the equation of indicatrix must be expressed in terms of generalized quadratic electro-optic tensor γ_{ijk}, as given by

$$\sum_{i,j,k} \left(\frac{1}{n_{ij}^2} + \gamma_{ijk} E_k \right) x_i x_j = 1 \tag{10.24}$$

By the usual contraction, $\gamma_{ijk} \leftrightarrow \gamma_{lm}$, where $l = 1, \ldots, 6$, and $m = 1, 2, 3$. i, j are related to l as follows: $1, 1 \leftrightarrow 1$; $2, 2 \leftrightarrow 2$; $3, 3 \leftrightarrow 3$; $2, 3 \leftrightarrow 4$; $3, 1 \leftrightarrow 5$; and $1, 2 \leftrightarrow 6$. Therefore, in general there exist a total of 18 electro-optic coefficients. Because of crystal symmetry, the number of electro-optic coefficients can be greatly reduced. For KDP (tetragonal class $\overline{4}2m$), all coefficients are zero with the exception of γ_{41}, γ_{52}, and γ_{63}. In the case of cubic crystals such as GaAs and CdTe ($\overline{4}3m$), the nonvanishing coefficients are $\gamma_{41} = \gamma_{52} = \gamma_{63}$. For LiNbO$_3$ and LiTaO$_3$ (trigonal class, 3m), the electro-optic tensor is more complex and has a total of eight nonvanishing tensor components: $\gamma_{11} = -\gamma_{12} = -\gamma_{62}$, $\gamma_{51} = \gamma_{42}$, $\gamma_{13} = \gamma_{23}$, and γ_{33}. Table 10.1 gives values of refractive indices and nonvanishing electro-optic coefficients of some crystals. For KDP we obtain the equation of indicatrix in the presence of an E field, $\mathbf{E} = E_1 i + E_2 j + E_3 k_3$, as

$$\frac{x_1^2}{n_0^2} + \frac{x_2^2}{n_0^2} + \frac{x_3^2}{n_e^2} + 2\gamma_{41}E_1 x_2 x_3 + 2\gamma_{41}E_2 x_1 x_3 + 2\gamma_{61}E_3 x_1 x_2 = 1 \tag{10.25}$$

where $n_0 = n_1 = n_2$ and $n_e = n_3$ are the ordinary and the extraordinary indices of this uniaxial crystal. Equation (10.25) can be simplified by a proper choice of the coordinate system (x_1', x_2', x_3') for crystal orientation such that the cross terms $x_i x_j$ vanish. For simplicity, if we apply the E field along x_3 axis, Equation (10.25) reduces to

$$\frac{x_1^2 + x_2^2}{n_0^2} + \frac{x_3^2}{n_e^2} + 2E\gamma_{63}x_1 x_2 = 1 \tag{10.26}$$

By a 45° rotation of the plane formed by x_1 and x_2, we obtain the new coordinate system (x_1', x_2', x_3) using the following transformation:

$$x_1 = \frac{\sqrt{2}}{2}(x_1' - x_2')$$

$$x_2 = \frac{\sqrt{2}}{2}(x_1' + x_2') \tag{10.27}$$

**TABLE 10.1   Refractive Indices and Electro-optic Coefficients
of Some Crystals**

Material	λ (μm)	n_e	n_0	γ_{ij} (10^{-12} m/V)
KDP	0.5	1.472	1.514	$\gamma_{41} = \gamma_{52} = 8.6$, $\gamma_{63} = 9.5$
LiNbO$_3$	0.5	2.245	2.344	$\gamma_{13} = \gamma_{23} = 9.0$, $\gamma_{22} = -\gamma_{12} = -\gamma_{61} = 6.6$
				$\gamma_{42} = \gamma_{51} = \gamma_{33} = 30$
GaAs	0.8–10	3.6–3.3		$\gamma_{41} = \gamma_{52} = \gamma_{63} = 1.2$
Quartz	0.6	1.553	1.544	$\gamma_{11} = -\gamma_{21} = -\gamma_{62} = -0.47$
				$\gamma_{41} = -\gamma_{52} = -0.2$

Substituting Equation (10.27) into (10.26), we obtain a simple equation of indicatrix as

$$\frac{x_1'^2}{n_1'^2} + \frac{x_2'^2}{n_2'^2} + \frac{x_3^2}{n_e^2} = 1 \tag{10.28}$$

which is an ellipsoid in its principal axes in the presence of the E field, provided that

$$\frac{1}{n_1'^2} = \frac{1}{n_0^2} + \gamma_{63}E \tag{10.29}$$

$$\frac{1}{n_2'^2} = \frac{1}{n_0^2} - \gamma_{63}E \tag{10.30}$$

We can rewrite Equation (10.29) as follows:

$$\begin{aligned} n_1' &= n_0(1 + n_0^2\gamma_{63}E)^{-1/2} \\ &= n_0[1 - \tfrac{1}{2} n_0^2\gamma_{63}E + O(\gamma_{63}^2)] \end{aligned} \tag{10.31}$$

Therefore, the change of refractive indexes Δn_1 can be approximated by

$$\Delta n_1 = -\tfrac{1}{2} n_0^3\gamma_{63}E \tag{10.32}$$

and similarly, we get

$$\Delta n_2 = \tfrac{1}{2} n_0^3\gamma_{63}E \tag{10.33}$$

Figure 10.9 shows amplitude modulators that use in part (a) a KDP and in part (b) a GaAs electro-optic crystal. In both cases, the amplitude modulator consists of a properly oriented crystal placed between two crossed polarizers, with the E field applied along the x_3 axis.

When a linearly polarized optical field enters the modulator as shown in Figure 10.9(a), it resolves into two orthogonal components along x_1' and x_3. Because of the difference in refractive index between these two components, a phase difference ϕ is developed as they propagate along the length L of the modulator. This phase difference contains two terms as given by

$$\phi = \frac{2\pi}{\lambda} L \left[(n_0 - n_e) - \frac{n_0^3}{2} \gamma_{63} \frac{V}{d} \right] \tag{10.34}$$

in which the first term $(n_0 - n_e)$ is independent of the E field and is due to the natural birefringence of the crystal, and the second terms is E-field dependent. The depth of modulation increases with increasing L and decreasing d. As these two components emerge from the modulator, they will interfere with each other because of a phase difference ϕ existing between them. If we let

$$E_1 = E_0 \quad \text{and} \quad E_2 = E_0 e^{-i\phi}$$

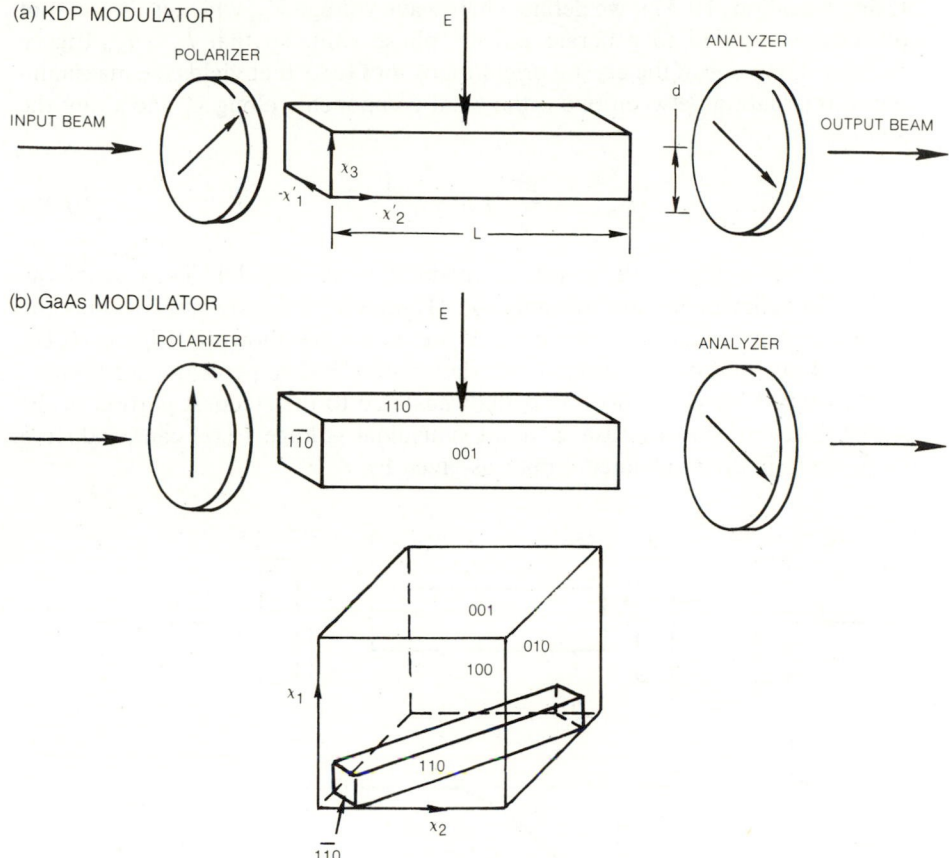

Figure 10.9   Electro-optic amplitude modulators using (a) a KDP crystal and (b) a GaAs crystal.

the resultant of these two complex fields is

$$E = \frac{E_0}{\sqrt{2}} (e^{-i\phi} - 1) \tag{10.35}$$

and the corresponding output density is

$$I_m \propto EE^* = 2E_0^2 \sin^2 \frac{\phi}{2} \tag{10.36}$$

The fractional intensity transmitted through the modulator is

$$\frac{I_m}{I_0} = \sin^2 \frac{\phi}{2} \tag{10.37}$$

Using Equation (10.37), we define a half-wave voltage V_π, which is the amount of voltage required to generate a 180° phase shift, so that $I_m = I_0$. Figure 10.9(b) shows one of the crystal orientations for GaAs that yields the maximum phase retardation between two orthogonal components along x_1' and x_2' by the amount

$$\phi_{max} = \frac{2\pi}{\lambda} L n^3 \gamma_{41} \frac{V}{d} \tag{10.38}$$

At low voltages, the depth of modulation is very low because of the $\sin^2(\phi/2)$ behavior for the transmission. Therefore, it is advisable to bias the modulator with a quarter-wave plate at the output, as shown in Figure 10.10. This $\lambda/4$ plate introduces a fixed retardation of $\pi/2$. The percent transmission, as shown in Figure 10.10(b), is being translated to the steepest portion of the $\sin^2(\phi/2)$ curve. In this case, a small sinusoidal voltage $V_m(t)$ can produce a large sinusoidally modulated output as given by

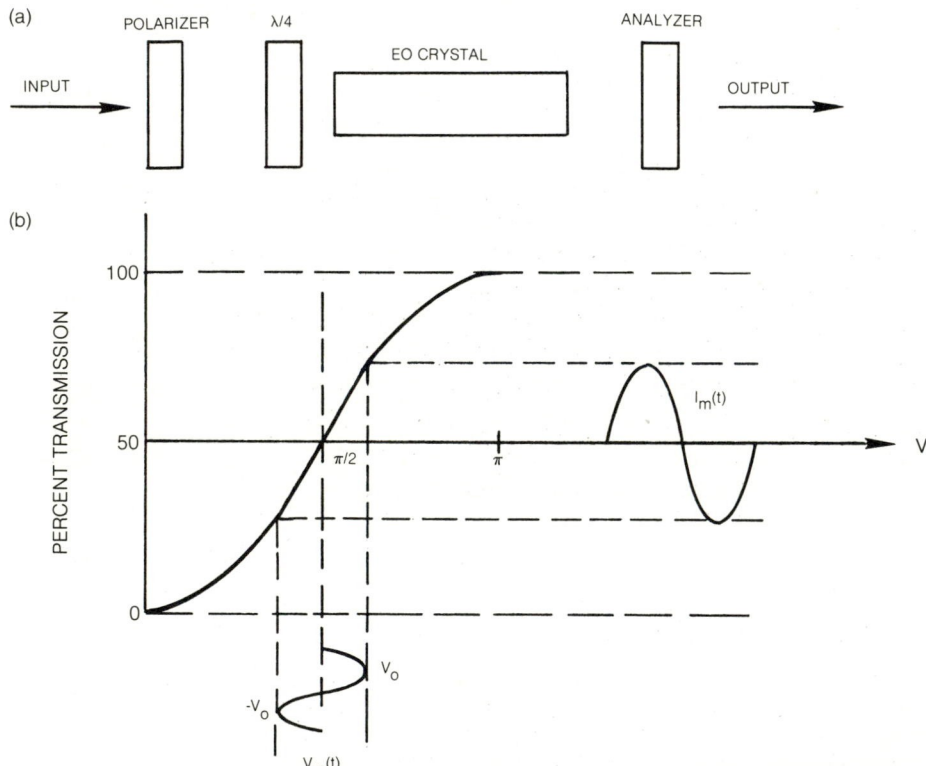

Figure 10.10 (a) Electro-optic amplitude modulator biased with a $\lambda/4$ plate; (b) percent transmission versus applied voltage for an electro-optic modulator biased with a $\pi/2$ phase retardation.

$$\frac{I_m}{I_0} = \sin^2\left(\frac{\pi}{4} + V_0 \sin \omega_m t\right)$$

$$= \tfrac{1}{2} + \tfrac{1}{2} \sin[V_0 \sin \omega_m t] \tag{10.39}$$

For small V_m, Equation (10.39) can be approximated by

$$\frac{I_m}{I_0} \simeq \tfrac{1}{2} + \tfrac{1}{2} V_0 \sin \omega_m t \tag{10.40}$$

To generate a phase shift in the light, it is only necessary to align the polarization vector along one of the induced birefringent axes, instead of producing two equal components as for the case of amplitude modulation. The phase shift $\Delta\phi$ is directly proportional to the change in refractive index Δn as

$$\Delta\phi = \frac{2\pi}{\lambda} L \Delta n \tag{10.41}$$

With a sinusoidal modulation $\Delta\phi \sin \omega_m t$, the output will contain both the upper and lower sidebands, which can be expressed in terms of Bessel functions of $\Delta\phi$ as

$$E_{\text{out}} = E_0 \cos(\omega t + \Delta\phi \sin \omega_m t)$$

$$= E_0 \left[J_0(\Delta\phi) \cos \omega t + \sum_{n=1}^{\infty} J_n(\Delta\phi) \cos(\omega + n\omega_m)t \right. \tag{10.42}$$

$$\left. + \sum_{n=1}^{\infty} J_n(\Delta\phi) \cos(\omega - n\omega_m)t \right]$$

The output spectrum is very complex and must be analyzed using either a tunable filter or a heterodyne receiver. If a circularly polarized light is used, it is possible to generate a single sideband with very high power conversion efficiency. The mathematics, however, is rather involved and is omitted.

For a reactive load, the power per unit bandwidth required to derive an electro-optic modulator is given by

$$\frac{P}{B} = \tfrac{1}{2} \left(\tfrac{1}{2} \varepsilon E_0^2\right) d^2 L \tag{10.43}$$

where the first factor ½ is a result of taking the mean-square energy of a harmonic field, the second factor (in parentheses) is a term representing the energy density of the peak field strength E_0, and $d^2 L$ is the volume of the modulator. We shall use the GaAs amplitude modulator as an example to calculate the power required for producing maximum phase retardation. Substituting Equation (10.38) into (10.43), we get

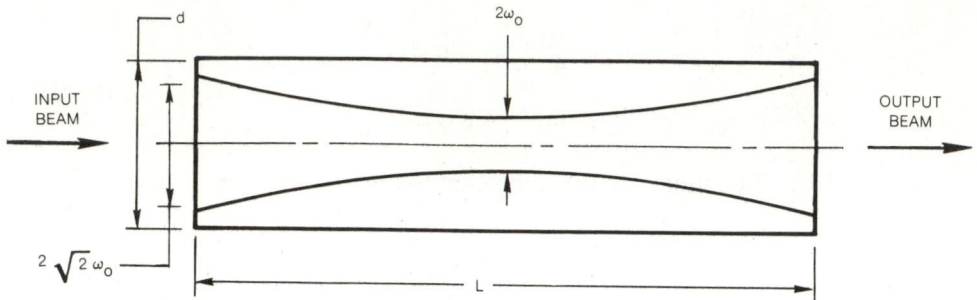

Figure 10.11 Optimum configuration for a Gaussian beam in an electro-optic modulator crystal.

$$\frac{P}{B} = \frac{\pi \varepsilon}{2} \frac{d^2 \phi_m^2 \lambda^2}{L n^6 \gamma_{41}^2} \tag{10.44}$$

To make the modulator more efficient, it is necessary to collimate the beam with two lenses such that the beam passes the modulator crystal in a confocal configuration, as shown in Figure 10.11. In this case a Gaussian beam diameter at the center of the modulator rod is $2\omega_0$ and at the ends is $2\sqrt{2}\,\omega_0$, where

$$\omega_0^2 = \frac{\lambda L}{2\pi n} \tag{10.45}$$

To minimize the difficulty of beam alignment, it is necessary to introduce a safety factor S such that

$$d = S\sqrt{8}\,\omega_0 \tag{10.46}$$

Combining Equations (10.45) and (10.46), we get

$$\frac{d^2}{L} = \frac{S^2 4\lambda}{n\pi} \tag{10.47}$$

Substituting Equation (10.47) into (10.44), we write

$$\frac{P}{B} = \varepsilon \frac{S^2 \lambda^3 \phi_m^2}{n^2 \gamma_{41}^2} \tag{10.48}$$

Equation (10.48) indicates that the modulator driving power increases rapidly with increasing wavelength. For long-wavelength sources, the high power requirement for the modulator may be an important factor that limits the system performance.

10.6 NOISE CHARACTERISTICS

In Section 10.4 we introduced several types of transmitter waveform distortions that can affect the fidelity of a communication system. In this section we discuss several noise sources that, combined with the waveform distortion, set the fundamental limitation on the information-carrying capability of the transmitter. The noise sources to be treated here are (1) the intrinsic noise of the laser source, (2) the partition noise, and (3) the noise resulting from interaction between the source and the fiber. Because the noise phenomena are random in nature, statistical methods must be used to deal with communication problems, which will be introduced in Section 10.7.

The intrinsic noise of semiconductor lasers originates from quantum statistical processes inside the cavity, most of which have been discussed previously. One dominant noise source, known as the shot noise, is caused by the injection current. Other noises are due to the randomness of the spontaneous recombination of carriers within the active layer and the statistical nature of absorption, scattering, and stimulated emission processes (see Chapter 7).

The measured intrinsic signal-to-noise (S/N) ratios are shown in Figure 10.12 as a function of injection current at $\omega = 50$ MHz and with a bandwidth $B = 10$ MHz for an index-guided CSP laser and a gain-guided V-groove laser. The S/N ratio reaches a minimum value at a point slightly above the threshold, indicating that this minimum point represents the state of maximum noise occurring at or near the threshold of laser oscillation. The smoother the transition from the nonlasing to the lasing state, the lower is the noise at threshold. For this reason, the noise is lower in a gain-guided laser than in an index-guided laser near the threshold. For injection currents sufficiently above the threshold ($I/I_{th} > 1.2$), the index-guided laser exhibits a better S/N ratio than does the gain-guided laser. In either case S/N is better than 70 dB for a noise bandwidth at 10 MHz. In actual systems, S/N is lower by a certain amount, which is determined by the modulation index.

Besides the intrinsic noise and waveform distortions, there are other problems, such as partition noise and power fluctuation, due to the interaction between the laser and the fiber. The partition noise is due to fluctuations among different lasing modes. Clearly, there would be no partition noise if the laser oscillates only in a single longitudinal mode. If the laser has two or three modes, the partition noise is greatest, because the total photon density is shared among these modes, and a slight perturbation can cause a significant amount of energy transfer between modes. On the other hand, if the laser output contains a large number of longitudinal modes, the partition photon density is relatively small in each mode. In this case, the spontaneous emission factor for each mode becomes relatively large, and can provide a natural damping of the modal fluctuation. However, a situation could occur in which a single-mode laser, when modulated by injection currents, could be forced to oscillate in multimodes, thus introducing partition noise.

Noise can also be induced by feedback as a result of reflection from the end face of a fiber. This problem can be treated by analyzing two coupled cavities. One of the two cavities is the laser itself and the other is the external cavity formed between the output facet of the laser and the end face of the fiber. The interaction between these two cavities can cause a considerable change in the emission spectra. A distinction must be made between the reflections from the near end and those from the far end. In the case of near-end reflection, a low-frequency noise usually occurs and is caused by fluctuations of either the laser cavity or the external cavity. The power spectrum of these fluctuations typically extends up to several kilohertz. In the case of far-end reflection, the submode spacing corresponding to the long external cavity is very narrow, so that a single longitudinal mode of the laser could break up into a number of submodes, which is a phenomenon commonly known as mode locking. Due to this self-modulation, the spectral envelope of the submodes will be considerably

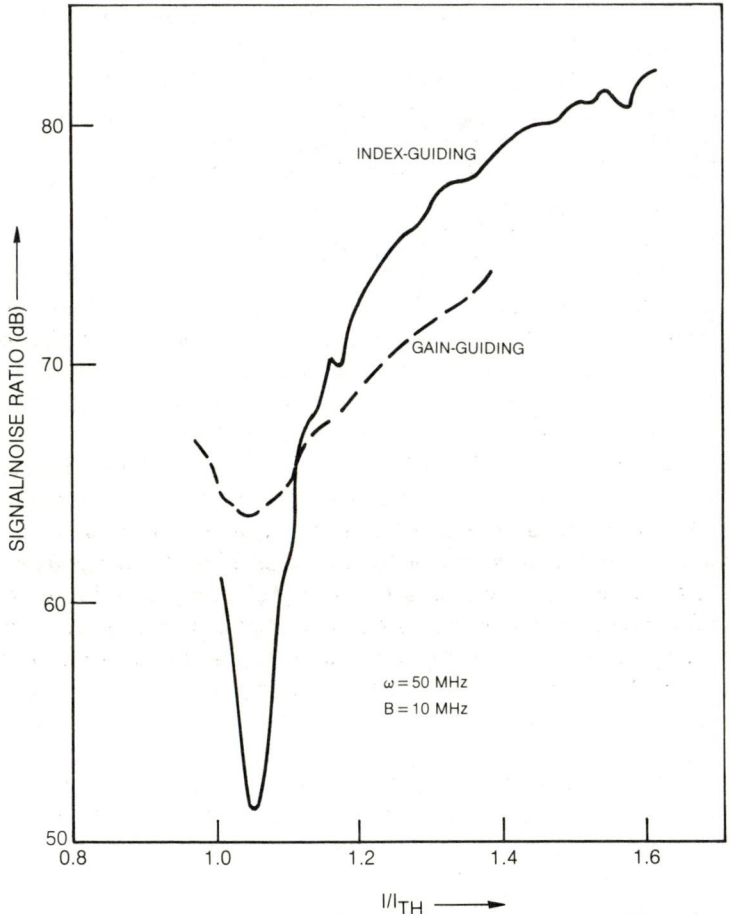

Figure 10.12 Measured signal-to-noise ratio (S/N) as a function of normalized injection current I/I_{th} for both index-guided and gain-guided semiconductor lasers. (From Ref. 10.2. Reprinted with permission of IEEE, © 1982.)

broadened. Self-pulsation can also occur if the inverse round-trip time is approximately equal to the relaxation frequency ν_0. The amount of reflection that the system can tolerate depends critically on the degree of coherence of the laser. For a laser cavity of 400-μm length, the round-trip time is of the order of 10 to 15 ps. If the front end of the fiber is located very close to the laser output facet, such that the cavity mode of a single-longitudinal-mode laser coincides with the external cavity mode, the natural line width could be significantly narrowed. However, this is not a very stable situation, because a slight change in cavity length of one of the cavities can result in a significant change in the wavelength of the laser. In both cases, interaction occurs if the end reflection R is greater than the amplitude of the spontaneous emission within the laser cavity. A typical value for the power reflection coefficient R to produce laser instability is about 10^{-4} for an index-guided laser and about 3×10^{-3} for a gain-guided laser. To avoid these complications, an optical isolator is required for high-data-rate optical communication systems.

10.7 ASPECTS OF COMMUNICATION THEORY

The inability to predict exact frequency and amplitude of an output from a receiving channel requires one to make only a best estimate of the message generated by the transmitter. Statistical averaging using probability theory is therefore a necessary process for many situations involving randomly varying phenomena such as random waveforms and noise problems in a communication system. In this section some basic conceptual aspects of communication theory are introduced. A more comprehensive discussion of this subject can be found in many texts (e.g., Ref. 10.3).

We shall first introduce the concept of a probability density function $p(x)$. In the real world we can obtain an average value by performing either a series of measurements on a system over a long period of time or a simultaneous set of measurements on a large number of identical systems. (If these experiments are performed carefully, they should lead to the same result.) The latter is known as the ensemble method and provides an elegant basis for establishing a mathematical model. This avoids the extended time required to complete the experiment. To assure that the two processes noted above are equivalent, new statistics involving an ergodic process must be considered. This is a matter of theoretical interest and will not be treated here.

Let x represent a typical measured value that defines a point at a corresponding distance from a fixed reference point on a straight line. If we divide this line into many small equal intervals of length Δx and count the number of points that fall in each interval, we obtain a probability density function $p(x)$ as defined by

$$p(x) = \lim_{\Delta x \to 0} \frac{\text{number of points in } \Delta x \text{ at } x}{N \, \Delta x}$$

where N is the total number of points.

The probability that the value falls within a specified range, say x_1 to x_2, is

$$P(x_1 < x < x_2) = \int_{x_1}^{x_2} p(x) \, dx$$

The probability that the value is less than a specified value, x, is

$$P(x) = \int_{-\infty}^{x} p(x) \, dx \tag{10.49}$$

The average value of a physical quantity $F(x)$ that satisfies the distribution function $p(x)$ is

$$\langle F(x) \rangle = \int_{-\infty}^{\infty} F(x) p(x) \, dx \tag{10.50}$$

The average value of x^n, which is known as the nth moment of the distribution, is defined by the expression

$$\langle x^n \rangle = \int_{-\infty}^{\infty} x^n p(x) \, dx \tag{10.51}$$

There are many types of probability density functions that describe the distribution of the ensemble of interests. The two distributions used most frequently in noise theory are the Poisson and Gaussian distribution functions. The probability density function of a Gaussian distribution is of the form

$$p(x) = \frac{1}{\sigma \sqrt{2\pi}} \exp\left[-\frac{(x - x_0)^2}{2\sigma^2} \right] \tag{10.52}$$

In Equation (10.52), the parameters have been adjusted such that $P(x)$ is normalized to give

$$\int_{-\infty}^{\infty} p(x) \, dx = 1 \tag{10.53}$$

The mean value $\langle x \rangle$ is x_0 and the variance

$$\sigma^2 = \langle x^2 \rangle - x_0^2 \tag{10.54}$$

Figure 10.13(a) and (b) show the nature of Gaussian functions for $p(x)$ and $P(x)$, respectively. Substituting Equation (10.52) into (10.49), we have

$$P(x) = \tfrac{1}{2} [1 + F(z)] \tag{10.55}$$

where $F(z)$ is defined by

$$F(z) = \frac{2}{\sqrt{\pi}} \int_0^z e^{-t^2} \, dt \quad \text{and} \quad z = \frac{x - x_0}{\sqrt{2\sigma}} \tag{10.56}$$

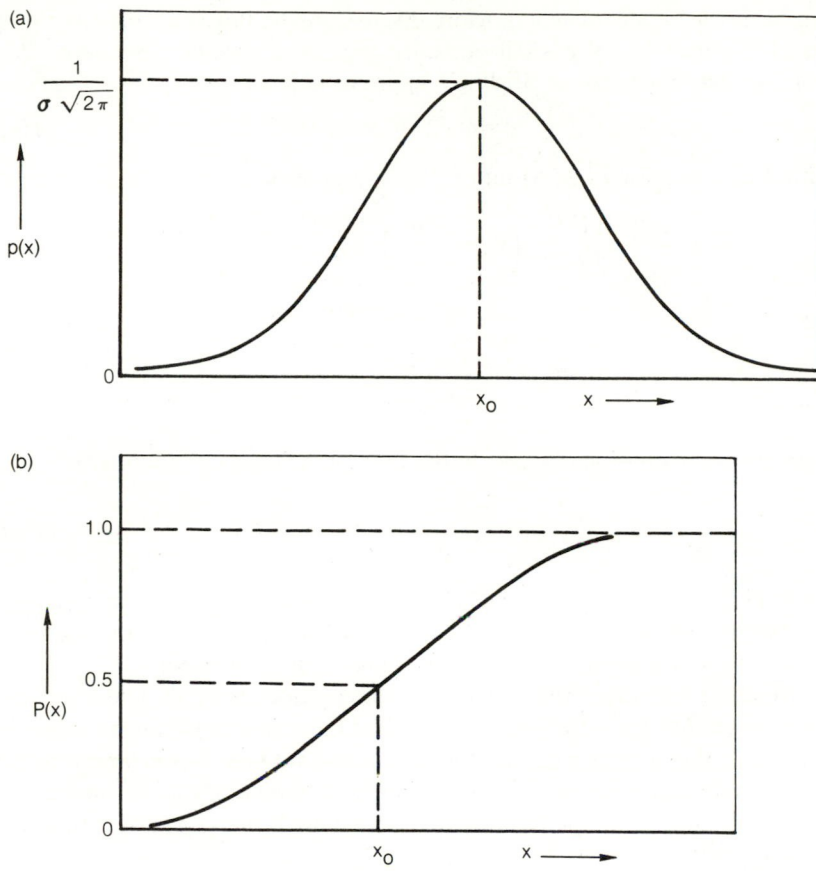

Figure 10.13 (a) Probability density function $p(x)$ of a Gaussian distribution; (b) probability function $P(x)$ of the corresponding distribution.

An important theorem in statistics, called the central limit theorem, shows in a very general way that the distribution of an ensemble of infinitely large numbers of independently and randomly distributed quantities must approach the Gaussian distribution. Almost all physical phenomena involving noise obey this theorem. A proof of this theorem is beyond the scope of this book.

The number of photons detected in a certain period of time is discrete, even though the light intensity is a continuous quantity. The probability $p(n)$ of detecting n photons is statistically independent of the number of photons previously detected and over a given time period obeys the Poisson distribution, which is given by the expression

$$p(n) = \frac{\langle n \rangle^n \exp(-\langle n \rangle)}{n!} \qquad (10.57)$$

where $\langle n \rangle$ is the mean value of n. More discussions on Poisson's distribution are given in Chapter 11. We shall consider only one special case here. When $n \gg 1$, we can make use of Stirling's approximation,

$$n! = \sqrt{2\pi n}\ n^n e^{-n} \tag{10.58}$$

Substituting Equation (10.58) into (10.57) and letting $n = \langle n \rangle + \delta n$, we get

$$p(n) = \frac{1}{\sqrt{2\pi\langle n \rangle}} \left(1 + \frac{\delta n}{\langle n \rangle} \right)^{-(\langle n \rangle + \delta n + 1/2)} e^{\delta n} \tag{10.59}$$

Since

$$\lim_{\langle n \rangle \to \infty} \left(1 + \frac{\delta n}{\langle n \rangle} \right)^{-(\langle n \rangle + \delta n + 1/2)} = e^{-\delta n - (\delta n)^2/2\langle n \rangle} \tag{10.60}$$

we can rewrite Equation (10.59) in the limit that $\langle n \rangle$ is very large as

$$p(\langle n \rangle) = \frac{1}{\sqrt{2\pi\langle n \rangle}}\ e^{-(\delta n)^2/2\langle n \rangle} \tag{10.61}$$

Equation (10.61) indicates that when the average number of photoelectrons is large, the fluctuations approach a Gaussian distribution about the mean with $\langle n \rangle = \sigma^2$. This is a good example of the central limit theorem.

Next, we introduce the concept of convolution integral. There are many physical situations for which we wish to relate the statistics of the measured sum of components in a system to the statistics of the individual contributions. If we let $p_1(x_1)$ and $p_2(x_2)$ be the probability density functions of two independent quantities x_1 and x_2, the probability density function of x_3, which is the sum of the two quantities (e.g., $x_3 = x_1 + x_2$), is

$$p_3(x_3) = \int_{-\infty}^{\infty} p_1(x_1)p_2(x_3 - x_1)\ dx_1 \tag{10.62}$$

Equation (10.62) is commonly known as the convolution integral, and has the familiar form of the Fourier integral in linear network analysis.

As shown in Chapter 6, the probability density function $p(x_3)$ for the sum of two independent variables is the inverse Fourier transform of the product of the Fourier transforms of the individual probability density functions. In statistics this function is called the characteristic function. If we let

$$P_1(t) = \int_{-\infty}^{\infty} p_1(x_1)e^{ix_1 t}\ dx_1 \tag{10.63}$$

and

$$P_2(t) = \int_{-\infty}^{\infty} p_2(x_2)e^{ix_2 t}\ dx_2 \tag{10.64}$$

then the characteristic function of the variable x_3 is equal to the product of the two Fourier transforms as

$$P_3(t) = P_1(t)P_2(t) \tag{10.65}$$

Extending this process to the sum of n variables, we write for the characteristic function of x_{n+1},

$$P_{n+1}(t) = P_1(t)P_2(t) \cdot \cdot \cdot P_n(t) \tag{10.66}$$

The inverse of P_{n+1} is

$$p_{n+1}(x_{n+1}) = \frac{1}{2\pi} \int_{-\infty}^{\infty} P_1(t)P_2(t) \cdot \cdot \cdot P_n(t)e^{-it(x_1+ \cdots +x_n)} \, dt \tag{10.67}$$

An example of the result above is to identify the sum of noise caused by the superposition of n independently occurring sinusoidal distributions, such as the outputs of n nonsynchronous oscillators which can be represented by the zeroth-order Bessel function of the first kind, J_0. The probability density function for the sum of n independent sinusoidal sources with peak value $\alpha_1 \cdot \cdot \cdot \alpha_n$ can be written in accordance with Equation (10.67) as

$$P_{n+1}(x_{n+1}) = \frac{1}{2\pi} \int_{-\infty}^{\infty} J_0(\alpha_1 t)J_0(\alpha_2 t) \cdot \cdot \cdot J_0(\alpha_n t)e^{-ix_{n+1}t} \, dt \tag{10.68}$$

For small $\alpha_n t$ we shall only retain the first two terms of the power-series expansion of $J_0(\alpha_n t)$ as

$$J_0(\alpha_n t) = 1 - \frac{(\alpha_n t)^2}{4} \tag{10.69}$$

Substituting Equation (10.69) into (10.68), we write

$$p_{n+1}(x_{n+1}) \simeq \int_{-\infty}^{\infty} e^{-ix_{n+1}t-(\alpha_1+\alpha_2+ \cdots +\alpha_n)t/4} \, dt = \frac{1}{\sigma \sqrt{2\pi}} e^{-x_n^2+1/2\sigma} \tag{10.70}$$

where

$$\sigma^2 = \tfrac{1}{2}(\alpha_1^2 + \cdot \cdot \cdot + \alpha_n^2) \tag{10.71}$$

The result above is rather remarkable and indicates that the resultant distribution is again in a Gaussian form with a standard deviation equal to the square root of the sum of the mean squares of each individual distribution. This is another example of the central limit theorem.

So far, we have dealt with the probability density function involving one variable. Similar techniques can also be applied to the analysis of physical quantities involving two or more random variables. Expressions for multivariate probability functions are rather complex and are beyond the scope of this text. However, an important application of bivariate statistics is the autocorrelation of two measurements which can be made for either a single noise source at two specified instances of time or two noise sources. As for the single-noise-source case, we let the two measured values be $x = f(t_1)$ and $y = f(t_1 + \tau)$. The autocorrelation function $R(\tau)$ is defined by the expression

$$R(\tau) = \lim_{T \to \infty} \frac{1}{T} \int_0^T f(t)f(t + \tau) \, dt$$

$$= \int_{-\infty}^{\infty} \int_{-\infty}^{\infty} xy p(x, y) \, dx \, dy \tag{10.72}$$

where $p(x, y)$ is a two-dimensional probability density function. The corresponding power spectrum $W(\omega)$ can be expressed in terms of $R(\tau)$ by Fourier transforms as

$$W(\omega) = \frac{1}{2\pi} \int_{-\infty}^{\infty} R(\tau)e^{i\omega\tau} \, d\tau$$

$$= \lim_{T \to \infty} \frac{F_x(-\omega)F_y(\omega)}{2\pi T} \tag{10.73}$$

where

$$F_x(\omega) = \int_0^T f(t)e^{-i\omega t} \, dt$$

$$F_y(\omega) = \int_0^T f(t + \tau)e^{-i\omega t} \, dt \tag{10.74}$$

We shall not discuss here the autocorrelation of two noise sources. The concepts and methods introduced above provide only the background necessary for us to treat the problems associated with the design of an optimum receiver, which are discussed in Chapter 11. In the rest of this section we introduce the most important concept concerning the information capacity of a transmitting channel. It was first formulated by Shannon (Ref. 10.4). The theorem of channel capacity states that there is a maximum information-carrying capacity C for any communication channel if it is constrained in power. To operate at a higher rate than C, the system is subjected to a higher probability of error. In the following, we state this theorem without offering the proof.

Theorem.   There exists a maximum capacity C given by

$$C = B \log_2 \left(1 + \frac{S}{N} \right) \tag{10.75}$$

for any communication channel having an information bandwidth B and a given signal-to-noise (S/N) ratio.

It follows from Equation (10.75) that in the limit as S/N approaches infinity, the information-carrying capacity also approaches infinity. Obviously, this is impossible, because every electronic system has noise which cannot be less than the fundamental limit set by the quantum noise. It has been shown by Gordon (Ref. 10.5) that in the quantum noise-limiting case, the classical channel capacity as given by Equation (10.75) is no longer valid. Gordon showed that for a single-mode system having an output power P given by

$$P = \langle n \rangle h\nu B \tag{10.76}$$

where $\langle n \rangle$ is the average number of photons per mode, the maximum information-carrying capacity C is given by

$$C = B \log\left(1 + \frac{S}{N + h\nu B}\right) + \frac{S + N}{h\nu} \log\left(1 + \frac{h\nu B}{S + N}\right)$$
$$- \frac{N}{h\nu} \log\left(1 + \frac{h\nu B}{N}\right) \tag{10.77}$$

As an example, we shall consider a resistor, having a thermal noise N, that can be expressed by the blackbody radiation formula at a temperature T as given by

$$N = h\nu B(e^{h\nu/kT} - 1)^{-1} \tag{10.78}$$

At room temperature, $N \gg h\nu B$. Substituting this condition into Equation (10.77), we obtain

$$C \simeq B\left[\log\left(1 + \frac{S}{N}\right) - \frac{h\nu B S}{2N(S + N)} \log e\right] \tag{10.79}$$

Since the second term in Equation (10.79) is very small compared with the first term in parentheses, we see that the quantum mechanical result approaches the classical result in the limit as $N \gg h\nu B$.

The capacity theory does not inform us how to achieve the maximum capacity. To achieve a reasonably high rate, the signal must be coded in such a way as to have a statistically predictable noise. Several basic coding schemes were introduced in previous sections. There are still more efficient coding schemes which are variations of the basic coding schemes. Readers are urged to pursue additional information from books specializing in this subject.

PROBLEMS

10.1. Using Equation (10.14), show that if $d \gg L_n$, the frequency response of the device is

$$\mathcal{R}(\omega) \simeq (1 + \omega^2 \tau^2)^{-1/2}$$

10.2. If the E field is applied along the (111) direction [e.g., $\mathbf{E} = E(i + j + k)/\sqrt{3}$], show that the equation of indicatrix of GaAs crystal is

$$\frac{1}{n^2}(x_1^2 + x_2^2 + x_3^2) + \frac{2}{\sqrt{3}}\gamma_{41}E(x_1x_2 + x_2x_3 + x_3x_1) = 1$$

10.3. Using the coordinate transformation

$$x_1' = \frac{\sqrt{2}}{2}(x_2 - x_3)$$

$$x_2' = \frac{\sqrt{6}}{6}(-2x_1 + x_2 + x_3)$$

$$x_3' = \frac{1}{\sqrt{3}}(x_1 + x_2 + x_3)$$

show that the equation of indicatrix as given by Problem 10.2 can be reduced to an ellipsoid in principal axes, with indices given by

$$n_{x_1'} = n_{x_2'} = n + \frac{1}{2\sqrt{3}}n^3\gamma_{41}E$$

$$n_{x_3'} = n - \frac{1}{\sqrt{3}}n^3\gamma_{41}E$$

10.4. Let x_1 and x_2 be two independent random variables with the probability density functions

$$p_1(x_1) = \begin{cases} (\pi\sqrt{1 - x_1^2})^{-1} & -1 \le x_1 \le 1 \\ 0 & \text{elsewhere} \end{cases}$$

$$p_2(x_2) = \begin{cases} x_2 e^{-x_2^2/2} & x_2 \ge 0 \\ 0 & \text{elsewhere} \end{cases}$$

Show that the product $x_3 = x_1 x_2$ has a Gaussian density function.

10.5. Let x be a random variable with the probability density function

$$p_1(x_1) = \tfrac{1}{2} e^{-|x_2|}$$

Establish the probability density function $p_2(x_2)$ of a random variable

$$x_2 = e^{x_1}$$

10.6. A random variable x which takes on only an integer value is said to be "Poisson" distributed if

$$p(n) = \delta(\langle n \rangle - n)\frac{\langle n \rangle^n e^{\langle n \rangle}}{n!}$$

(a) Plot $p(n)$ for $\langle n \rangle = 2$, $n \leq 5$.

(b) Find $\langle n \rangle$ and σ^2.

10.7. Let x be a sinusoidal function of t:

$$x(t) = A\cos(\omega t + \varphi)$$

where A, ω, and φ are statistically independent random variable.

(a) Show that the correlation function $R(\tau)$ is

$$R(\tau) = e^{-|\tau|}$$

(b) The power spectrum $F(\omega)$ is

$$F(\omega) = \frac{2}{1 + \omega^2}$$

REFERENCES

10.1. I. P. Kaminow and E. H. Turner, *Proc. IEEE, 54,* 1374 (1966).

10.2. K. Peterman and G. Arnold, *IEEE J. Quantum Electron., QE-18,* 543 (1982).

10.3. J. M. Wozencraft and I. M. Jacobs, *Principles of Communication Engineering,* John Wiley & Sons, Inc., New York, 1965.

10.4. C. E. Shannon, *Proc. IRE, 37,* 10 (1949).

10.5. J. P. Gordon, *Proc. IRE, 50,* 1898 (1962).

11

Photodetectors and Optical Receivers

11.1 INTRODUCTION

Photodetectors are devices that convert optical signals into identical electrical waveforms. There are many types of detectors operating on the basis of pyroelectric, thermoelectric, or photoelectric effects. Detectors for fiber optical systems usually are photodiodes which are photoelectric devices. Photodiodes convert incoming photons into electron–hole pairs in a regeneration time on the order of 10^{-10} s. Physical processes responsible for these devices are exactly the reverse of light emission in a semiconductor and have already been discussed in Chapter 7. In this chapter we first discuss the device aspect of photodiodes and study the structural configurations. Then we analyze the device responsivity and quantum gain. It is important to have a working knowledge of their operating mechanisms. In most cases, the output from the photodiode is usually very weak and often requires amplification and signal processing. Therefore a typical optical receiver consists of a photodiode, an amplifier, and some sort of filtering circuits. Because of the statistical nature of the pair-production and multiplication processes, these devices are inherently noisy. We examine the sources of noise. By minimizing the noise sources, it is possible to achieve a high-fidelity fiber transmission system at a low bit error rate (BER) with an optimum signal-to-noise (S/N) ratio. The gain versus bandwidth and S/N versus BER are just some of the important trade-off parameters in the design of a receiver for an optical fiber system.

11.2 *pn* AND *pin* PHOTODIODES

Photodiodes are made of semiconductor *pn* junctions under reverse bias. When a photodiode is illuminated by photons of a given frequency, electrons are excited from the valence band into the conduction band. Thus electron–hole pairs are generated within the depletion region density. These electron-hole pairs will drift toward the *n* and *p* regions, respectively, under the influence of the internal field, which builds up from the internal space-charge density as a result of the difference in Fermi level between the *n* and *p* materials. At zero bias, this drifting current flowing through the junction is balanced by an opposite current due to diffusion of majority carriers. If an external reverse bias is applied across the junction, as shown in Figure 11.1, the diffusion of majority carriers will be greatly reduced, resulting in a net current flow. The photoexcited carriers can migrate across the junction. If we ignore the loss due to recombination in the depletion region, the photocurrent can be written as

$$I = \frac{e \eta P_0}{h \nu} \tag{11.1}$$

where e, P_0, and $h\nu$ are the electronic charge, optical power, and photon energy, respectively. η is the conversion efficiency. If electron–hole pairs are produced outside the depletion region, the probability for them to recombine in the diffusion region is very great, so that the conversion efficiency will be reduced.

Let ρ_n and ρ_p be the charge densities and W_n and W_p be the depletion widths of the *n* and *p* materials, respectively. From Gauss's law, we obtain an expression for the field E as given by

$$E = \frac{\rho_n W_n}{\varepsilon} = \frac{\rho_p W_p}{\varepsilon} \tag{11.2}$$

The bias voltage V_b across the junction can be expressed as

$$V_b = \frac{\rho_n W_n^2 + \rho_p W_p^2}{2\varepsilon} \tag{11.3}$$

If $\rho_p \ll \rho_n$, $W_n \ll W_p$, so that

$$W_p \simeq \sqrt{\frac{2\varepsilon V_b}{\rho_p}} \tag{11.4}$$

Substituting Equation (11.4) into (11.2), we obtain an approximate expression for the field at or near the junction:

$$E \simeq \sqrt{\frac{2V_b \rho_p}{\varepsilon}} \tag{11.5}$$

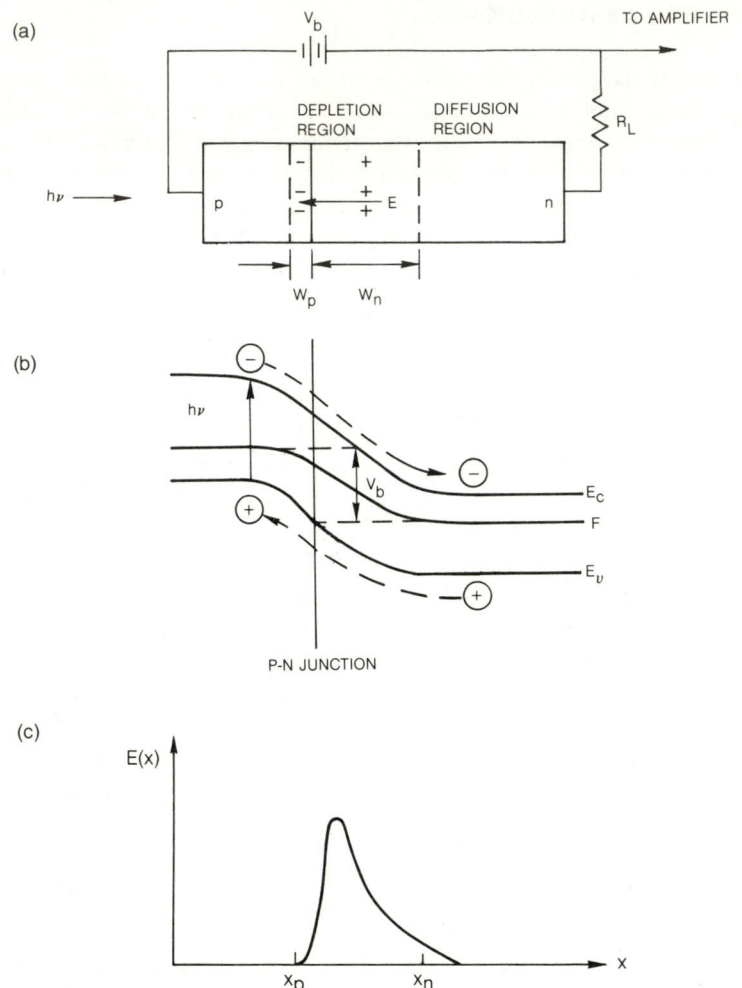

Figure 11.1   (a) External reverse-biased *pn* photodiode showing the depletion regions in *p* and *n* materials; (b) energy-band diagram of a reverse-biased *pn* junction; (c) electric field $E(x)$ in a reverse-biased *pn* photodiode.

The quantum efficiency of a simple *pn*-junction device is usually very low because a large fraction of optical power is not efficiently utilized. In the case of a silicon *pn*-junction photodiode, the depletion width of the junction is much shorter than the absorption length because the optical absorption coefficient is very small, so that most of the optical power is absorbed inside the diffusion region. Only a very small fraction of the carriers can be generated in the depletion region to provide a displacement current. Clearly, such a device is not only inefficient, but also has a relatively slow response time, as a result of a random diffusion process. These difficulties can be avoided by adding an

(a) **pin** PHOTODIODE CONFIGURATION

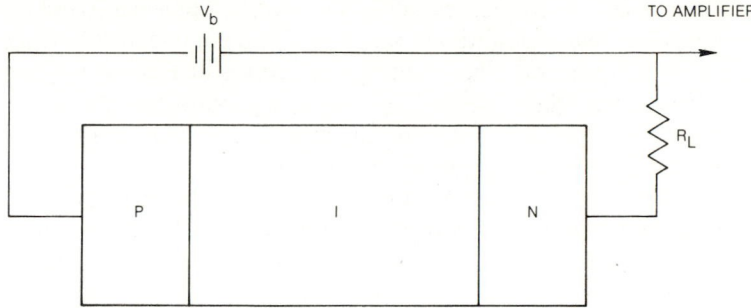

(b) E FIELD IN THE DEPLETION REGION

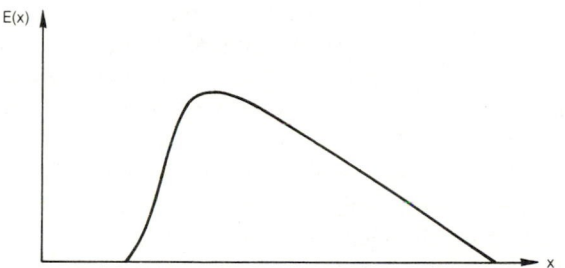

Figure 11.2   (a) External reverse-biased *pin* photodiode; (b) *E* field in the depletion region of a *pin* photodiode. [From *Fundamentals of Optical Fiber Communications, 2e*, Academic Press, Inc. (1981) Ed. by M. Barnoski.]

intrinsic (undoped) and high-resistivity semi-insulating layer within the *pn* junction to form a *pin* structure, as shown in Figure 11.2. The width of the *i* layer must be made many times wider than the absorption length. Because the field strength is high in the *i* layer, electron–hole pairs can be quickly swept out of the *i* toward the *n* or *p* region, respectively. The wider the thickness of the *i* layer, the higher is the quantum efficiency. However, one takes a penalty of reducing the speed of response by using a very thick insulating layer between the junction. For relatively low-data-rate systems, pin diodes have been widely used. The typical quantum efficiency of pin diodes lies in the range 0.5 to 0.9 and the responsivity varies from 0.4 to 0.65 A/W. For Si-based pin diodes, the response time is typically of the order of several nanoseconds and the spectral response covers a range from ultraviolet to near infrared with a responsivity peak at around 0.9 μm. The typical range for the bias voltage is between 20 and 100 V.

Materials that are commonly used to fabricate photodiodes are Si, Ge, GaAs, InAs, and InGaAs. In general, indirect bandgap materials, such as Si and Ge, are preferred over direct bandgap materials, primarily because the surface recombination in a direct bandgap material can lead to a substantial loss of carriers as a result of trapping by surface states, without producing photocurrents of significant magnitude. On the other hand, indirect absorption requires the help of a third body such as a phonon to conserve the momentum in the transfer process. This requirement makes the transition probability less likely to occur than the process involving only two bodies in a collision. However, it is possible to avoid this difficulty by making a pin configuration to provide a longer absorption length. Figure 11.3 shows the absorption coefficients of Si and Ge as a function of wavelength. For comparison purposes, the absorption coefficient of GaAs is also presented. The absorption begins at the band edge and increases rapidly with increasing photon energy; however, the increase is more gradual for indirect materials than for direct materials. The threshold for direct absorption in silicon occurs at 4.1 eV, which lies in the ultraviolet, whereas in GaAs, direct absorption occurs at 1.45 eV. The results shown in Figure 11.3 indicate that an absorption length of about 50 μm is required to absorb all the light at 0.83 μm in Si, whereas in GaAs only 1 μm is required. For wavelengths below 1 μm, Si photodiodes are considered to be ideal for use in optical fiber systems, but they are not very sensitive when the wavelength exceeds 1 μm. For wavelengths above 1 μm, Ge photodiodes are often used. At 1.5 μm a rapid increase in the absorption coefficient occurs in Ge, as shown in Figure 11.3. This corresponds to a direct transition in Ge from its band structure. As mentioned before, the surface recombination becomes a problem at this wavelength. A maximum efficiency of 50% can be obtained by

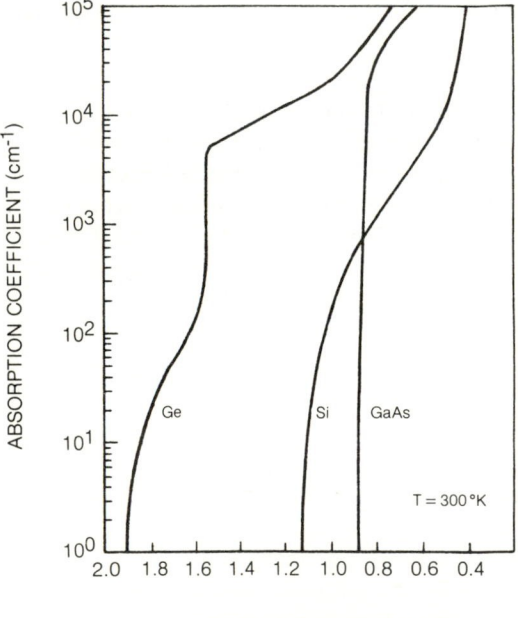

Figure 11.3 Absorption coefficient $\alpha(\text{cm}^{-1})$ versus optical wavelength for Ge, Si, and GaAs semiconductors. (From Ref. 11.2. Reprinted with permission of John Wiley and Sons, Inc., New York, © 1980.)

using a Ge photodiode and 30% for an InAs diode at about $1.3\ \mu$m. There are other photodiodes, such as AlGaSb and InGaAsP, which are made by an alloy process of binary into ternary or quaternary III–V compounds. High quantum efficiency and high-speed photodiodes for detection of light at longer wavelengths could become commercially available using these alloys. Besides, its material dependence, quantum efficiency is also dependent on the reflectivities and geometry of the device, the absorption coefficient $\alpha(\lambda)$, and the operating temperature. At room temperature, the quantum efficiency of a Si pin diode can be as high as 90% at $0.9\ \mu$m. The value decreases rapidly with increasing λ and reduces to about 20% at $1.06\ \mu$m. At longer wavelengths, materials other than Si must be considered (Ref. 11.1). More recently results (Ref. 11.8) indicate that η values as high as 80% can be obtained for wavelengths ranging from 0.9 to $1.3\ \mu$m by using Al_xGa_{1-x} AsSb photodiodes.

11.3 AVALANCHE PHOTODIODES

Another way to improve the performance of a *pn*-junction photodiode is to produce an avalanche gain in these devices by means of impact ionization. The avalanche condition can be created by increasing the reverse-biased voltage just slightly below the breakdown value of the semiconductor so that a very high electric field ($>10^5$ V/cm) is established across the junction. Electrons and holes traversing under such a high field can acquire sufficient kinetic energy to produce additional electron–hole pairs through inelastic collisions. The energy transfer in this process can be sufficient to bring an electron from the valence band into the conduction band. This process can be multiplied rapidly with a multiplication factor M that is usually represented by an exponential function of the bias voltage.

Figure 11.4(a) shows a typical avalanche photodiode structure which separates the depletion region into two different regions. The first region is a wide drift i region, in which photons are absorbed, and the second one is a narrow p region, in which the photogenerated carriers are multiplied. When a sufficiently high reverse-biased voltage is applied such that the depletion layer of the diode just reaches through into the low-concentration π region, the electric field at the junction is about 5 to 10% less than that required to cause avalanche breakdown. A slight increase in applied voltage could cause the depletion layer to extend rapidly to the p^+ contact. The resistivity of the π region of the device is typically about 5000 Ω-cm and its thickness can be as much as 200 μm at a bias voltage of less than 100 V over and above that required to achieve reach-through. Usually, this device is operating in a fully depleted mode so that all carriers are collected by drift alone.

The electric E-field profile for this diode is shown in Figure 11.4(b). Avalanche multiplication occurs in the high-field region near the pn junction with a typical width of about 2 μm. The field in the π region is much lower but

a) AVALANCHE PHOTODIODE CONFIGURATION

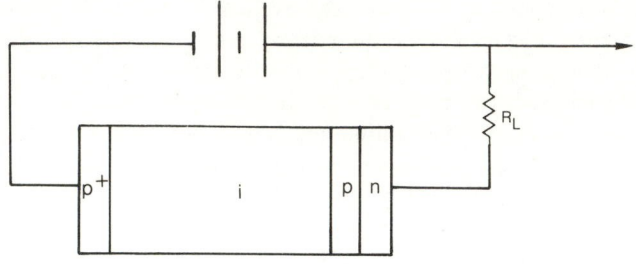

b) ELECTRIC FIELD PROFILE CROSSING THE DEVICE

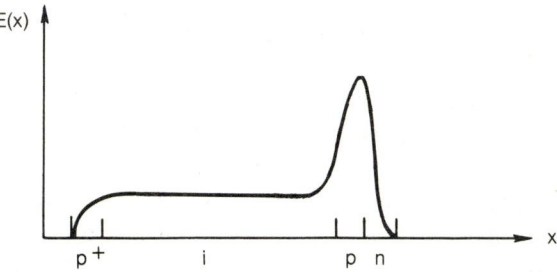

Figure 11.4 (a) External reverse-biased avalanche photodiode; (b) E-field profile extended across the device configuration.

sufficient to assure that the carriers gain enough kinetic energy to reach saturation. The transition time required for reach-through is in the nanosecond range, which is the typical response time of the diode. It can be estimated by the expression W/v_e, where v_e is the drift velocity of the electrons. The transit-time multiplication is much shorter than the time required for electrons to go across the depletion layer and can be considered to occur instantaneously.

Both the avalanche gain and excess noise are related to an important parameter k, which is the ratio of the ionization rates α and β for electrons and holes, respectively. Clearly, the rates at which electrons and holes are created in the semiconductor are different and also structurally dependent. Figure 11.5 shows the measured α and β values for silicon. The results of Figure 11.5 are only qualitative, because the information on the orientation of the electric field for these measurements is unknown. At low fields, the ionization rates differ significantly and usually are structurally dependent. At high fields, the rates become relatively independent of field orientation and crystal structure, because the number of collisions is so large that the ionization rates become isotropic. It is reasonable to expect that at very high fields the k value approaches unity.

The multiplication factor M and the excess noise factor F of avalanche photodiodes have been analyzed in great detail by McIntyre and co-workers (Ref. 11.3). We shall examine only a few special cases that can provide a reasonable understanding of the operating mechanism of typical devices.

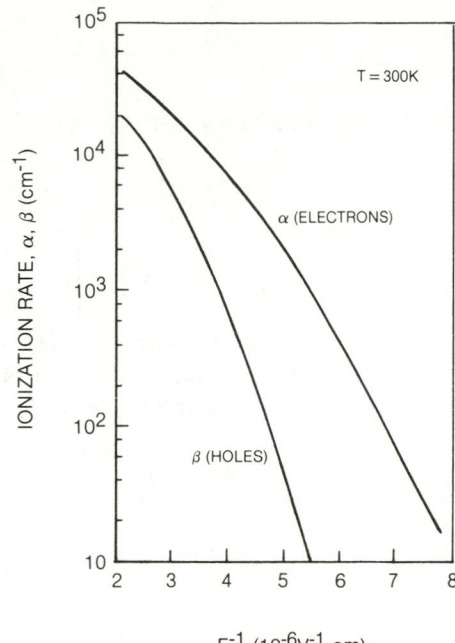

Figure 11.5 Ionization rates for electrons α and holes β in silicon as a function of $1/E$ field at 300°K. (From Ref. 11.2. Reprinted with permission of John Wiley and Sons, Inc., New York, © 1980.)

Consider that additional electron–hole pairs are generated within the depletion layer at x under a sufficiently high field. In traversing a distance dx, the electron and the hole will produce, on the average, $\alpha\, dx$ and $\beta\, dx$ ionizing collisions, respectively. These additional carriers, in turn, can gain enough energy from the field to cause further ionization, until an avalanche of carriers has been created.

Over a long path W, the multiplication $M(x)$ can be expressed by the expression

$$M(x) = 1 + \int_0^x \alpha(x)M(x)\, dx + \int_x^W \beta(x)M(x)\, dx \qquad (11.6)$$

Differentiating Equation (11.6), we get

$$\frac{dM(x)}{dx} = (\alpha - \beta)M(x) \qquad (11.7)$$

Equation (11.7) has a solution of the form

$$M(x) = M(0) \exp\left[\int_0^x (\alpha - \beta)\, dx \right] = M(W) \exp\left[-\int_x^W (\alpha - \beta)\, dx \right]$$

$$(11.8)$$

Substituting Equation (11.8) into (11.6), we get

$$\frac{1}{M(W)} = 1 - \int_0^W \alpha \, dx \exp\left[-\int_x^W (\alpha - \beta) \, dx \right] \tag{11.9}$$

Substituting Equation (11.8) into (11.9), we get

$$M(x) = \frac{\exp\left[-\int_x^W (\alpha - \beta) \, dx \right]}{1 - \int_0^W \alpha \, dx \exp\left[-\int_x^W (\alpha - \beta) \, dx \right]} \tag{11.10}$$

For simplicity we consider the case for which the electron and hole ionization coefficients are equal, e.g., $\alpha = \beta$. Equation (11.10) indicates that M is no longer a function of x and is given by

$$\frac{1}{M} = 1 - \int_0^W \alpha \, dx = 1 - \delta \tag{11.11}$$

In reality, $\alpha \neq \beta$. We shall define the average of ionization rate ratio

$$k = \frac{\beta}{\alpha} \tag{11.12}$$

In terms of k, Equation (11.10) can be approximated by

$$M = \frac{1 - k}{\exp[-(1 - k)\delta] - k} \tag{11.13}$$

If $\beta \ll \alpha$ or k approaches zero, then Equation (11.13) becomes

$$M \simeq e^\delta \tag{11.14}$$

In this case, the carrier multiplication increases exponentially with δ; however, it remains finite as long as $\beta = 0$. Thus such a diode would never reach the avalanche breakdown. Figure 11.6 shows the carrier multiplication as a function of electric field and δ for various values of k in an avalanche photodiode. Results of Figure 11.6 indicate that for k values other than zero, the multiplication goes to infinity at some finite values of electric field. Therefore, the choise of k values is very important in the design of a fiber optic receiver system.

Another important consideration in choosing an avalanche photodiode is the excess noise factor F. Assuming that the diode is shot-noise limited, the noise current dI generated in dx is determined by the mean square of the fluctuation in photocurrent generated in the same space. It is given by

$$\langle (dI - \langle dI \rangle)^2 \rangle = 2e \, dI \, BM^2 \tag{11.15}$$

where B is the detector bandwidth. The result of Equation (11.15) will be

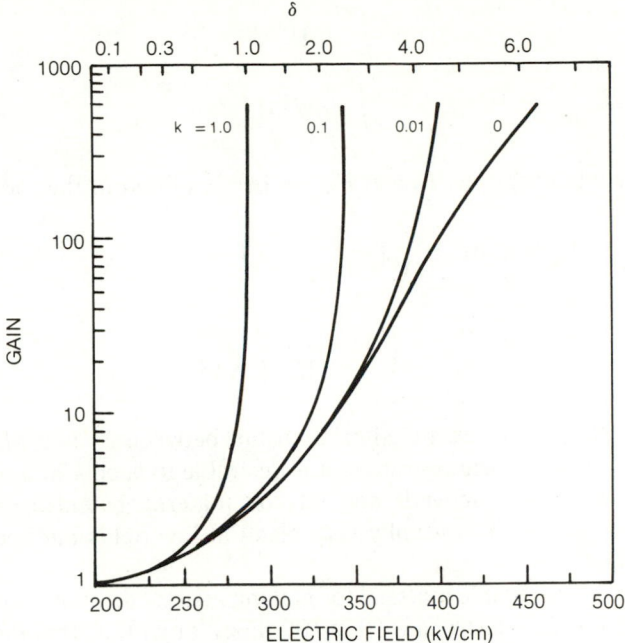

Figure 11.6   Electronic gain in an avalanche photodiode as a function of E field at various k values. (From Ref. 11.3.)

derived in Section 11.4. Integrating this element of noise current, we obtain the total noise spectral density as given by

$$N = 2\,eI_0BM^2F \qquad (11.16)$$

where I_0 is the average current of the photodiode, and F is the ratio of the actual noise to the noise of an ideal device if the multiplication process is noiseless. Expressions for the excess noise factor have been derived (Ref. 11.3) for both electrons and holes as given by

$$F_e = \frac{k_2 - k_1^2}{1 - k_2}\,M_e + 2\left[1 - \frac{k_1(1 - k_1)}{1 - k_2}\right] - \frac{(1 - k_1)^2}{M_e(1 - k_2)} \qquad (11.17)$$

$$F_h = \frac{k_2 - k_1^2}{k_1^2(1 - k_2)}\,M_h - 2\left[\frac{k_2(1 - k_1)}{k_1^2(1 - k_1)} - 1\right] + \frac{(1 - k_1)^2 k_2}{k_1^2(1 - k_2)M_h} \qquad (11.18)$$

where k_1 and k_2 are two different weighted averages of k as defined by

$$k_1 = \frac{\displaystyle\int_x^W \beta M\,dx}{\displaystyle\int_0^W \alpha M\,dx} \qquad (11.19)$$

$$k_2 = \frac{\displaystyle\int_x^W \beta M^2 \, dx}{\displaystyle\int_0^x \alpha M^2 \, dx} \tag{11.20}$$

For all practical purposes, F_e and F_h can be simplified in the forms

$$F_e = k_2 M_e + \left(2 - \frac{1}{M_e}\right)(1 - k_2) \tag{11.21}$$

$$F_h = \frac{k_2 M_h}{k_1^2} - \left(2 - \frac{1}{M_h}\right)\left(\frac{k_2}{k_1^2} - 1\right) \tag{11.22}$$

Figure 11.7 shows the functional relationship between F_e and M_e for various values of k. For low-noise operation, it is desirable to keep k at a relatively low value. The value of k depends not only on material but also on the device structure. For silicon, k is usually very small at low fields but becomes larger with increasing field strength.

Silicon is a favorable material for making avalanche photodiodes, because its ionization rate for electrons is 10 to 100 times larger than that for holes in the spectral region 0.8 to 0.9 μm. In the first-generation optical fiber systems, Si APDs have been used extensively and have provided excellent performance, with a quantum efficiency reaching nearly 100% and a response time around

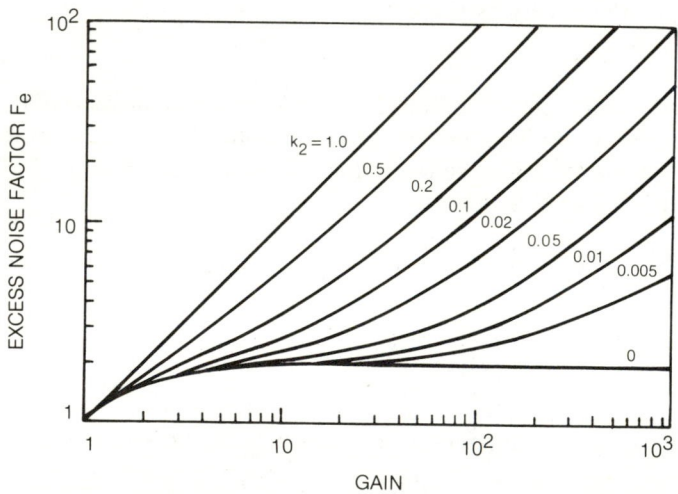

Figure 11.7 Excess electronic noise factor of an avalanche photodiode as a function of gain at various k values. (From Ref. 11.3.)

1 ns. These devices usually provide a current gain of 100 with an excess noise factor of about 5. In the longer-wavelength region (1 to 1.6 μm), silicon becomes transparent, so that other materials with narrower bandgaps must be used. Again similar to the case of pin photodiodes, germanium can be used for making APDs. However, the dark current in Ge is much larger than in Si (about 10^{-7} A), and nearly equal carrier ionization rates result in high excess noise. In spite of these shortcomings, recently developed Ge APDs have exhibited an excess noise factor of 7 at a gain of 10, a bandwidth of 500 MHz, and a quantum efficiency of ~ 0.8, at $\lambda = 1.3$ μm.

Other materials, such as InGaAsP and InGaAs, have also been used to make APDs sensitive at longer wavelengths. Figure 11.8 depicts the construction of two types of APDs employing a narrow-bandgap InGaAs or InGaAsP layer for absorption and a wide-bandgap InP layer for multiplication. This scheme has been shown to minimize the tunneling dark current. For example, the mesa-type APD has a primary dark current of 3 nA and an excess noise factor $F = 10$ at a gain of 20. The planar structure, on the other hand, usually is more noisy than the mesa type because of the dark current problem.

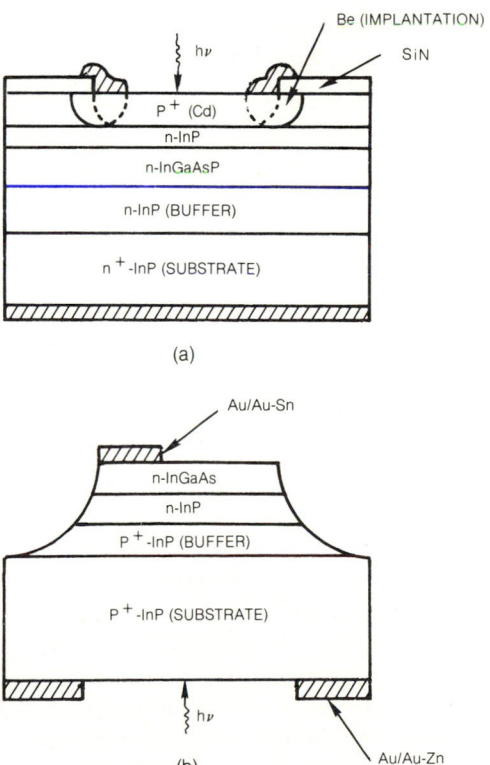

Figure 11.8 InGaAs/InP avalanche photodiodes having separate regions for absorption (InGaAs) and multiplication (InP): (a) planar type; (b) mesa type. (From Ref. 11.8. Reprinted with permission of IEEE, © 1983.)

11.4 NOISE IN PHOTODIODES

Photoexcited carriers in a diode contain both signal and noise. We have discussed the concept of a random event. The noise generated in a diode is of a statistical nature and signal dependent in generating electron–hole pairs. Both the number n of the electron–hole pairs and the generation time t_n are random quantities and satisfy Poisson statistics. The total number n of electron–hole pairs generated during a time period τ obeys the probability distribution

$$p(n) = \frac{\Lambda^n e^{-\Lambda}}{n!} \qquad (11.23)$$

where Λ is a time integral of the probability that either 1 or 0 electron–hole pairs will be created in a time interval over the entire period. We can write that

$$\Lambda = \int_t^{t+\tau} \frac{\eta}{h\nu} P_0(t) \, dt \qquad (11.24)$$

where $P_0(t)$ is the optical power in watts, η is the detector quantum efficiency, and $h\nu$ is the photon energy. This expression represents simply the average number of detected photons incident during the time period τ. Therefore, we shall write

$$p(n) = \frac{\langle n \rangle^n e^{-\langle n \rangle}}{n!} \qquad (11.25)$$

The fluctuation about the mean δn is $n - \langle n \rangle$, and the mean square of the fluctuation is

$$\langle (\delta n)^2 \rangle = \langle n^2 - 2n\langle n \rangle + \langle n \rangle^2 \rangle = \langle n^2 \rangle - \langle n \rangle^2 \qquad (11.26)$$

However, the quantity $\langle n^2 \rangle$ is the second moment of $p(n)$ and can be calculated as follows:

$$\langle n^2 \rangle = \Sigma \, n^2 p(n) = \langle n \rangle + \langle n \rangle^2 \qquad (11.27)$$

Substituting Equation (11.27) into Equation (11.26), we get

$$\langle (\delta n)^2 \rangle = \langle n \rangle \qquad (11.28)$$

We obtain one of the remarkable results of Poisson distribution. It shows that the mean-square fluctuation of this distribution is equal to its mean value. Using this result, we now calculate the shot noise of photodiodes. The shot noise N_{sh} is primarily due to the statistical nature of electron–hole generation in the diode. During the photoexcitation process, there are two noise sources: One is attributed to the randomness of photocurrent generation; the other is related to the randomness of the generation rate. For simplicity we shall assume that the transit time for these charges to reach the electrodes is negligible, so that the detector's response is fast compared with the repetition rate of the optical pulses. Under these circumstances, we pass these short pulses through a low-pass filter with a transmission characteristic given by $(\sin x)/x$. The output current, I, through this filter at any instant is, then,

$$I = \frac{en}{\tau_p} \tag{11.29}$$

where τ_p is the average width of the pulse. The average current I_0 is

$$I_0 = \frac{e\langle n\rangle}{\tau_p} \tag{11.30}$$

From Equations (11.29) and (11.30) we can write the fluctuation in current that obeys the Poisson statistics, as

$$\langle(\delta I)^2\rangle = \frac{e^2}{\tau_p^2}\langle(\delta n)^2\rangle = \frac{e^2}{\tau_p^2}\langle n\rangle = \frac{eI_0}{\tau_p} \tag{11.31}$$

To calculate τ_p, we see that the frequency response of the filter is

$$F(\omega) = \frac{\sin \omega t}{\omega t} e^{-i\omega t} \tag{11.32}$$

At each frequency, the noise power is filtered by a factor $|F(\omega)|^2$. Hence the effective bandwidth is

$$B = \int_0^\infty |F(\omega)|^2 \, d\omega = \frac{1}{2\tau_p} \tag{11.33}$$

Substituting $1/2B$ for τ_p in Equation (11.31), we obtain

$$N_{sh} = \langle(\delta I)^2\rangle = 2eI_0B \tag{11.34}$$

The expression (11.34) has been commonly used for estimating the shot noise of a photodiode having a bandwidth B. It shows that the shot noise is directly proportional to the photocurrent. For APD, the higher the multiplication factor, the higher is the detector's noise, as indicated by Equation (11.16).

In addition to shot noise, there are a number of other noise sources encountered by the receiver. The most common ones are thermal background noise, quantum noise, and amplifier noise. The level of the thermal noise N_T is often given in terms of kT watts per hertz, where k is Boltzmann's constant and T is the absolute temperature. The familiar expression for the thermal noise is given by

$$N_T = \frac{h\nu}{\exp(h\nu/kT) - 1} \tag{11.35}$$

At optical frequency (e.g., $\lambda = 1$ μm), $h\nu \gg kT$ and the photon energy has a value on the order of 2×10^{-19} J, as compared to the kT value of 4×10^{-21} J at room temperature. Therefore, at optical frequency the thermal background noise is usually insignificant. Quantum noise, on the other hand, manifests itself in several possible ways, depending on the detection techniques. For direct detection as usually is the case for many optical fiber systems, quantum noise N_Q arises primarily from the statistical counting of photons. Substituting Equation (11.1) into (11.34), we can write an expression for the mean-square fluctuation in photocurrents due to quantum noise as

$$N_Q = \langle (\delta I)^2 \rangle = 2eB \left(\eta \, \frac{eP_0}{h\nu} \right) \tag{11.36}$$

As an example, we consider a digital communications system operating at a data rate B bits per seconds. If the detector is relatively free of dark current noise, there will be no error in the absence of an optical pulse. The probability of an error in the presence of an optical pulse of energy E can be estimated by the expression

$$P(E) = e^{-E/h\nu} \tag{11.37}$$

Using Equation (11.37), we see that in the quantum-noise-limited case, the energy E must be greater than $21 h\nu$ for an error probability less than 1 in a billion (10^{-9}). In other words, only in 1 out of a billion events will there be no electron–hole pair generation, if the number of photons in a pulse is greater than 21 for quantum-noised-limited detection. Expressing this in terms of the minimum detectable power for a 10^{-9} error rate, we write

$$P_{\min} \simeq \frac{21 h\nu B}{2} \tag{11.38}$$

In Equation (11.38), a factor of 2 is introduced because of the use of a binary code. As an example, for a system bandwidth of 10 Mb/s, the $P_{\min}$ is about 2×10^{-11} W or -77 dBm for an incident light wave at 1 μm. This estimate assumes an error rate of 10^{-9} and the receiver is only quantum-noise-limited.

The output from a photodiode is usually fed into an amplifier which is an integral part of the receiver system. Figure 11.9 shows a simple diagram of the detector–amplifier configuration and its equivalent circuit for the purpose of analysis. For simplicity, we shall assume that amplifier noise is dominated by Johnson noise that is generated in the 50-Ω input resistance. This is a reasonable assumption because the admittance of the combined diode capacitance C_d and the amplifier input capacitance C_a is usually very small compared to that of the 50-Ω resistance. A resistor R at a temperature T emits noise power according to Equation (11.35). We can rewrite Equation (11.35) as

$$N_T = \frac{h\nu}{1 + h\nu/kT + \frac{1}{2}(h\nu/kT)^2 + \cdots - 1}$$

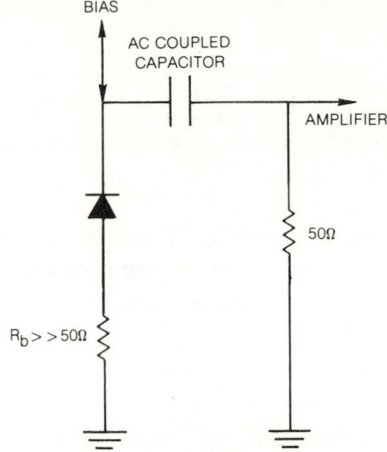

(a) A SIMPLE PHOTODIODE-AMPLIFIER CIRCUIT

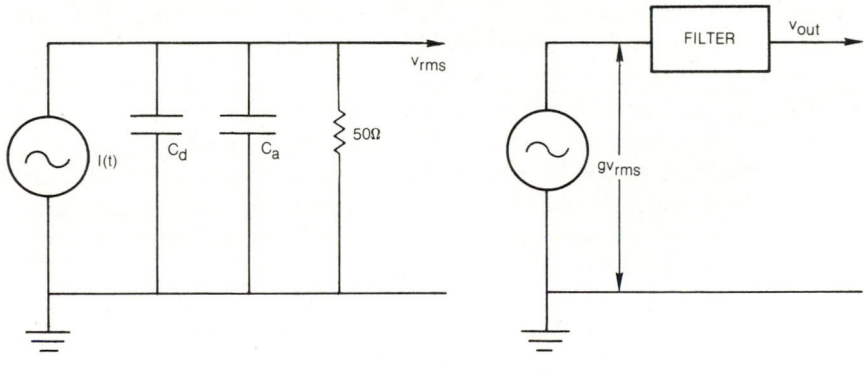

(b) THE EQUIVALENT CIRCUIT

Figure 11.9   (a) Simple photodiode–amplifier circuit; (b) equivalent circuit for the photodiode–amplifier combination.

At radio frequencies (e.g., $h\nu \ll kT$), we have

$$N_T \simeq kT$$

The mean-square fluctuation of the photocurrent can be calculated as

$$\langle (\delta I)^2 \rangle = \frac{2N_T}{R} \int_{-\infty}^{\infty} |F(\omega)|^2 \, d\omega$$

where $|F(\omega)|$ is the Fourier transform of the impulse response of the RC network. For a total of two pass bands, the integral yields a value of $2B$, where B is the filter bandwidth. Therefore, the mean-square fluctuation of the thermal noise current or the Johnson noise in the load resistor R can be expressed as

$$\langle (\delta I)^2 \rangle \simeq \frac{4kTB}{R} \tag{11.39}$$

Or the mean-square fluctuation in voltage can be expressed as

$$\langle (\delta V)^2 \rangle = 4kTBR \tag{11.40}$$

where B is the bandwidth of the amplifier. The expressions above indicate that the amplifier thermal noise from a resistor can contribute significantly to the excess noise power. We introduce a parameter Z:

$$Z = \frac{\sqrt{4kTB/R}}{eB} \tag{11.41}$$

that is a commonly used parameter to represent a figure of merit for the amplifier. Since photodiodes are modeled as a current source in parallel with a capacitor as shown in Figure 11.9(b), there exists essentially no thermal noise associated with these sources. All thermal noises come from the input impedance of the amplifier. For a simple 50-Ω input amplifier, Z is about 35,000 at $B = 10$ MHz, and varies with $B^{-1/2}$. In this case, the amplifier noise is 35,000 times as high as the signal produced by a single electron–hole pair. Therefore, it is important to choose amplifiers having a low Z value.

    For frequencies below 25 MHz, it is possible to reduce Z values by using a low-capacitance diode bonded to a substrate containing either a Si or GaAs FET amplifier of comparable capacitance. In this way, a Z value as low as 1000 can be obtained. For frequencies above 25 MHz, the gain of FET approaches unity. Therefore, it is necessary to use a bipolar transistor whose Z value is relatively independent of B. Operating at high signal levels, transimpedance amplifiers are often used. For more details, readers should consult books on amplifiers (e.g., Ref. 11.4).

11.5 FREQUENCY RESPONSE

The frequency response of a photodiode is determined primarily by three factors: (1) the carrier collection time τ_c, which is the time required for the electric field to sweep out the photoexcited carriers within the depletion region; (2) the RC time τ_{RC}, which is the time required to discharge the junction capacitance C_d through a combination of internal and external resistances; and (3) the diffusion time τ_d, which is the time required for these carriers to diffuse into the depletion region. The total photodiode rise time τ_r, defined as the time of response from 10% to 90% of a pulse height is essentially equal to the largest of the three as mentioned above. To a good approximation, we write

$$\tau_r = (\tau_c^2 + \tau_{RC}^2 + \tau_d^2)^{1/2} \tag{11.42}$$

The expression for τ_c is given by

$$\tau_c \simeq \frac{W}{2} \frac{W}{\mu V_b} \qquad (11.43)$$

where W, μ, and V_b are the width of the depletion region, carrier mobility, and reverse bias voltage, respectively. For silicon photodiodes, τ_c can be estimated to a good approximation by using the expression

$$\left[\tau_c \simeq \frac{\rho_n}{400} \quad \text{or} \quad \frac{\rho_p}{1000} \right] \text{ns} \qquad (11.44)$$

where ρ_n and ρ_p are the resistivities in ohm-cm of n-type or p-type materials. The circuit defining the RC time constant of a silicon photodiode is shown in Figure 11.10, in which R_s is the internal resistance, R_j the junction resistance, R_L the load resistance, C_j the junction capacitance, and I the photocurrent generated by the incident light. R_s is comprised of three parts: the resistance of the undepleted region of the silicon chip, the contact resistance, and the collection resistance associated with the resistivity of the front surface generated by diffusion. The total capacitance of the diode could exceed the C_j value by an additional capacitance associated with the metalized surface for external packaging and wiring connections. Taking these factors into account, we write

$$\tau_{RC} = 2.2(R_S + R_L)(C_j + \Sigma\, C_n) \qquad (11.45)$$

Typical R_c values range from a few ohms to a few hundred ohms and C_j can be estimated by

$$C_j = \frac{19{,}200A}{\rho_n^{1/2} V_b^{1/2}} \quad \text{pF} \qquad (11.46)$$

where A is the photodiode active area. The C_j value could be increased by several picofarads due to additional external capacitance.

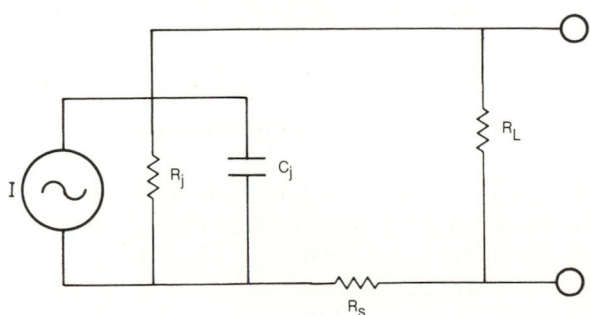

Figure 11.10   Equivalent circuit for a photodiode.

The diffusion time depends critically on the photon absorption depth, which is a strong function of λ. If the absorption depth exceeds the depletion depth, the diffusion time can be estimated for the p-on-n structure by the expression

$$\tau_d \simeq \frac{1}{13}\left(\frac{3}{\alpha} - 0.54\rho_n^{1/2}V_b^{1/2} \times 10^{-4}\right)^2 \tag{11.47}$$

where α is the absorption coefficient, and for the n-on-p structure by

$$\tau_d = \frac{1}{36}\left(\frac{3}{\alpha} - 0.32\rho_p^{1/2}V_b^{1/2} \times 10^{-4}\right)^2 \tag{11.48}$$

Risetime values for commercially available pin silicon photodiodes range from about 0.5 ns for small-area devices up to several microseconds for very large-area devices operating in a nonbias mode. For the latter case, the expressions above involving V_b must be modified by replacing V_b with V_0, which is the self-depletion bias voltage. Fall-time values are usually longer than τ_r for partially depleted devices but are about equal to τ_r for fully depleted structures. The τ_r for avalanche silicon photodiodes is significantly shorter than that of pin diodes; however, APDs require very careful control of operating bias voltage and temperatures and they are available only in very small sizes.

Assuming that τ_c and τ_d can be greatly reduced by optimizing the device configuration, the upper cutoff frequency f_c is then determined by the RC time constant alone as

$$f_c = \frac{1}{2\pi RC} \tag{11.49}$$

Combining Equations (11.49) and (11.42) we can relate the 10% to 90% rise time τ_r to the cutoff frequency f_c by the expression

$$\tau_r \simeq \frac{0.35}{f_c} \tag{11.50}$$

where the cutoff frequency f_c is often regarded as the bandwidth B of the system.

11.6 SIGNAL TO NOISE AND ERROR PROBABILITY

The signal-to-noise (S/N) ratio and the bit error rate, (BER) are major parameters for determining the fidelity of an optical receiver system. S/N is defined by the ratio of the mean square of the signal current to the sum of the mean-square fluctuation of the currents. For an avalanche photodiode having an internal gain M and an external load R_L, the signal-to-noise ratio at the output of a direct detection system can be written as

$$\frac{S}{N} = \frac{(mI_0M)^2/2}{2eI_0M^2FB + 4kT\,B/R_L} \tag{11.51}$$

where m is the modulation index of the light with an average power P_0. In the denominator, only the shot noise and the amplifier noise are considered. Other noise sources, such as thermal background, dark current, leakage current, and quantum fluctuation, should also be included, if any of them represents a significant portion of the total system noise. One meaningful quantity that determines the detector's figure of merit is the noise equivalent power (NEP), which is defined as the amount of light power impinging on the active detector area which will produce an output signal power that is equivalent to the noise output of the detector. NEP represents the minimum detectable power of a detector and is a function of the bandwidth. Typical NEP values for silicon pin diodes are in the range 1 to 10×10^{-14} W/Hz$^{1/2}$. More discussion on minimum detectable power is given in the next section.

The signal-to-noise ratio in general increases with increasing P_0. For high-fidelity communication systems, $S/N \gg 1$. The higher the S/N, the lower is the probability of errors in the system. In this section we examine the causes of errors and calculate the probability of bit error rate. Exact calculation is very difficult. Only a simple approximation is introduced to estimate the bit error rate for a series of digital Gaussian pulses incident on a photodiode. Because of pulse spreading, adjacent pulses can overlap, as shown in Figure 11.11. This problem is commonly referred to as intersymbol interference. Figure 11.11(a) shows a series of three incident Gaussian pulses having $\sigma = 0.25\tau$, where τ is the spacing between pulses. If $\sigma > 0.25\tau$, intersymbol interference becomes significant. As an example, we shall consider a case for which $\sigma = 0.5\tau$, as shown in Figure 11.11(b). The amplitude of a normalized Gaussian pulse is $1/\sqrt{2\pi}\,\sigma$. The intersymbol interference is represented by the eye formation. The eye opening ε is a measure of the interference. In this case $\varepsilon = 0.8\tau(1 - 0.27) = 0.584\tau$. To reduce the overlap, an equalizer can be used as a filter. One must be cautioned that in the implementation of an equalizer, more noise could be introduced into the system; therefore, great care must be exercised in choosing a filter that can match the input pulse shape precisely.

A simpler analysis of error probability for a digital communication system can be found in the book by Pratt (Ref. 11.5). For more comprehensive analysis, readers should consult the work by Personick (Ref. 11.6). We shall look at this problem with a simplified approach without going through great mathematical detail. Considering a digital system in which a series of short pulses is incident on the photodiode, the signal power can be written as

$$P_0(t) = \Sigma\,\delta h_p(t - n\tau) \tag{11.52}$$

where δ can be either zero or one and h_p represents the pulse shape. τ is the time

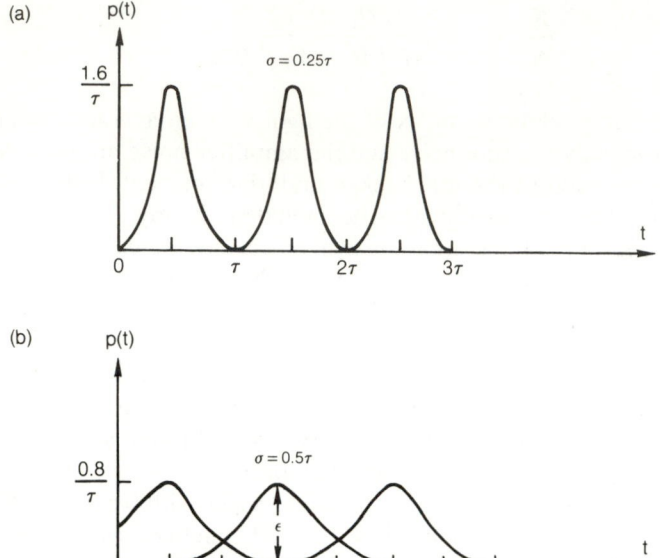

Figure 11.11 Series of three normalized Gaussian light pulses showing the effect of intersymbol interference with (a) $\sigma = 0.25\tau$; (b) $\sigma = 0.5\tau$. (From Ref. 11.4. Reprinted with permission of Academic Press, Inc., New York.)

between pulses or the time slot for one bit. The mean current generated from the photodiode at time t is given by

$$I(t) = \frac{e\eta}{h\nu} MP_0(t) \tag{11.53}$$

The mean voltage $\langle V \rangle$ at the output of an equalizer filter can be written as

$$\langle V \rangle = \frac{e\eta}{h\nu} MP_0(t) * h_d(t) * h_{eq}(t) \tag{11.54}$$

where $h_d(t)$ and $h_{eq}(t)$ are the impulse responses of the detector and the equalizer networks, and $*$ represents the convolution integral. As an example, we shall consider a simple digital PCM intensity modulation system employing direct detection. We shall derive the system probability of detection error for both shot-noise and thermal-noise-limited cases. The carrier signals of a PCM/IM system are represented by ones and zeros. The transmitted signals with a given information bandwidth can be corrected and filtered at the receiver terminal. For example, a bit decision can be made as to whether the photodetector output exceeds a decision threshold during a bit period. Within each bit interval, two probabilities of detection must be evaluated in order to determine the probability of system detection error. The first one is the probability P_S that the decision threshold will be exceeded by the transmitter signal and detector noise. The second one is the probability P_N that the threshold will be exceeded by the noise alone. The probability of making an error, P_E, depends on the probability that the signal and noise do not exceed the threshold when the transmitter signal is present, and that the noise alone exceeds the threshold when the transmitter signal is absent. For a system transmitting "one" or "zero" bits with a probability P, we write

$$P_E = P(1 - P_S) + (1 - P)P_N \qquad (11.55)$$

For equal likelihood (e.g., $P = \frac{1}{2}$) we have

$$P_E = \frac{1}{2} 1(1 - P_S + P_N) \qquad (11.56)$$

For shot-noise-limited operation, the threshold decision is based on the actual counting of the photoelectrons. The number of counts, n, during a bit period must be compared with the threshold number n_{th}. If n is equal to or larger than n_{th}, a "one" bit is assumed to be registered. If n is less than n_{th}, a "zero" bit is registered. Because of extremely low level detection, the detection probabilities are assumed to have a Poisson distribution. We therefore write

$$P_S = \sum_{n=n_{th}}^{\infty} (\langle n_S \rangle + \langle n_N \rangle)^n \, \frac{\exp(-\langle n_S \rangle - \langle n_N \rangle)}{n!} \qquad (11.57)$$

and

$$P_N = \sum_{n=n_{th}}^{\infty} \langle n_N \rangle^n \, \frac{\exp(-\langle n_N \rangle)}{n!} \qquad (11.58)$$

where $\langle n_S \rangle$ is the average number of photoelectrons per bit due to the signal and $\langle n_N \rangle$ is the average number of photoelectrons per bit due to the noise.

The decision rule for judging the presence of a signal is usually followed by a ratio test:

$$\frac{P_S}{P_N} \geq \frac{1 - P}{P} \qquad (11.59)$$

where P is the a priori probability that a signal is transmitted. Solving for the threshold value n_{th} for which equality exists in Equation (11.59) (e.g., $P = \frac{1}{2}$) yields

$$n_{th} = \frac{\langle n_S \rangle}{\ln(1 + \langle n_S \rangle / \langle n_N \rangle)} \qquad (11.60)$$

Figure 11.12 is a plot of the likelihood ratio tests threshold n_{th} as a function of the average number of signal and shot-noise photoelectron counts for a pulse-code intensity modulation system. Figure 11.13 is a plot of the probability of detection error for the shot-noise-limited decision threshold based on the

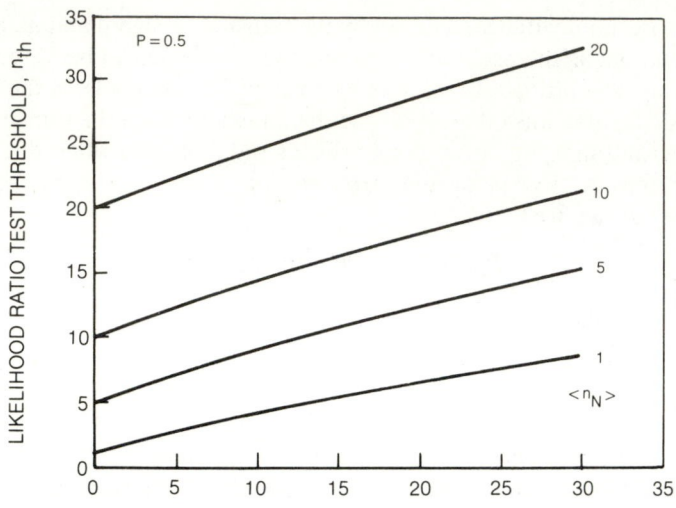

Figure 11.12 Threshold n_{th} as a function of $\langle n_S \rangle$ and $\langle n_N \rangle$ for a PCM/1M system with $P = \frac{1}{2}$. (From Ref. 11.5. Reprinted with permission of John Wiley and Sons, Inc., New York ,© 1969.)

likelihood ratio test. The cusps in the curves are due to integer changes in the detection threshold as $\langle n_S \rangle$ increases. For the case $\langle n_N \rangle = 0$, a value for P_E reaches 10^{-9} when $\langle n_S \rangle = 21$, consistent with the quantum noise-limited case discussed before.

As the system noise increases, the optical power must be increased accordingly. If photoelectrons contained in one bit become very large, Gaussian statistics can be used to describe the detection probabilities. For most optical fiber systems, this turns out to be the case because system operation is usually thermal noise limited. We shall write

$$P_S = \int_{I_{th}}^{\infty} \frac{1}{\sqrt{2\pi}\,\sigma} \exp\left[-\left(I - \frac{e}{\tau}\langle n_S \rangle\right)^2 / 2\sigma^2\right] dI \qquad (11.61)$$

and

$$P_N = \int_{I_{th}}^{\infty} \frac{1}{\sqrt{2\pi}\,\sigma} \exp(-I^2/2\sigma^2)\,dI \qquad (11.62)$$

where I_{th} is the decision threshold current, τ the bit period, and σ^2 the thermal noise current variance. By the ratio test it can be shown that for $P = \frac{1}{2}$,

$$I_{th} = \frac{e}{2\tau}\langle n_S \rangle \qquad (11.63)$$

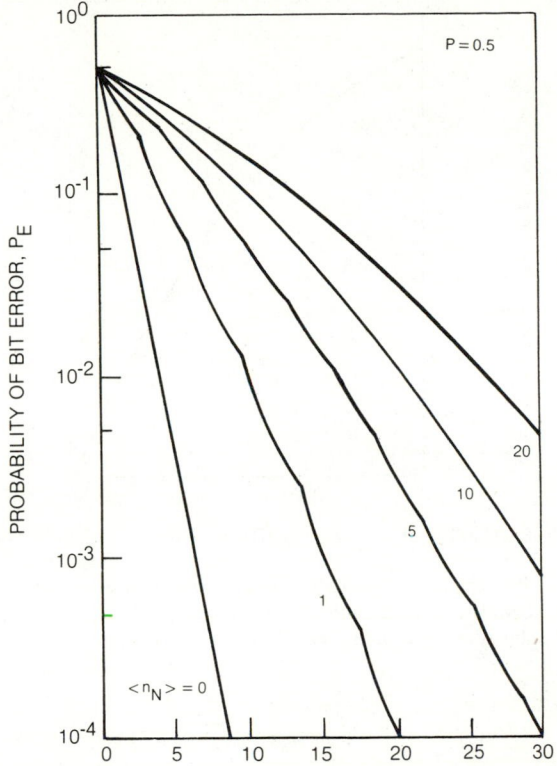

AVERAGE NUMBER OF SIGNAL PHOTOELECTRONS PER BIT, $<n_S>$

Figure 11.13 Probability of detection error as a function of $\langle n_S \rangle$ for a shot-noise-limited PCM/1M system with $P = \frac{1}{2}$. (From Ref. 11.5. Reprinted with permission of John Wiley and Sons, Inc. New York, © 1969.)

The probability of detection error can be expressed in terms of the error function as

$$P_E = \frac{1}{2}\left[1 - \mathrm{erf}\left(\frac{e\langle n_S \rangle}{2\sqrt{2}\,\tau\sigma} \right) \right] \tag{11.64}$$

where

$$\mathrm{erf}(Q) \equiv \frac{2}{\sqrt{2\pi}} \int_Q^\infty \exp\left(-\frac{x^2}{2} \right) dx \tag{11.65}$$

Substituting Equation (11.39) for the thermal noise current variance into Equation (11.64), we obtain

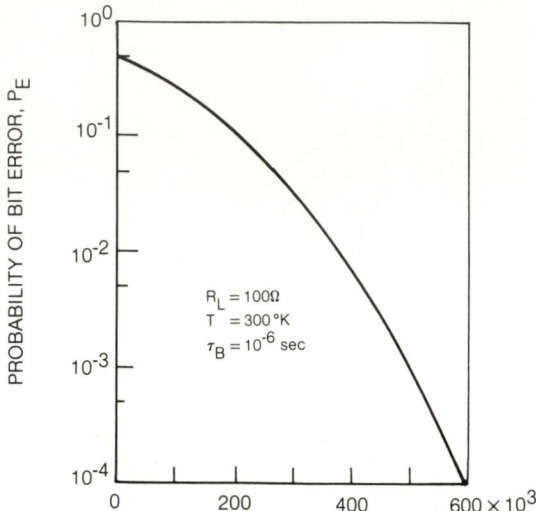

AVERAGE NUMBER OF SIGNAL PHOTOELECTRONS PER BIT, $<n_S>$

Figure 11.14 Probability of detection error as a function of $\langle n_S \rangle$ for a thermal-noise-limited PCM/1M system with $P = \frac{1}{2}$. (From Ref. 11.5. Reprinted with permission of John Wiley and Sons, Inc. New York, © 1969.)

$$P_E = \frac{1}{2}\left[1 - \mathrm{erf}\left(\frac{e\langle n_S\rangle}{4\sqrt{2}} \sqrt{\frac{RB}{kT}} \right) \right] \tag{11.66}$$

Figure 11.14 is a plot of the probability of detection error for a PCM/IM direct detection communication system under a thermal-noise-limited operation as a function of the average number of photoelectrons during each bit period. In this calculation, we have assumed that the transmitter signal is only slightly above the noise. To reduce the probability of error, one must significantly increase the signal power. In the Gaussian approximation, the bit error rate (BER) can be expressed as

$$\mathrm{BER} = \tfrac{1}{2}\,\mathrm{erf}(Q) \tag{11.67}$$

where Q is just one-half of the S/N ratio. Figure 11.15 is a plot of the error function versus Q. For an error probability of 10^{-9}, we obtain from Figure 11.15 a corresponding value for Q to be 6, or a corresponding S/N value for the optical power of 12 or 10.8 dB. This corresponds to an electrical S/N of 144 or 21.6 dB.

The analysis above is oversimplified. In general, receivers contain inter-symbol interference, and detectors such as APD have excess multiplication noise. For these reasons, the Gaussian approximation is not completely satisfactory. However, experimental results indicate that the Gaussian approxi-

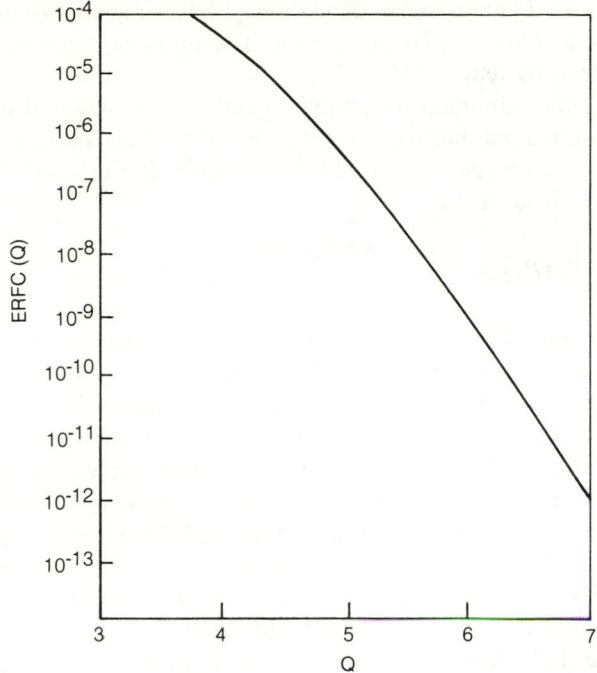

Figure 11.15   Plot of the error function.

mation for calculating receiver sensitivity is good to within 1 dB. The short-coming lies in the fact that the optimum threshold is closer to the center of the "eye" than the Gaussian approximation predicts. This is a consequence of the highly skewed nature of the probability distribution of randomly multiplied shot noise.

11.7 MINIMUM DETECTABLE POWER

When designing a receiver, it is important to know the minimum power P_{min} required to obtain a reliable detection by maintaining a desired level of fidelity for a given information bandwidth. The value of P_{min} can be determined from the information on the performance characteristics of detectors and amplifiers. Ideally, the highest receiver sensitivity is reached when the minimum detectable signal is not limited by the receiver, but rather by fluctuations in the signal current alone. In this case, the minimum detectable power becomes

$$(mP_0)_{min} = \frac{4h\nu}{m\eta} FB \frac{S}{N} \qquad (11.68)$$

where m is the modulation index. Equation (11.68) represents the quantum-noise-limited sensitivity of direct detection. The noise equivalent power NEP can be determined by letting $S/N = 1$.

In reality, the minimum detectable signal of a broadband direct photo-detection system without internal current gain ($M = 1$) is usually limited by the thermal noise of the detector and the load resistance. In this case, the minimum detectable power is given by

$$(mP_0)_{min} = \frac{h\nu}{e} \left(\frac{8kTB}{\eta^2 R_L} \right)^{1/2} \left(\frac{S}{N} \right)^{1/2} \tag{11.69}$$

At room temperature, these values can be significant especially for very large B value. It is possible to reduce the noise by using avalanche diodes with internal gain. However, it is difficult to reach the ideal case because of the excess noise factor $F(M)$. The best that one can do is to increase the gain until the shot noise reaches a level comparable to the thermal noise. At optimum gain, the minimum detectable signal can be reduced by approximately a factor of $2/M$.

Smith and Garrett (Ref. 11.7) have analyzed this problem by calculating the optimum gain required to reach a desired S/N level with a minimum optical power in terms of the number of noise-equivalent photoelectrons. This analysis is rather complicated and involves many parameters including various noise sources, intersymbol interference, and nonzero extinction ratio. One of the most significant results emerging from this analysis indicates that the optimum gain depends critically on the amplifier noise. If one can eliminate the amplifier noise, the optimum gain will occur at very low value (≤ 10) and the receiver becomes relatively insensitive to intersymbol interference. It is possible to reduce the amplifier noise by using a FET in the input stage of the amplifier with a hybrid circuit technology. This approach makes the pin diode very attractive compared to an APD especially from the economic point of view. Because pin diodes do not require high bias voltage, there are no special requirements for gain control as for the case of APD, which requires special stabilizing circuitry against temperature fluctuation. However, the penalty that one pays for using pin diodes is the FET bandwidth limitation. As already discussed, the Z value for the case of FET receivers varies as $B^{1/2}$. Therefore, for a system with large bandwidth, it is necessary to use a bipolar amplifier because its Z value is independent of B.

The performance of typical receivers that use FET or bipolar front ends is shown in Figure 11.16. These sensitivity curves have been calculated by using Gaussian approximation and assuming an error rate of 10^{-9}. For example, at 100 Mb/s, a receiver sensitivity of -55 dBm of optical power can be achieved by using a bipolar amplifier with an APD. Similar sensitivity can also be achieved by using a FET amplifier and a pin photodiode without gain; however, the bit rate is severely limited to below 10 Mb/s. Figure 11.16 indicates that extremely sensitive optical receivers can be built for both high-data-rate digital or wide-bandwidth analog transmission signals. For low-data-rate systems, a

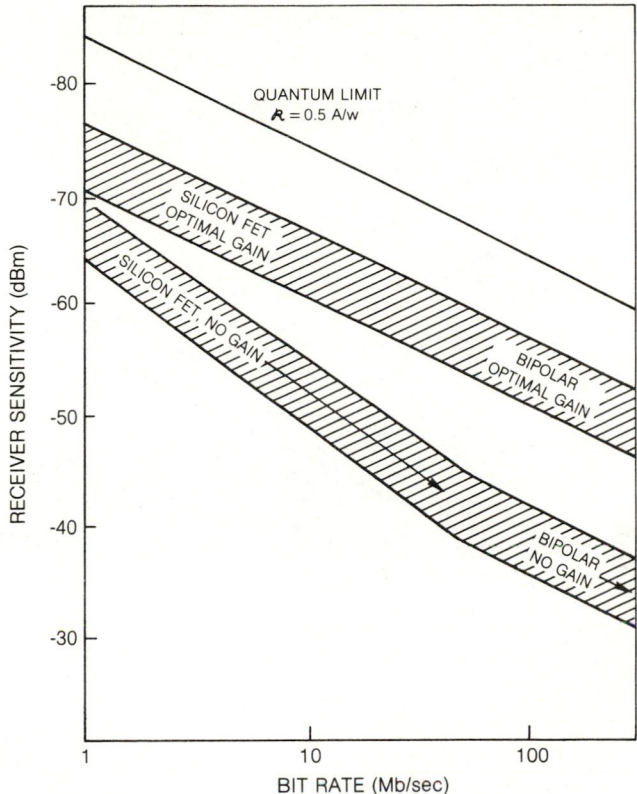

Figure 11.16 Performance of typical photoreceivers with and without gain. (From Ref. 11.4. Reprinted with permission of Academic Press, Inc., New York.)

simple silicon FET front-end amplifier with a pin photodiode is sufficient. At higher bit rates ($>$50 Mb/s), the use of APD detectors and bipolar amplifiers with automatic gain control circuitry become necessary. The design of analog systems is more complex than that of digital systems because in digital systems distorted signals can easily be regenerated by using readily available electronic circuits. On the other hand, if the linearity of the source and coherent transmission can be achieved without too much difficulty, analog systems should certainly be considered.

Recent advances (Ref. 11.8) in long-wavelength devices have brought a new perspective for future optical fiber systems. Figure 11.17 summarizes the state of the art of the performance of optical digital repeaters operating in the spectral region 1.2 to 1.6 μm. The ordinate is the average number of signal photoelectrons required in a bit interval to achieve an error probability of 10^{-9}, assuming an equal probability of zeros and ones. This number is proportional to the average optical energy per pulse. The abscissa is the bit rate. The solid

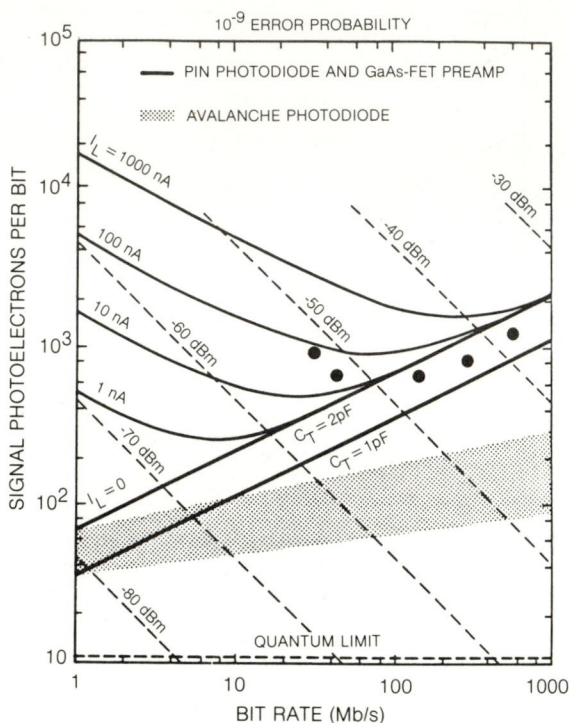

Figure 11.17 Long-wavelength receiver sensitivity as a function of the bit rate for $P_E = 10^{-9}$. (From Ref. 11.8. Reprinted with permission of IEEE, © 1983.)

curves represent calculated results for the range of parameters pertinent to currently available InGaAs pin photodiodes and GaAs FETs. It is interesting to note that the sensitivity, when measured in number of photoelectrons per bit, is independent of the wavelength. Since the optical power is proportional to the number of photoelectrons per bit, it is possible to superimpose on Figure 11.17 with dashed diagonal lines, which correspond to the minimum detectable power at $\lambda = 1.3$ μm and a quantum efficiency of 0.7 for the photodiode. The quantum-noise-limited detection is indicated by the dashed horizontal line, located at $\langle n_S \rangle = 11$. The dots represent the best experimental values presently available. It is anticipated that the performance of long-wavelength components will continue to improve rapidly. The two solid straight lines with slopes proportional to $B^{1/2}$ represent the receiver performance of a pin photodiode and a GaAs FET preamplifier having a capacitance at two different values (1 and 2 pF), a transconductance of 50 mA/V, and negligible leakage current. As leakage current increases, receiver performance degrades sharply at low bit rates, but is relatively independent of the signal intensity I_L at high bit rates, as shown in the figure by the confluence of the curves for various values of I_L. The dotted band represents projected performance for future avalanche devices. In conclusion, long-wavelength repeaters operating at or above 1 Gb/s rates with sensitivities of about 12 to 15 dB above the quantum noise are very realistic projections in the not-too-distant future. The results of Figure 11.16 can be used by designers of present systems and the charts in Figure 11.17 are useful for future system planning purposes.

PROBLEMS

11.1. With 50 V of reverse-biased voltage, calculate the doping level ρ_p required to fully deplete the photoexcited carriers within a width $W = 0.1$ mm. What is the maximum electric field inside this device?

11.2. Show that the second moment of Poisson's distribution is

$$\langle n^2 \rangle = \langle n \rangle + \langle n \rangle^2$$

11.3. Calculate the Johnson noise of a 5000-Ω resistor for a bandwidth $B = 10$ MHz.

11.4. If the diffusion time is negligible, calculate the rise time of a silicon photodiode having an active area of 0.1 cm^2 and a thickness of 0.04 cm. The resistivity is 400 Ω-cm. The device is reverse biased at 50 V and is connected to an external load of 50 Ω.

11.5. The absorption depth of a silicon photodiode at 0.9 μm is about 75 μm. Assuming the same parameters as in Problem 11.4, what would be the depletion depth of the diode? Calculate the diffusion time if the reverse-biased voltage of the diode is reduced to 5 V.

11.6. Using the Gaussian pulse-shape function, (a) plot the curves for $\sigma = 0.25\tau$ and $\sigma = 0.5\tau$, and (b) determine the eye-opening parameter formed by the zero and one pulses with a $\sigma = 0.5\tau$.

11.7. Calculate the minimum optical power required for a 50 Mb/s communication system that has a total system loss of 40 dB.

REFERENCES

11.1. Special Issue on Quarternary Compounds Semi-conductor Materials and Devices—Sources and Detectors. *IEEE J. Quantum Electron., QE-17,* (1981).

11.2. T. Pearsall, *Optical Fibre Communications,* ed. M. J. Howes and D. V. Morgan, John Wiley & Sons, Inc., New York (1980).

11.3. R. J. McIntyre, *IEEE Trans. Electron Devices, ED-13,* 164 (1966), and *ED-19,* 703 (1972); P. P. Webb, R. T. McIntyre, and J. Conradi, *RCA Rev., 35,* 234 (1974); J. Conradi, *IEEE Trans. Electron Devices, ED-19,* 713 (1972).

11.4. S. D. Personick, *Fundamentals of Optical Fiber Communications,* ed. M. K. Barnoski, Academic Press, New York (1976).

11.5. W. K. Pratt, *Laser Communication Systems,* John Wiley & Sons, Inc., New York, 1969.

11.6. S. D. Personick, *Bell Syst. Tech. J., 52,* 843 (1973).

11.7. R. D. Smith and I. Garrett, *Opt. Quantum Electron., 10,* 211 (1978).

11.8. T. Li, *IEEE J. Select. Areas Commun., SAC-1,* 356 (1983).

12

Optical Fiber Systems

12.1 INTRODUCTION

With the accumulated knowledge of various optical fiber components, we are now in a position to analyze optical fiber systems for which each component has a given set of specifications to be considered for various applications. Based on system requirements and constraints, a designer must first choose the most suitable components and then evaluate overall performance in terms of the total frequency response and power budget of the system. An optimum system is one in which the power margin is at a minimum while the system response is faster than that required for the system bandwidth and fidelity. To achieve this, one must perform a trade-off analysis among various possible component selections. This type of design analysis usually involves an iterative process for which computer-aided design (CAD) is often necessary. In this section we introduce various techniques for selecting suitable components. To assure system reliability, an allowance of component degradation is often made. Parameters involved in component and system optimization are summarized. Examples are given to illustrate the procedure for making this type of analysis. Descriptions on several practical optical fiber systems are also given.

12.2 PRELIMINARY DESIGN GUIDE

Four basic inputs required for the design of a communication link are (1) the data rate or bandwidth; (2) the fidelity, indicated by either S/N or BER; (3) the link length and number of terminals; and (4) the type of data and signal waveforms to be transmitted. Once these requirements are specified by the user, the designer can begin the iterative process for system optimization. It is customary to start the analysis by examining the total dispersion bandwidth of the fiber and the source combination for a given terminal spacing. If the pin photodiode is sufficient, there is no reason to use the more sophisticated circuit for operating APD. If the initial choice for a fiber–source combination cannot meet the required data rate, either an upgrading of the combination or a repeater may be considered without sacrificing the transmission fidelity. Since the optical fiber industry is still in a stage of rapid growth and expansion, both the quality and cost of various components will undergo several significant transitions. Therefore, it is desirable to install the best commercially available fibers into systems because the replacement cost of this item is far greater than that of terminal components such as sources and detectors. It is expected that considerably more reliable sources and efficient couplers will become available at a lower cost in the future. If necessary, the system can easily be upgraded by only replacing these terminal components. It is advisable to allow for possible system expansion in data-handling capacity even though the initial investment cost may be higher.

Table 12.1 serves as a guide for the initial choice of various components based on the communication length and data-rate requirements. The initial selection is not important because through an iterative procedure these components will be optimized in terms of the total system performance, cost, and availability. There are many companies in the United States, Canada, Japan, and in Europe, where commercial product lines are well characterized and performing reliably in accordance with their specifications.

For data transmission between short distances at rates below 2 Mb/s, very reliable systems can be made at very reasonable cost. For these systems the best choice of source is LED, because the cost is low and its performance is very reliable, with a typical lifetime greater than 10^6 h. Using large NA values, the coupling efficiency between a LED and a fiber bundle can be made as high as $1 - L_p$, where L_p is the packing fraction loss. For short communication links at low data rates, fiber dispersion usually does not present a problem and the information can be transmitted by using either a pulse code or IM with either a transistor logic or a simple IC circuit. A pin photodiode with an integrated FET preamplifier at the front end is sufficient to use as the receiver. The typical receiver noise for a pin receiver at 2 Mb/s is below -40 dBm for BER of 10^{-9}. In addition to the considerations above, environmental constraints, such as the operating temperature range, humidity, corrosion, radiation, and so on, must also be taken into account for the design of system packaging.

For system requirements falling in the medium range (i.e., $L \leq 1$ km and $B \leq 100$ Mb/s) the component trade-off and selection process become more involved, because there exist many options, each of which requires a detailed examination, if the system is to be optimized. In this case, a computer-aided

TABLE 12.1 Preliminary Choice of Components for Various System Requirements on Channel Length L and Bandwidth B

Component	$L < 100$m $B < 2$ Mb/s	$L < 1$ km $B < 30$ Mb/s	$L < 1$ km $B < 100$ Mb/s	$L < 1$ km $B < 100$ Mb/s
Source	LED (GaAs)	LED (AlGaAs) Ld (AlGaAs)	LD (AlGaAs)	LD (InGaAsP) LD (AlGaAs)
Fiber	Step-index fiber bundle	Step-index fiber or bundle	Graded-index (low loss)	Single mode or graded-index
Detector	pn or pin	Si pin	Si APD GaAs APD	Si APD Ge APD
Amplifier	FET	Trans-impedance	Temperature-compensated bipolar	Temperature-compensated bipolar
Driver	IC or transistor	Prebias	Pre-bias or EO modulator	EO modulator

design can be very helpful and will be outlined in the following section. For very high-data-rate and very long-distance communication channels, the choice of components is again limited. In this case one inevitably chooses the best in each category.

12.3 DESIGN ANALYSIS

For systems with intermediate bandwidth requirements, a detailed analysis must be performed to optimize the system by evaluating its performance with specifications for the transmitter, cable channels, and the receiver. In the case of the transmitter, the parameters that need to be considered include the optical power, wavelength, spectral width, beam size and shape, frequency response, output linearity, and modulation or coding format. Other parameters that also have an influence on the system performance are the lifetime, transmitter noise, and environmental factors such as temperature, vibration, and humidity. In the case of the fiber, critical parameters are the diameters for the core and the cladding, index profile, attenuation coefficient, modal and material dispersion, pulling strength, and microbending loss. In the case of the receiver, the noise equivalent power, frequency response, quantum efficiency, amplifier noise, and gain are important parameters that characterize the receiver performance. Even though the system trade-off analysis among these parameters can be rather complex, the system requirements in terms of data rate, distance between terminals, number of terminals, fidelity, and environmental constraints such as temperature range, humidity, vibration, space, and cost usually dictate the choice of certain key components that are the most likely candidates for the system. Once the initial choices are made, a routine iterative process can be followed for the optimization of each component. Figures 12.1, 12.2, and 12.3 are the flowcharts for design analysis of a transmitter, fiber, and a receiver, respectively. The analysis is complicated by the fact that some of the parameters in one subsystem are dependent on parameters in another subsystem. For example, the choice of spectral width and wavelength for the source depends on the dispersion characteristics of the fiber and also on the spectral response of the detector. For this reason, an optimum system may not be unique, and often the choices are made on the basis of availability and cost.

Once a system configuration is established, two computations must be carried out, one involving calculation of the total system response and the other related to the power budget. The former gives an estimate for the system information-handling capability and the latter gives an estimate on the system reliability, based on the available transmitter power. Both of these calculations are necessary and provide an indication of the fidelity of the system. The total system response can be expressed in terms of the rise time for the transmitter τ_t, the fiber τ_f, and the receiver τ_r as follows:

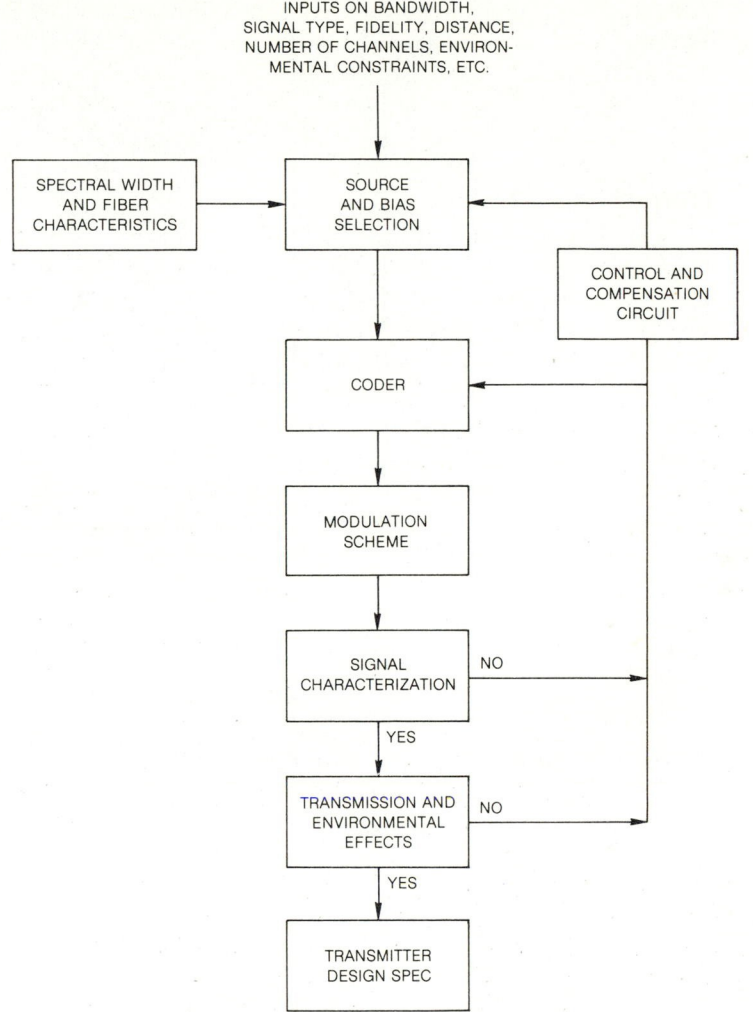

Figure 12.1   Design flowchart for optical fiber transmitter.

$$\tau_s \simeq \sqrt{\tau_t^2 + \tau_f^2 + \tau_r^2} \qquad (12.1)$$

On the other hand, the amount of information to be transmitted by the system with bandwidth B can also be expressed in terms of a time constant τ_B. Table 12.2 gives τ_B in terms of either the bit rate or the bandwidth B for various signal types. If $\tau_s < \tau_B$, the system response is considered to be adequate.

The power budget can be estimated by calculating (1) the minimum optical power P_d that can be detected by a chosen receiver at a given bandwidth and a signal level required by the S/N ratio, and (2) the maximum signal power

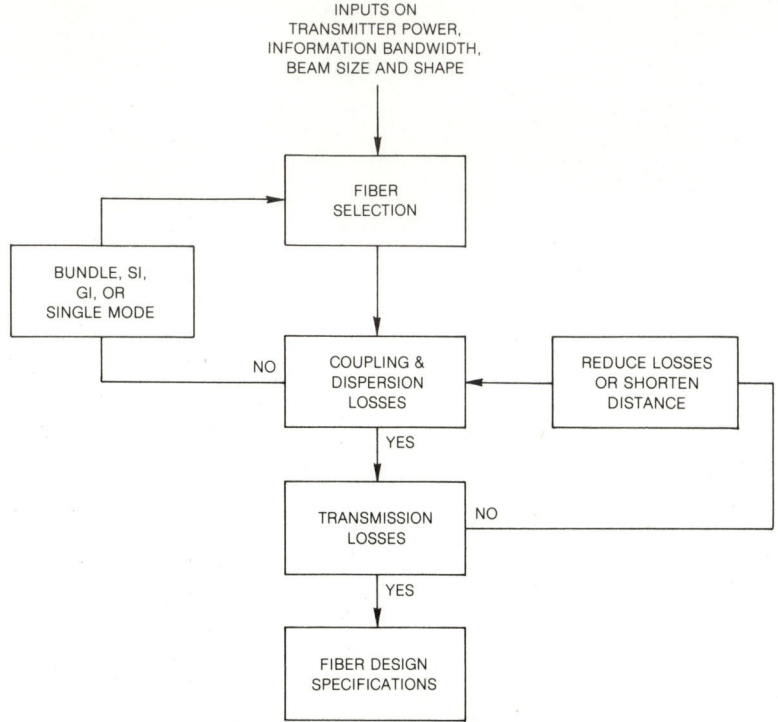

Figure 12.2   Design flowchart for optical fiber transmission channel.

P_s that can reach the receiver after allowing for all possible system losses. If $P_s > P_d$, the system so chosen is considered to be adequate. The power margin that is the difference between P_s and P_d should be at least 10 dB to allow for component degradation and other unexpected problems. Table 12.3 gives the results of an analysis for power budget and rise times of a multichannel color TV system. The system components consist of a AlGaAs DH laser, a graded-index fiber, and a star coupler, which form a distribution network to 10 receiving terminals at a distance of 0.5 km utilizing a APD receiver.

Another example, given below, describes a typical digital data link involving only one transmitter and one receiver at a bit rate of 20 Mb/s. The length of this link is assumed to be 8 km. Over this length, four graded-index fiber cables are used with a total of three splicings. A LED is chosen as the source with an average output at 3 dBm. An APD detector is used in this link and has a time constant of 3 ns. Table 12.4 gives the results of the power budget and system rise-time analyses for this data link. The calculated system rise time is about 14 ns, which is well within the time requirement (35 ns) specified by the system bandwidth. However, the power budget calculation indicates that the power margin is too small to allow for any excess loss. To increase the excess

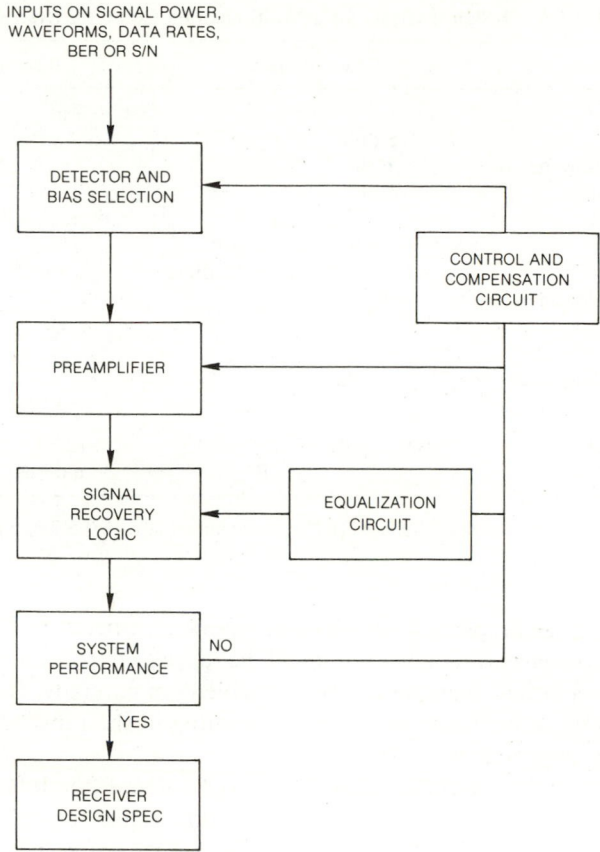

Figure 12.3   Design flowchart for optical fiber receiver.

power margin a lower-loss fiber with an attenuation coefficient of less than 5 dB/km must be considered. This example illustrates that a low-cost but reliable LED source is sufficient for use in a high-fidelity (BER = 10^{-9}) data transmission system operating at a bit rate of 20 Mb/s over a distance of 8 km without the need of a repeater. If a laser source is used, the length of this link can easily be increased by about a factor of 2 or more. A considerably longer

TABLE 12.2   Rise-Time Estimate for Various Signal Types

Signal	τ_B
NRZ	0.7/bit rate
RZ	0.7/2(bit rate)
IM	0.7/2B
PCM	1/(sampling rate)(bits/sample)(B)

TABLE 12.3   Design Analysis for a Multi-channel Color TV System[a]

Component	Power Budget	Risetime
Transmitter		$\tau_t = 5$ ns
AlGaAs laser	$P_l = 15$ dBm	
Degradation allowance	3 dB	
Source coupling loss	5 dB	
Fiber and coupler		$\tau_f = (\tau_{mod}^2 + \tau_{mat}^2 + \tau_c^2)^{1/2}$
GI (3 dB/km)	$a_f = 3 \times 0.5 \times 10 = 15$ dB	
Star coupler	$a_c = 10 \log 10 + 13 = 23$ dB	$= 1.6$ ns
Degradation allowance	3 dB	
Receiver		$\tau_r = 3$ ns
APD	$P_r = -55$ dBm	
Degradation allowance	5 dB	
Receiver coupling loss	10 dB	
System performance:	Power margin $= 7 - 41 + 40$	$\tau_s = (25 + 2.56 + 9)^{1/2}$
	$= 6$ dB	$= 6.05$ ns

[a]System requirements: $B = 6$ MHz, BER $= 10^{-9}$, 10 terminals at $L = 0.5$ km, $\tau_B \simeq 8$ ns.

data link can be accomplished by using components operating at 1.55 μm. Examples of a longer-wavelength system will be discussed in Section 12.6. The components used in this example are all available commercially. It is expected that the performance of these components operating in the range 0.82 to 0.85 μm will also be improved with time.

The last example given below describes an optical repeater. It basically consists of a photodiode, an amplifier, an equalizer, and a signal regenerator. Figure 12.4 is a block diagram of a typical PC, optical fiber repeater. The attentuated

TABLE 12.4   Design Analysis for a Digital Data Transmission System[a]

	Power Budget		Rise Time
Source: LED	3 dBm		6 ns
Signal: NRZ	-3 dBm		
Fiber: G1: $\alpha = 5$ dB/km	-40 dB	Dispersion	10 ns
Splicing loss (3)	1.5 dB		
Detector: APD	59 dBm		3 ns
Source/fiber coupling	10 dB		
Fiber/detector coupling	1 dB		
Temperature degradation allowance	1 dB		
Other allowance	5 dB		
System requirement: $40 + 1.5 + 10 + 7 =$	58 dB	$0.7/20$ Mb/s $= 35$ ns	
System performance: $3 - 3 + 59 =$	59 dB	$1.11\sqrt{6^2 + 3^2 + 10^2} = 13.4$ ns	

[a]System requirements: BR $= 20$ Mb/s, BER $= 10^{-9}$; two terminals, terminal spacing $= 8$ km, data coding: PCM, NRZ.

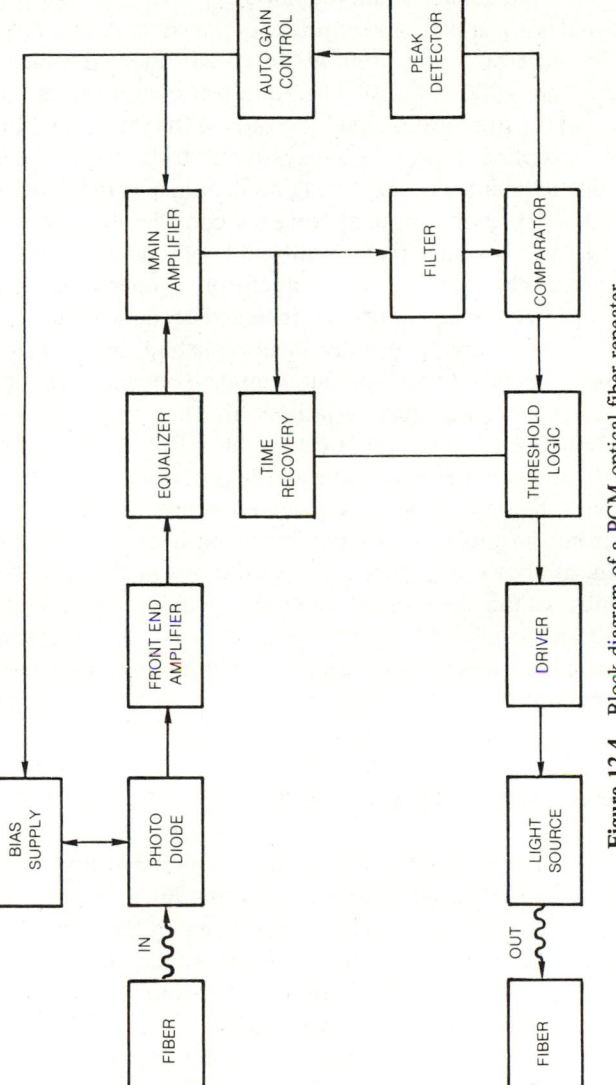

Figure 12.4 Block diagram of a PCM optical fiber repeater.

and distorted optical signal, after propagating along a long fiber, is collected by a photodiode, which converts the optical signal into an electrical signal. The amplifier, which is designed for a capacitive source typically of a few picofarads and having an input impedance equal to $[j\omega(C_{ampl} + C_{det}) + 1/R_{in}]^{-1}$, amplifies the received signal with as little added noise as possible. A low-noise front end of the amplifier must be very carefully designed. Typical values for the Z parameters vary from 1000 to 5000. The equalizer compensates for the effects of the detector and the fiber dispersion by reducing the intersymbol interference at a penalty of the optical power. If the system is truly dispersion limited, the repeater spacing can be increased by using an equalizer. On the other hand, the equalizer should not be used if optical power is considered to be at a premium. The signal regenerator is made up of a number of components, including a signal comparator, timing recovery circuit, a waveform regenerating logic, a driver, and a source. The regenerator must be designed to be able to reproduce the desired pulse shape. This is usually accomplished by using a feedback mechanism that responds to the output signal. The automatic gain control (AGC) circuit compensates for variations in the incoming signal levels, amplifier gain, and the temperature effect in the APD device. To reproduce a desired pulse stream, it is necessary to provide a periodic clock signal which is synchronized with the time slots of the received pulses. Such a clock signal can be generated from the receiver's output by using a phase-locked-loop timing recovery circuit, as shown in Figure 12.4. Variations in the pulse pattern could cause a phase jitter of the clock signal which can generate an accumulative error in the output. A trade-off between the amount of narrow-bandwidth filtering required in the phase-locked loop and the ability to lock onto the phase modulated component must be made in establishing a reliable phase-locked loop.

12.4 TELECOMMUNICATION SYSTEMS

Analog signal processing, in particular, the frequency-division-multiplexing technique, has been widely used in earlier telecommunication systems. In recent years, however, digital signal processing has been by far the fastest-growing technique, and will probably dominate the future telecommunication network. The reason for the rapid growth of digital transmission over the analog system has much to do with economics. The terminal cost for a FDM is considerably higher than that for a PCM, which requires a simpler circuitry, involving only an A/D converter and a digitizer. For each voice channel, the analog speech signal is limited to a frequency range of 300 to 3400 Hz. Each voice signal is usually sampled at a rate of 8 kHz and each sample is converted into a digital "byte" involving 8 bits. Usually, the first bit gives the polarity of the signal and the remaining 7 bits provide the magnitude of the signal on a logarithmic scale. Within this framework, each voice channel requires a transmission rate of 64 Kb/s. For a time-domain multiplex system containing 24 voice channels, a bit

TABLE 12.5   Standards for Digital Telecommunication Rates
in the United States and Europe

Class	Bell Systems		European Systems	
	Channel	Rates (Mb/s)	Channel	Rates (Mb/s)
T1	24	1.544	32	2.048
T2	96	6.312	120	8.448
T3	672	46.304	480	34.368
T4	4032	274	1920	139.364

stream of 1.54Mb/s is required. Table 12.5 shows the standard transmission rate in the United States and in Europe. The time-multiplexed channels are divided into consecutive time slots. Among these slots, there will be at least one slot used for the purpose of synchronization, redundancy, or transmitting the dialing number. Therefore, the total number of voice channels is slightly less than those allowed by the total bit rates of the system. These time-multiplexed blocks are classified by labeling them with T1, T2, T3, and so on, for U.S. telecommunication systems. For European networks, the standards are slightly different and correspond to the information rates shown in Table 12.5.

Transmission lines of present telecommunication systems utilize (1) twist wires, (2) coaxial cables, (3) terrestrial and satellite microwave, and (4) optical fibers. Each system has its advantages and disadvantages. This section presents a brief historical sketch of these systems. The first transcontinental telecommunication occurred in 1915 and was transmitted on an open-wire pair. By 1930, open-wire multiplex systems with up to 12 voice channels were in widespread use. A major breakthrough was the development of the first coaxial cable transmission, which was put into service in 1940. This was a 3-MHz system capable of transmitting 300 voice channels or a single TV channel. Since then the coaxial systems have evolved into a T4 (274 Mb/s) coaxial system, which was put into service in 1975.

Microwave links, unlike the radio waves that are reflected by the ionosphere, propagate in a line of sight. At microwave frequencies, much wider transmission bandwidth becomes available. The first microwave system operating at 4 GHz, which linked New York and Boston, was put into service in 1948. It has been the workhorse of the T2 Bell systems serving coast to coast since 1951. The latest system introduced in 1973 has a capacity of carrying 12 two-way communication channels, each with approximately 1800 voice circuits.

After the breakthrough of low-loss glass fibers in 1970, the first major fiber optic T3 system was put into field trial at Atlanta in 1976 (Ref. 12.1). Since then many operating systems at rates ranging from T1 to T4 have been installed. Details on these systems are given in the following section. To make a

TABLE 12.6 Comparison between Coaxial, Microwave, and Fiber Optic Communication Systems at a 45-Mb/s Bit Rate

System	Format	Gain (dB)	Loss (dB)	Length (km)
Coaxial (9.5-mm coax)	4B3T	90	85	7
Microwave (11 GHz)	16 QAM	109	100	50
Fiber optic (1.55 μm)	NRZ	51	45	50

comparison between coaxial, microwave, and fiber optics systems, we shall arbitrarily select a 45-Mb/s digital communication system and list typical performance characteristics expected from each of the foregoing three technologies. Table 12.6 summarizes system performance characteristics for each of the three using the state-of-the-art components. For comparison purposes the most important parameter is the maximum transmission length for each system without the need of a repeater. As shown in Table 12.6, the coaxial system is by far the shortest of the three. The optical transmission line can be made as long as the microwave link, provided that long-wavelength components ($\lambda = 1.55$ μm) are used. Looking toward the future, both the optical and the microwave systems will likely continue to grow. However, it is anticipated that the growth in light-wave technology will be more significant because optical systems are considerably simpler and can be manufactured more economically in the long run. Microwave technology is now a mature field in which most microwave systems are operating near their fundamental noise-limited capacity. Further improvements for these nearly ideal systems will be more difficult to achieve than for the newly emerging technology of fiber optics.

12.5 IN-SERVICE OPTICAL COMMUNICATION SYSTEMS

There exist many in-service optical communication systems both in the United States and overseas (Ref. 12.1). In the United States, American Telephone and Telegraph Corporation has installed two operating systems: one is an interoffice trunk that transmits at a rate of 44.7 Mb/s, located within the Chicago metropolitan area; the other is an intercity transmission line service between New York City and Washington, D.C., at a rate of 90 Mb/s. In Japan, one of the earlier projects is the Higashi-ikoma Optical Visual Information System (Hi-Ovis), which began to operate in 1978, after completing a successful field trial in 1976 (Ref. 12.2). The Hi-Ovis project is sponsored by the Japanese government; its objective is to provide a two-way interactive CATV service in the model town of Higashi-ikoma, a suburb of Osaka, Japan.

Since 1977, many optical communication systems have been installed in the United States, Japan, and Europe, covering a wide range of bit rates, environmental conditions, and applications. In Europe, in-service systems are being developed at 8, 34, and 140 Mb/s. Among them, the 34-Mb/s system is especially attractive because it interfaces well with many existing networks and offers a high enough bit rate to make it cost effective. A brief description of the Chicago project, the Hi-Ovis project, and the British Post Office project will be given below. These projects represent a major part of the applications which will probably occupy a large portion of the fiber optics market in the future.

The Chicago project, the first major optical fiber communication system put into service (in 1979), was designed to evaluate the optical fiber technology for a wide range of Bell system service conditions. The system connects Illinois Bell customers in the Brunswick Building to the Franklin central office at a length of about 0.9 km, and also connects that office to the Wabash central office at a length of about 1.6 km. This system carries a variety of commercial traffic, including telephone channels, interoffice trunks at T3 rate, a picture phone, and a 4-MHz standard black-and-white video conference service. System parameters are listed in Table 12.7 and are very similar to those previously used in a field-trial experiment conducted in Atlanta, Georgia. The only exception is that the fibers used in the Chicago project have a lower attenuation coefficient ($\sim$4 dB/km) than those used in the Atlanta experiment. These cables were installed over a route having a total length of 2.5 km by pulling them through about 32 manholes with a total of five splices in the congested Chicago metropolitan area. Some of the manholes are partially filled with water, so this installation provides a real test for this new technology. A

**TABLE 12.7   System Parameters Used
in the Chicago Project**

System	
Bit rate:	44.7 Mb/s
Total length:	2.5 km
Repeater spacing:	7 km
Channels:	144
Fiber	
Type:	Graded-index and ribbon structure
NA:	0.23
Dimension:	55-μm core, 110-μm cladding
Length:	0.66 km
Loss:	4 dB/km
Dispersion:	1.3 ns/$\sqrt{\text{km}}$
Splicing loss:	0.8 dB
Source	
Type:	AlGaAs DH laser
Wavelength:	0.82 μm
Power:	-3 dBm
Detector	
Type:	Si APD
Sensitivity:	-54 dBm

polyethylene duct was installed in the underground conduit for these fibers to maintain a controlled environment. Cables were pulled through the duct by personnel without any special training, but the five splices made in the manhole were done by specially trained personnel. Results indicated that the system loss in the Chicago project is actually lower than that of the Atlanta experiment. This was attributed to a decrease in both the fiber attenuation coefficient and microbending loss. The Chicago project has been carrying commercial traffic since May 1977 without any outage. The performance has been outstanding, as indicated by a carefully monitored record. The most useful information emerging from this Chicago project is that, in dealing with the real world, the repeater spacing cannot be derived simply from the known parameters, such as transmitter power, fiber loss, and receiver sensitivity. The added unknown losses in cable pulling and splicing are dictated by the route congestion, access restriction, and locations. These factors also play an important role in the design of a real system. Designers must allow sufficient margin to account for extremely hostile environments, especially in metropolitan areas, where the underground ducts and conduits are often shared with other services, so that these telephone transmission lines will be able to withstand extremely rough treatment. In addition, water, soil, stream, hydrocarbon, and corrosive chemicals may be present in the underground conduits and manholes. Another important result that emerged from this Chicago project is the fact that the measured transmission loss of the installed system is very low. It is reassuring that very long repeater spacing is possible. The repeater is a very costly item and the number of repeaters, especially those located in the manholes, must be reduced as much as possible. With low-loss fiber transmission, repeaters can either be eliminated or located in exchange centers instead of underground facilities. This is a very important trade-off consideration in selecting new systems for future investment.

The Hi-Ovis project is a good representation of analog video CATV systems. This is a rather complex system which provides computer-controlled video service to about 160 home subscribers and 8 local studio terminals located at public premises such as the city hall, schools, police and fire stations, and medical centers. The system uses approximately 400-km fibers. It consists of a 36-channel distribution network, each of which has a length of 6 km. In addition, there are 24 fibers 400 m in length, 18 fibers 500 m in length, 6 fibers 5.5 km in length, 4 fibers 1.5 km in length, and 2 fibers 31 km in length. The longest transmission distance for analog video and FM audio signals is about 4 km between a UHF receiving station located on top of Ikoma mountain and the control station in Higashi-ikoma. The system does not make use of repeaters.

The major feature of this system is the interactive programming between the users and the central station. At the station, a video switching network contains 32 inputs and 168 outputs. Two out of 32 inputs are used for video signal transmission from the home terminal video cameras to the station and the remaining 30 inputs are used for video signal transmission from the center to TV monitors of the subscribers. There are 11 cables distributed from the station, each of which can serve 16 subscribers with a two-way communication linkage involving two fibers. The components used in the Hi-Ovis project are: LED sources with an average output of 1 mW, plastic-clad fibers with a NA of 0.25 and a loss of 16 dB/km, pin photodiodes with a capacitance of 2.5 pF, a quantum efficiency of 0.8, a time constant of 10 ns, and a S/N ratio of 56 dB for a video signal transmitted over a distance of 0.5 km. Figure 12.5 is a block diagram of various components involved in the Hi-Ovis system. It contains basically three subsystems. The first is the front-end equipment, which includes a broadcast network and a minicomputer control electronics to handle service requests of various subscribers. Services include alphanumeric data, movies, educational programs, local news, community events, and shopping information. The second is the information transmission subsystem, which consists of a video switching network designed to handle various services being offered. A shared frequency-division-multiplexed architecture is employed. To increase the system bandwidth capability, a large number of fibers are used. The third subsystem is the subscriber equipment, which consists of a terminal controller, a keyboard and a TV receiver, a video camera, and a microphone. The drop cable

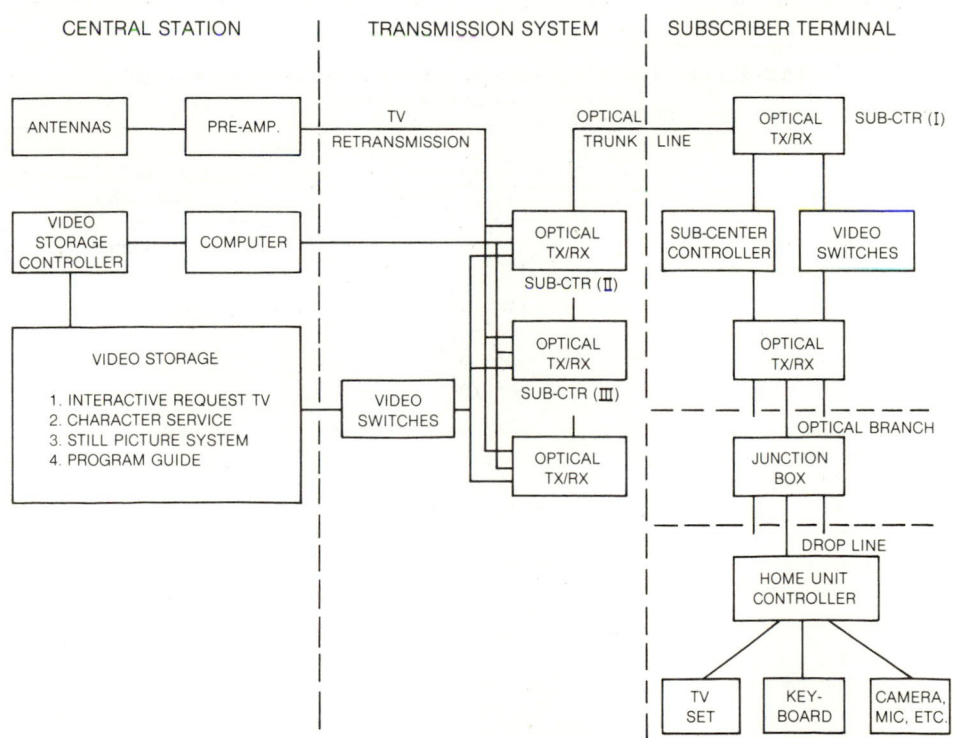

Figure 12.5   Block diagram of Hi-Ovis project.

to a subscriber's home consists of two optical fibers which connect directly to the distribution box. The optical transmitter and receiver were installed in a splicing box. The commercial market for interactive service at home is still in the distant future because the installation cost of the home terminal subsystem is very high. This situation could change when mass production of these items occurs.

A field-trial experiment conducted by the British Post Office (Ref. 12.3) involved two 8-Mb/s optical fiber transmission lines over a 13-km route using LED and LD sources, and a 140-Mb/s optical fiber transmission line over a 8-km route using laser diodes. These routes are between the BPO Research Center at Martlesham Heath and the Ipswich Telephone exchange in Suffolk, England. The system parameters are listed in Table 12.8. The receiver sensitivity of the 8-Mb/s and the 140-Mb/s systems for an error rate of 10^{-9} is -60 dBm and -43 dBm, respectively. For the 8-Mb/s system where LED is used, a repeater is placed in the middle of the 13-km path. No repeater is needed for both systems where LD is used. At 140 Mb/s, results indicated that repeater spacing of at least 8 km is possible with the commercially available Corning graded-index fibers. The spacing can be extended by using lower-loss fibers and can be significantly extended by using components operating at 1.55 μm.

TABLE 12.8   System Components and Parameters of British Post Office Field-Trial Experiments

	Bit Rate	
	8 Mb/s	140 Mb/s
System length	13 km	8 km
Telephone channel	120	1920
Source	LED	LD
Wavelength	0.82 μm	0.84 μm
Injection current	300 mA	30 mA
Output	65 μW	1 mW
Photodiode	APD	APD
Amplifier	Si J-FET	GaAs FET
Capacitance	10 pF	6 pF
Gain	40	80
AGC	25 dB	29 dB
Logic gate	Dual D-type bistable	D-type bistable
Receiver sensitivity	-59.7 dBm	-43 dBm
Fiber	Corning graded index	Corning graded index
Loss	4.5 dB/km	4.5 dB/km
Installation	Existing ducts	Existing ducts
Splicing	V-groove	V-groove

12.6 LONG-HAUL SYSTEMS

Design considerations for long-distance communication systems are different from those for interoffice and intracity trunks. The long-haul fiber optical communication systems require repeaters at regular spacings which are determined primarily by the transmitter power, receiver sensitivity, bandwidth, fidelity, and system losses. Their values are derived from parameters involving fiber attenuation and dispersion characteristics, system noise, signal distortion, and intersymbol interference. Figure 1.5 gives the ranges in which repeater spacing for various types of systems can be realized. The bit error rate for a typical telecommunication system is 10^{-9}. For systems involving a large number of repeaters, the accumulated BER can be very high.

As an example, we shall describe an optical repeater utilizing the state-of-the-art components for a long-haul and high-data-rate ($>$1-Gb/s) communication link to be used in the transoceanic transmission system. Several systems are presently under development both in this country and overseas (see Table 12.9). Of most importance is the system reliability for the transoceanic operation. This requires very careful selection of fiber optical components with proven performance and operating life. To achieve a maximum repeater spacing, it is necessary to select a very low dispersive single-mode fiber operating at 1.3 μm and having a loss figure at $\lesssim$0.5 dB/km. At this wavelength, the fiber dispersion can be kept to less than a few ps/km-nm. An InGaAsP semiconductor laser, which is driven by a GaAs FET, must be used as the transmitter. The laser can be directly modulated in the PCM format by a pulse signal current superimposed on a dc pre-bias current. The modulated output is then coupled into the single-mode fiber. Special cabling technology is required for transoceanic operation, where cable is under tremendous tension ($\gtrsim$10 tons) during the process of laying or raising from a depth of about 6000 m. It is important to design the cable with a strength that can survive the undersea

TABLE 12.9   System Design Parameters of Undersea Optical Fiber Systems

	United States (Bell System)	Japan (NTT)
Transmission length	8000 km	1000–10,000 km
Depth	7.5 m	8 km
Bit rate	274 Mb/s	260–400 Mb/s
Wavelength	1.3 μm	1.3 μm
Reliability	8 years	10 years
Repeater spacing	25–50 km	25–50 km
Power consumption	4 W	3–5 W
Fiber loss	$<$1 dB/km	$<$1 dB/km
Fiber strength	8 tons	7.5–10 tons

environment without deformation and breakage. The 1.3-μm detectors for the receiver must be either Ge APD or InGaAs APD, followed by an amplifier having a low-noise FET front end. InGaAs APD is expected to have a higher sensitivity than Ge APD; however, InGaAs APD is still in the research and development stage and not available commercially at present.

Present experimental results (Ref. 12.4) show that at a BER of 10^{-9}, the Ge APD receiver sensitivity is -31.9 dBm, which is lower than the InGaAs APD sensitivity for a 2-Gb/s RZ system over a 44.3-km transmission length. The RZ signal waveform in general results in about 1 dB improvement in the receiver sensitivity compared to NRZ signals at the same data rate. To extend the repeater spacing of a 2-Gb/s communication system beyond 45 km, it is necessary to reduce the fiber loss further, to below 0.3 dB/km, and also to decrease the dispersion loss to less than 3 ps/km. To reach this level of performance, a zero-dispersive fiber with extremely low loss (≤ 0.2 dB/km) must be used. Such a low-loss transmission line has been demonstrated by using a laser diode operating at 1.55 μm (Ref. 12.5). Fiber optic components operating at 1.55 μm are considered to be most attractive for long-haul and high-data-rate systems; however, the technology at this wavelength has to be developed, especially in the detector area, because the performance of Ge APD degrades rapidly beyond 1.3 μm.

12.7 MULTITERMINAL CONTROL AND DATA DISTRIBUTION

Fiber optics can offer many services other than telecommunication. Systems that transmit and distribute data within a building or a group of buildings can now make use of optical fiber technology to its best advantages. Other applications include electric power monitor and control, energy and load management, data bus for aircraft and traffic control and command, and so on. This section describes briefly some of these applications.

The intrafacility network within a building consists of an interface unit between a high-data-rate optical fiber transmission link and the subscriber's equipment, which is comprised of video terminal, telephone, computer and facsimile, and so on. Information and data can be distributed to a number of local fiber optical loops by means of a central control processor. Each loop contains a subscriber interface unit that supports several terminals. Time-division multiple access is a common technique to be used in this network, where the time is divided into many slots that can be made available to various terminals. In this scheme, several protocals can be considered; for example, time slots within a recurring time frame can be dedicated to users, or the time-slot access can be operated in contention among various users, who compete for the allocation of a time slot.

Optical fiber transmission lines have been utilized by the electric power utility industry in its power monitor and control systems. In a high-voltage

(>200 kV) transmission network, there are many noise and electromagnetic interference problems associated with the conventional communication circuits that cause difficulties in supervising, protecting, and controlling power delivery from the generating plant to a control center located in a metropolitan area. A system suitable for this purpose must be reliable and must have long-haul capability at bit rates as high as 30 Mb/s. From the control center, a multiterminal data distribution system is needed to communicate with a number of division offices and substations. A combination of technologies in microprocessors and fiber optics has provided the power utility industry with an improved load management. One of the most important aspects is the distribution automation, which provides all communication and load control functions, including time-of-day metering, remote meter reading, feeder switching, capacitor bank control, transformer temperature monitoring, fault location, and isolation.

In conclusion, fiber optic technology will continue to grow and will capture a major share of the market within this decade in telecommunication, computer, and power utility industries, as well as other services, such as education, entertainment, home protection, banking, medical and health, business trade, and many military systems. It will also be used in a variety of instrumentation for automobiles, ships, and aircraft. As the technology matures, the costs of various components will be reduced in a way similar to that typified by the solid-state electronics industry. When this occurs, optical fiber technology will become extremely competitive with already established radio-frequency and microwave technologies. An attractive application is to introduce services such as high speed and interactive video in major population centers, which can be connected by long-haul fiber systems. Time-division and wavelength-division multiplexing can be used to relieve congestion as traffic grows. The potential for growth of this technology appears immense. Not only can optical fiber perform better and more economically than conventional transmission media in most applications, but it will also help to bring new services that are not possible or practical with transmission media available today.

PROBLEMS

12.1. Calculate the system response and power budget for a digital link with the following specifications: BR = 50 Mb/s, BER = 10^{-9}, signal format is NRZ, terminal length is 10 km, and number of terminals is 2.

12.2. Design an optimum two-way telecommunication system, capable of transmitting a bit rate of 400 Mb/s at an error rate of 10^{-9} over a distance of 10 km.

12.3. Design a distribution system which serves 100 parallel terminals each of which has a transmitter and a receiver and is located at a distance about 100 m from a star coupler. Calculate the total system loss.

12.4. For the system defined in Problem 12.3, what is the required transmitter power, assuming that 5-dB/km fibers with 0.2 NA are used?

12.5. Design a serial distribution system which serves 100 subscribers along a linear path of a total length 10 km. Calculate the total system loss.

REFERENCES

12.1. Third International Conference on Integrated Optics and Optical Fiber Communication, San Francisco, April 27–29, 1981, Tech. Digest.

12.2. First International Conference on Integrated Optics and Optical Fiber Communication, Tokyo, July 18–20, 1977, Tech. Digest.

12.3. R. W. Berry, D. J. Brace, and L. A. Ravenscroft, *IEEE Trans. Commun., COM-26,* 1020 (1978).

12.4. J. Yamada and T. Kimura, *IEEE J. Quantum Electron, QE-18,* 718 (1972).

12.5. T. Miya, Y. Terumuma, T. Hosaka, and T. Miyashita, *Electron. Lett., 15,* 106 (1979).

Index

Index